Natural Science in Archaeology

Series Editors
Bernd Herrmann
Günther A. Wagner
Christopher E. Miller

For further volumes:
http://www.springer.com/series/3703

Cristina Corsi • Božidar Slapšak
Frank Vermeulen
Editors

Good Practice in Archaeological Diagnostics

Non-invasive Survey of Complex Archaeological Sites

Springer

Editors
Cristina Corsi
Department of Humanities
University of Cassino
Cassino
Italy

Frank Vermeulen
Department of Archaeology
Universiteit Gent
Ghent
Belgium

Božidar Slapšak
Department of Archaeology
University of Ljubljana
Ljubljana
Slovenia

ISSN 1613-9712
ISBN 978-3-319-01783-9 ISBN 978-3-319-01784-6 (eBook)
DOI 10.1007/978-3-319-01784-6
Springer Cham Heidelberg New York Dordrecht London

Library of Congress Control Number: 2013956503

Printed on acid-free paper

Springer is part of Springer Science+Business Media (www.springer.com)

Contents

Good Practice in Archaeological Diagnostics: An Introduction

1

Cristina Corsi

Contents

By the term 'archaeological diagnostics', we mean a whole array of methodologies and approaches to the survey of archaeological sites, mainly referring to those that do not imply excavations or at least only very limited ones. 'Non-destructive' or 'non-invasive' approaches to the investigation of ancient landscapes have always been available to archaeologists, since the first methodological definitions of the discipline were drafted (e.g. Bradford 1957; Clarke 1977, 1990; Pasquinucci and Trément 2000; Renfrew and Bahn 2000). Among the most traditional methods, the collection of historical documentation and the field survey are undoubtedly the best developed, but aerial archaeology also provided a boost as soon as photography and flying machines came into use (Chap. 2 by Ceraudo, in this volume). The use of ancient sources, archive research and toponymy for the investigation of cultural landscapes is so rooted that it has not been possible here to devote specific chapters to these branches of archaeological research. Robust new GIS processing of historical cartography could surely have merited special attention (some reviews of recent case studies are in Corsi and Vermeulen 2007 and Börner et al. 2012). However, we have decided to limit the already wide spectrum of this volume to the newest technological developments achieved in remote sensing and geophysical surveying and to the most recent methodological innovations that have been introduced to the broad approach of the archaeological survey of greenfield sites.

A very important section of this volume deals with aspects related to the visualisation of survey

C. Corsi
Department of Humanities, University of Cassino, Cassino, Italy
e-mail: c.corsi@unicas.it

C. Corsi et al. (eds.), *Good Practice in Archaeological Diagnostics*, Natural Science in Archaeology,
DOI 10.1007/978-3-319-01784-6_1, © Springer International Publishing Switzerland 2013

data and their valorisation. In the field of digital technologies for virtual reconstruction and data visualisation, recent years have seen not only spectacular developments but also a growing awareness of the need for 'regulation' and the delineation of standards and guidelines. In this regard, we present here the up-to-date results of the international debates that have produced the indispensible 'charters' of London and Seville.

When we attempted this enterprise, we were obviously aware of the fact that the themes encompassed would be very extensive and that this collective work could not hope to be comprehensive, neither in the range of topics nor in the technicality of the contributions. Our intention was, and still is, primarily to report on the intensive exchange and collaboration carried out in recent years and secondly to offer an up-to-the-minute hint for further discussion.

Not least, our intention is also to provide an instrument to young researchers and students as a starting point for the framing these nowadays very popular subjects of discussion and training and to offer them the possibility of deepening their knowledge of the aspects that they feel are closest to their interests and suitable to their talents.

We have taken care to avoid overlaps with the much more technical manuals about specific techniques, such as the very popular Seeing the Unseen: Geophysics and Landscape Archaeology, edited by Salvatore Piro and Stefano Campana (Campana and Piro 2009) and the manual by Armin Schmidt entitled *Geophysical Data in Archaeology: A Guide to Good Practice* (Schmidt 2001a, 2013; see also Schmidt and Ernenwein 2013) or the *Arts and Humanities Data Services Guides to Good Practice* (2004; available online at: http://www.ahds.ac.uk/guides/). These indeed cover several fields, but there is no framing of the disciplines, only guidelines for good practice in archiving and data management. In the most recent book by Imma Ollich-Castanyer, *Archaeology, New Approaches in Theory and Techniques* (Ollich-Castanyer 2012), non-destructive approaches and the visualisation and valorisation of complex sites do not get any attention. About 'cyberarchaeology' and virtual reconstruction, we can only list the books edited by Maurizio Forte (Forte 1997; Forte and Williams 2002), Juan Barcelo et al. (2000), Gary Lock (2003) and Mark Greengrass and Lora Hughes (2008), just to mention a handful.

Instead, we have tried to cover the two main spheres of our work: research and valorisation. We have sought to provide a good coverage of the different methods of data capture, all possibly included in the definition of 'remote sensing', and we have provided an insight into the different approaches to data integration. At the same time, by mainly examining the aspects related to the interpretation and visualisation of data and via the discussion of specific management plans for the valorisation of this peculiar category of site where most evidence is 'invisible' to visitors, we intend to lay the ground for a discussion about the essential aspects of cultural heritage management.

The discussion about the basic principles of the digital reconstructions has been extended on several occasions to specialists and the wider public. We are, of course, aware of the dangers, but at the same time we want to stress how much not only interdisciplinary teamwork but also 3D visualisation has enhanced our comprehension of spatial phenomena and relationships. Surely a 3D digital reconstruction is more effective and 'convincing' than a plan reconstructing the layout of a town; however, it is not necessarily more 'inventive' or less scientific.

1.1 Making a 'Radiography' of the Past

This book constitutes the final and possibly most durable 'deliverable' of the project Radio-Past (www.radiopast.eu), the Marie Curie/People Industry and Academy Partnerships and Pathways (IAPP) project entitled, 'Radiography of the past: Integrated non-destructive approaches to understand and valorise complex archaeological sites' that has aimed to join together different resources and skills to improve, refine and validate intensive archaeological surveys on complex sites, with a special focus on abandoned ancient urban sites in the Mediterranean.

To fulfil the objectives of the programme, the consortium of seven partners was composed of academic institutions, University of Évora (P), Ghent University (B), University of Ljubljana (SL) and the British School at Rome (UK), and private companies, 7Reasons Media Agency (A), Past2Present (NL) and Eastern Atlas (D).

The Radio-Past project has sought to integrate different methodologies in the widely developed field of non-destructive survey technologies as applied to archaeology, and it has also pursued the validation of the results through innovative methods of visualisation and the development of strategies for the efficient management of the cultural heritage sites studied. One of the main objectives of this project was to allow a multiplicity of methods and research approaches and to generate methodological guidelines for archaeological diagnostics. The idea was to develop a standard set of survey approaches, based on a series of already widely used methods as well as more innovative methods such as active low-altitude aerial photography, geophysical prospection, light detection and ranging (LiDAR) surveys and geomorphological observations, which can in the future be efficiently used in a comparable and integrated way on a wide range of complex sites in Europe.

Furthermore, the project was also concurrently targeting the development of effective scientific systems for the dissemination of survey results. In particular, the combination of high-resolution fieldwork with computer-based means of mapping and data visualisation allows the virtual reconstruction of buried towns or large settlements within a relatively short space of time, as opposed to the more traditional excavation-centred approach where it can take generations before a broader view of the site becomes available.

With these aims in mind, a link-up was pursued between the project and the EU policies for cultural heritage and landscape management. The core field research done within the framework of Radio-Past is fully compliant with Article 3 (ib) of the European Convention on the Protection of Archaeological Heritage, better known as the Treaty of La Valletta 1992, where it is stated that 'to preserve the archaeological heritage and guarantee the scientific significance of archaeological research work, each Party undertakes: … to ensure … that non-destructive methods of investigation are applied wherever possible'. Cultural heritage management authorities will benefit widely from this approach as such integrated surveys of complex sites will provide them with a very effective tool for gauging the degree of archaeological survival on sites in their care and for choosing appropriate conservation strategies.

The operative strategy that the consortium decided to apply is the creation of 'open laboratories', that is, archaeological sites where fieldwork was ongoing over several years, if not decades, and where the partners were involved at different levels. These sites are spread over the Mediterranean, including the Atlantic Lusitania (*Ammaia* in Portugal), the Tyrrhenian coasts (*Mariana* in Corsica and *Portus* at the mouth of the River Tiber), Adriatic Italy (Potenza Valley), the Aegean Sea (Boeotia) and reaching beyond the Alps to *Carnuntum* along the Danube (Austria). The idea was to test and validate methodologies and strategies and discuss results and interpretations. The Roman town of *Ammaia* was the most important 'open lab' of the project; here, all the teams gathered periodically for survey campaigns and carried out processing, interpretation and visualisation and even training activities. Strategies for the validation of the results were developed in all the partner institutions, while dissemination activities were conducted regularly at all levels.

For this reason, the *Ammaia* case study has played a key role in some of the papers collected here. However, we have always been concerned to develop standards and guidelines for good practice that can be extended to every type of 'complex archaeological site'. It is undoubtedly true that all archaeological sites are complex, but we would like to stress here that by this definition we mean large settlements where structures, buildings and infrastructures are developed and where a long occupation has possibly brought with it transformations and overlapping changes.

The authors of the contributions have been selected from among the research 'staff' of the project but they also include internationally known specialists who were involved as speakers at the two international events organised in the framework of the project (the Valle Giulia Colloquium of Rome, 2009; the Colloquium of Ghent, 2013) and the three Specialisation Fora, the high formation training activities organised in 2010, 2011 and 2012.

In this way, this volume offers contributions on different aspects of the full research process (data capture, data management, data elaboration, data visualisation, site management, dissemination and communication and even data presentation), setting out the most up-to-date and state-of-the-art guidelines for good practice in each field.

1.2 Data Acquisition Versus Understanding

However, the intention of this collective work is to go beyond the aspects of 'archaeological diagnostics' that have already been carefully explored in depth. We have sought, indeed, to bid for the deeper disclosure of possibilities offered by the integration of these different survey techniques, going beyond the data capture procedures to penetrate the most important aspects of interpretation and understanding.

Too often in fact we are confronted with 'revolutionary discoveries' that are instead just puzzles of data without any historical in-depth or methodological criticism. Knowledge is very different from data collection, and aspects related to interpretation should be carefully and openly discussed.

A recent essay by Martin Millett offers a very good résumé of the contribution of geophysical surveys to the understanding of complex sites and specifically of Roman towns. Yet, when describing the methodological framework, he mentions only the surface collection and topographical survey as approaches used 'to investigate a full range of Roman urban centres' in central Italy. These approaches are considered 'antagonists' of the previous archaeological work focused on excavation and study of above ground architectural remains, which produced 'high-resolution' data about very limited parts of the settlements. Within this framework, I consider that the contribution of the full panoply of non-invasive instruments of research should not be underestimated, which is – to resume our comparison with the medical diagnostic – the anamnesis part of it. Neglecting historical sources and historical cartography and ignoring previous research do affect our understanding of the settlements and of the human beings who populated them, of the social structures which animated them and of the processes and the activities which took place there.

So, if it is true that remote sensing and geophysical surveys in archaeology are undergoing increasingly sophisticated technological development and achieving increasingly reliable results and that the rapidity of the process of acquisition and analysis of data have achieved unprecedented quality and unparalleled resolution, it is also true that the methodology of research cannot only be inspired by the objective of the 'maximization of data collection across as broad an area as possible' (Millett 2012, p. 26): historical criticism and the distinction between data acquisition and the generation of knowledge have always to be kept in mind. The methodological framework for this type of research still has to be considered, and much more theoretical elaboration is needed when (and if) we formulate the scientific questions behind our research.

It is intended that this volume should make ripples in the stagnant pond and stake out the ground for further discussion.

1.3 A Question of Integration

We fully agree with the warning by Keay et al. (2009, pp. 154–155) that it is simplistic and superficial to think that applying only a single technique of geophysical surveying can enhance our 'understanding' of a certain ancient site, providing a bi-dimensional 'accurate and high-resolution representation of archaeological and geomorphological features'. Only an integrated survey 'can furnish the researcher with a variety of data sets and provide a more nuanced and complex representation of a site'.

In this sense, it is possibly more proper to refer to data fusion and to the problems related to it (an issue relatively recently discussed by Armin Schmidt 2001b). Data fusion, mostly undertaken when data capture has been carried out using the same parameters and resolution, has proved to be very efficient at enhancing the quantity of detected features and the quality of their interpretation.

However, our idea of integration goes deeper into the complexity of the reconstruction of an 'invisible' or yet almost completely buried settlement, starting from understanding the reasons for its birth and for its abandonment, grasping its three-dimensional characteristics, going through its rise and decline and its changes and transformations and understanding the material culture and the daily life of its inhabitants.

For these reasons, when carrying out fieldwork, we have sought to apply the widest range of approaches, and in this volume we assembled papers from specialists in different disciplines. The underlying philosophy is that only a real and strong integration of approaches and techniques can bring about the understanding of a complex site, and for these reasons, in the project Radio-Past and in this book, we have pulled together researchers from disparate fields (archaeology, geophysics, geology, geomorphology, ICTs, CAD and virtual reality, cultural heritage management, chemistry, archaeometry, etc.).

This extended concept of 'integration' brings out a theme raised by Jeroen Poblome during the closing panel of the Radio-Past Colloquium, held in Ghent in January 2013. I share the worry that the increasing level of specialisation and technological mastery is promoting the idea that a 'standard' research requires all these branches of science to be mastered at the highest level and that no decent project can be carried out without deploying a full array of expensive techniques, which require the most specialised know-how and the state-of-the-art instrumentarium. This is obviously not true, but it is undeniable that multidisciplinary teamwork and the integration of staff and resources are necessary to tackle research agendas in a well-designed and well-managed project.

Related to this aspect of research agenda and guidelines, some words have to be said about the structuring of 'workflows'. We intend this term in its widest meaning, as planning the full process from data collection and fieldwork to the archiving and data processing phases, from the visualisation of results to the communication and dissemination to all kinds of audiences, until the valorisation and management of the sites.

When drafting the proposal for the project Radio-Past, this aspect of the valorisation of these very peculiar sets of survey data was specifically taken care of. Archaeology cannot postpone anymore the urge for society of playing a 'social' role, for sustainable development and for the valorisation of our cultural heritage and of our historical landscapes. The public widely perceives archaeology is synonymous only with excavations, and people interpret the mission of archaeologists as only having as its aim a 'finding', and they do not see it as a process of 'understanding' our past. The elaboration of targeted management plans for sites where most of the archaeological evidence is 'invisible' is a first step toward the sustainable integration of archaeology into the social and economic texture of smaller and wider regions.

1.4 Size Matters

When attempting the survey of a complex site, we have to face the matter of the 'scale'. Townscapes and landscapes require different ranges of resolution, but whatever the case, high resolution of smaller fields or lower resolution of larger extents is unavoidably related to huge amounts of data, bringing with it troubles with data archiving, processing and retrieving. Technological developments, starting with digital cameras and the diffusion of low-altitude flying devices like drones for remote sensing and real-time kinematic (RTK) automatic or mechanical sensors for geophysical surveys, have exponentially increased the amount and the resolution of data available for each site. Most ICTs and GIS processing has also dramatically enhanced the quantity and quality of information that we can retrieve from 'traditional' sources, like historical cartography and pictorial urban and country landscapes views.

The relationship between the time invested in the survey and the extent of the surface area surveyed or the resolution of each unit of surface has increased enormously, which means that in the same time span, we can now investigate much wider terrains or in the same time span obtain a much higher resolution. But the processing time has not been cut down at the same rate, and data management is becoming a higher priority in our workflow. This raises the issue of complexity: the availability of more data can surely enhance the quality of the result but can at the same time raise more questions for interpretation.

1.5 The Fourth Dimension

The term 'complexity' occurs very frequently in our discussion, and it can be applied to most facets of the general interpretation of sites investigated mainly by means of non-destructive approaches. With rapidly increasing experience and know-how, the availability of very extensive comparative research and the technical and technological improvements in hardware and software have made the interpretation of individual features faster and easier; but we have to admit that we are still too often powerless when we have to understand the fourth dimension.

Diachronic evolution can be in most cases be snatched by collecting surface artefacts, and relative chronology can be glimpsed with the help of a ground-penetrating radar (GPR) survey. But absolute chronology is still the exclusive domain of traditional excavations, and therefore it is limited to the few 'windows' that we can open to it in a complex and large site.

This consideration does not have to spoil the enthusiastic atmosphere that is at the moment animating the teams working on these themes, but there is a challenge at stake here.

For the time being, there will be continue to be a topic of discussion about whether it makes sense to search for the standardisation of procedures in archaeological surveys. We are aware that geographical and cultural peculiarities, which have been shaped by the elapsed centuries of different types of land use, make each archaeological site a case study in itself, but we are keen to prosecute the delineation of guidelines for good practice in archaeological diagnostics.

If archaeological diagnostics aspires to be considered a science in all respects, this is a process we have to endure. It has worked with the stratigraphic excavation methodology – why not with survey?

References

Arts and Humanities Data Services (2004) Guides to good practice. Available online at http://www.ahds.ac.uk/guides/

Barcelo JA, Forte M, Sanders D (eds) (2000) Virtual reality in archaeology, BAR international series S 843. Archaeopress, Oxford

Börner W, Uhlirz S, Dollhofer L (eds) (2012) Proceedings of the 16th international conference on cultural heritage and new technologies (CHNT 16, Vienna, Austria, 14–16 November 2011). Museen der Stadt Wien – Stadtarchäologie, Vienna

Bradford J (1957) Ancient landscapes: studies in field archaeology. G. Bell and Sons, London

Campana S, Piro S (eds) (2009) Seeing the unseen. Geophysics and landscape archaeology. Taylor and Francis, London

Clark AJ (1990) Seeing beneath the soil. Prospecting methods in archaeology. B.T. Batsford, London

Clarke DJ (ed) (1977) Spatial archaeology. Academic, New York

Corsi C, Vermeulen F (2007) Digital vs. historical cartography for studying archaeological sites: a case study from Corsica. In: Proceedings of the 11 workshop "Archäologie und Computer. Kulturelles Erbe und Neue Technologien", Vienna, 18–20 Oktober 2006), 1–19. Abstract: http://www.stadtarchaeologie.at/tagung/archiv/ws11/abstracts/vermeulen.htm)

Forte M (ed) (1997) Virtual archaeology. Thames & Hudson Ltd., London

Forte M, Williams RP (eds) (2002) The reconstruction of archaeological landscapes through digital technologies (Boston, Massachusetts, USA, 1–3 November 2001). Archaeopress, Oxford

Greengrass M, Hughes L (eds) (2008) The virtual representation of the past. Ashgate, Farnham/Burlington

Keay S, Earl G, Hay S, Kay S, Oogden J, Strutt KD (2009). The role of integrated geophysical survey methods in

the assessment of archaeological landscapes: the case of Portus. Archaeol Prospect 16:154–166. Published online 23 July 2009, Wiley Inter Science (www.interscience.wiley.com). doi:10.1002/arp.358
Lock G (2003) Using computers in archaeology: towards virtual pasts. Routledge, London
Millett M (2012) Understanding Roman towns in Italy: reflections on the role of geophysical survey. In: Johnson PS, Martin M (eds) Archaeological survey and the city, vol 2, University of Cambridge Museum of Classical Archaeology monographs. Oxbow, Oxford, pp 24–44
Ollich-Castanyer I (ed) (2012) Archaeology, new approaches in theory and techniques. Intech. Open Access Book. http://www.intechopen.com/books/archaeology-new-approaches-in-theory-and-techniques
Pasquinucci M, Trément F (eds) (2000) Non-destructive techniques applied to landscape archaeology. Oxbow, Oxford
Renfrew C, Bahn PG (2000) Archaeology: theories, methods and practice. Thames and Hudson, London
Schmidt A (2001a) Geophysical data in archaeology: a guide to good practice. Oxbow, Oxford
Schmidt A (2001b) Visualisation of multi-source archaeological geophysics data. In: Mauro C, Paola C (eds) Quaderni Prospezioni Archeologiche (50th anniversary issue). Fondazione Ing. Carlo Maurilio Lerici/Politecnico di Milano, Rome, pp 149–160
Schmidt A (2013) Geophysical Data in Archaeology: A Guide to Good Practice (2nd, fully revised edition). Oxbow, Oxford and Oakville
Schmidt A, Ernenwein E (2013) Guide to Good Practice: Geophysical Data in Archaeology. 2nd edition. Available at: http://guides.archaeologydataservice.ac.uk/g2gp/Geophysics_Toc. Accessed October 2013

Part I

Remote Sensing

Aerial Photography in Archaeology

2

Giuseppe Ceraudo

Contents

G. Ceraudo
Dipartimento di Beni Culturali,
Università del Salento, Lecce, Italy
e-mail: giuseppe.ceraudo@unisalento.it

2.1 Historical Overview and Assessment

The interest in aerial photography on the part of those working in our sector of study dates back to the beginnings of photography, with the first aerial photograph taken during flight in 1858 by the Frenchman Gaspard-Felix Tournachon (known as "Nadar"), who photographed the Avenue du Bois de Boulogne in Paris from a hot-air balloon (Fig. 2.1).

In the archaeological field, this technique for obtaining images from above was employed by the German Friedrich Stoltze as early as 1879, to document the state of the excavations in *Persepolis*. However, a leading role in the development of the technique in this phase was played by Italy. Indeed, the first flight undertaken for archaeological purposes in Europe took place in Rome in early June 1899, organised by the archaeologist Giacomo Boni. To document the excavations then in progress in the Roman Forum, in 1899 photographs were taken from a tethered balloon belonging to the Special Brigade of Military Engineers (Figs. 2.2 and 2.3). A few years later in England, in the summer of 1906, R. H. Sharpe took pictures of Stonehenge from a military hot-air balloon (Fig. 2.4).

Despite its pioneering application in the documentation of the excavations of the Roman Forum by Boni and others in subsequent years (Ostia, Pompei, Porto: Fig. 2.5), this study technique in Italy did not experience the development

C. Corsi et al. (eds.), *Good Practice in Archaeological Diagnostics*, Natural Science in Archaeology,
DOI 10.1007/978-3-319-01784-6_2, © Springer International Publishing Switzerland 2013

Fig. 2.1 Oblique aerial photograph of Paris taken by Nadar in 1858 (Piccarreta and Ceraudo 2000)

and widespread adoption that might have been expected. In other countries like the UK and Germany, instead, developments of instruments (cameras and aircrafts) and know-how (photo-interpretation techniques) were progressively achieved.

With the outbreak of the First World War, aerial photography became a key tool in military reconnaissance, and consequently the procedures useful for the reading and interpretation of photographic images began to be codified and refined. From the large quantity of aerial photographs taken for military purposes in those years, lessons were learned that would also be of use to studies of archaeological topography (Fig. 2.6). In Italy however with the exception of a few attempts by Giuseppe Lugli, effective and rigorous applications of this tool began to be seen only after the Second World War with the fundamental work of Ferdinando Castagnoli, John Bradford, Giulio Schmiedt, Dinu Adamesteanu, Nereo Alfieri and others.

In the interwar period, the use of aerial photographs for archaeological purposes saw significant development, including on a theoretical level.

Between 1925 and 1932, important research was conducted at the behest of Father Antoine Poidebard, particularly in Syria (Fig. 2.7). This soldier and clergyman, nicknamed the "Flying Priest", established the foundations of archaeological photo-interpretation and provided valuable insight concerning the timing and the techniques required in order to ensure the appearance of certain archaeological features in the photographic images.

By then, the utility of aerial photography in desert contexts, where the continuity of settlement had been interrupted, was well established. In contrast there remained much doubt about its potential for areas that are still inhabited and cultivated today, where it was assumed that successive human transformations must have obliterated any trace of their most ancient phases. However, the studies by O.G.S. Crawford conducted in Great Britain from 1922 onwards demonstrated the extensive applicability of the method even in areas characterised by long-standing continuity of settlement. In several European countries, and in many of the lands included in their expanding colonial domains, aerial photography for archaeology was applied by amateur pilots but also in the framework of governmental-supported aerial reconnaissance programs.

The start of the Second World War led to the interruption of the research, but it also provided

Fig. 2.2 The captive balloon of Brigata Specialisti of the Military Engineers of Italian Army inside the central nave of the Basilica of Maxentius (early 1900s) (Ceraudo 2004)

researchers with an enormous quantity of photographic material that had been acquired for military reasons (Fig. 2.8). The result was a considerable boost for this type of study, which by then was well past its pioneering stage. Indeed, a substantial quantity of images from that time is held by a number of important aerial photography archives throughout Europe.

The numerous images acquired in those years today provide us with historic testimony concerning the organisation of territories before the extensive urbanisation and infrastructure building that was to profoundly alter the agrarian landscape of Italy and Europe as a whole in the post-war period. Paradoxically, these images were in some ways more representative of the

Fig. 2.3 Excavation campaign in the central part of the forum (area of Comizio and of *Niger Lapis*) recorded by G. Boni on a captive balloon of Brigata Specialisti of the Military Engineers of Italian Army (Ceraudo 2004)

ancient layout of places than of the modern situation.

In the subsequent period, from 1960 onwards, Europe saw growing interest in the various techniques used in aerial photography as applied to archaeology. In Italy, however, such images were mostly limited to vertical photos of the military type, more suitable for an overall reading of the terrain. This was a direct consequence of a restrictive law dating back to 1939 which banned private companies and organisations from freely taking aerial photographs at low altitudes.

In contrast, in some European countries (Great Britain, France, Belgium, Germany), there was a tendency for systematic aerial reconnaissance to be conducted by private aviators (two famous names in this regard are Roger Agache and Otto Braasch) or by specially created research centres.

In addition the period saw many important events which provided an occasion for cultural exchange, including the 8th International Congress of Classical Archaeology in Paris 1963; the 10th Congress of the International Society of Photogrammetry, Lisbon 1964 and the 2nd

Fig. 2.4 Stonehenge from an Army balloon (Sharpe 1906) (Bewley 2004)

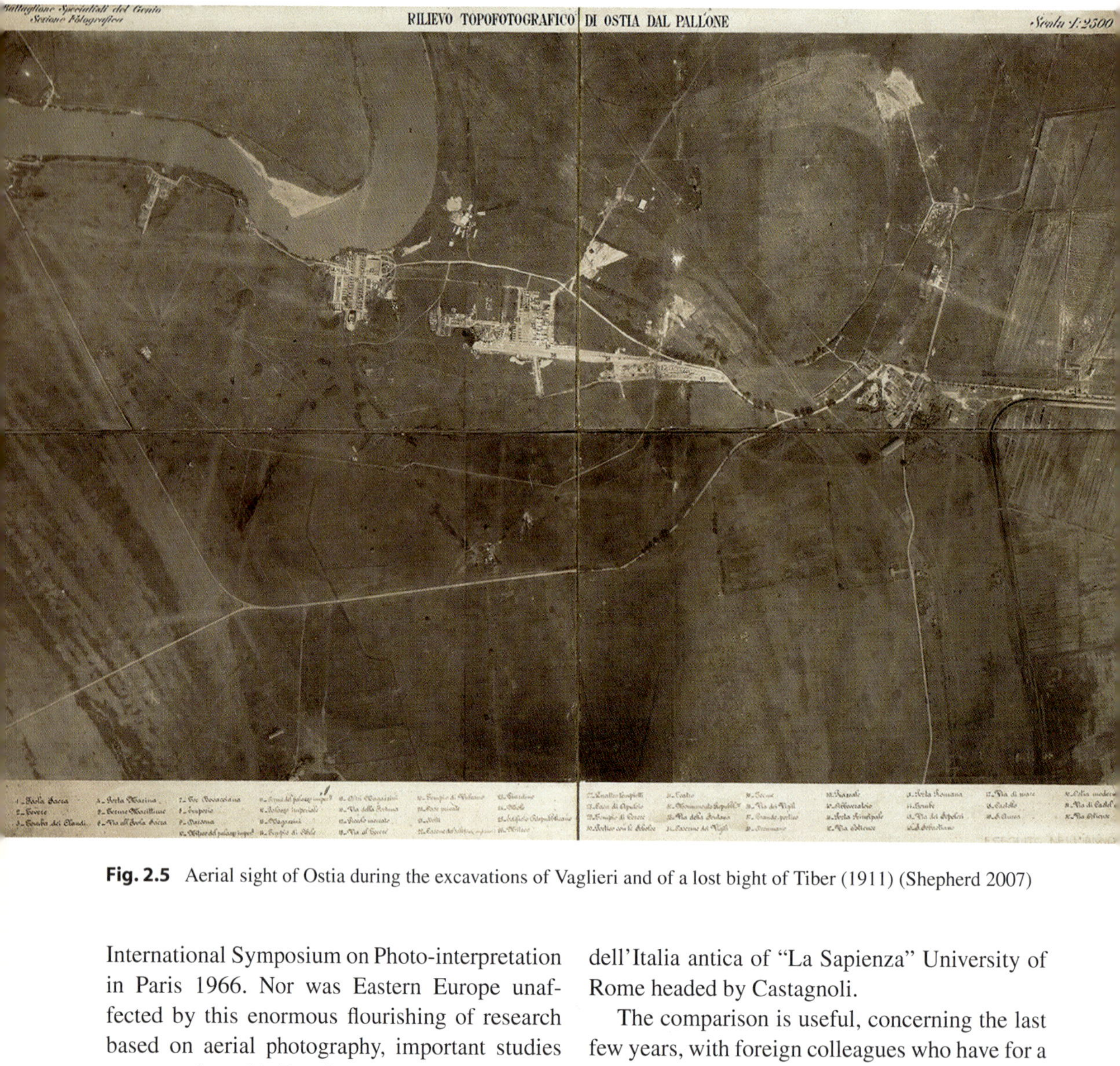

Fig. 2.5 Aerial sight of Ostia during the excavations of Vaglieri and of a lost bight of Tiber (1911) (Shepherd 2007)

International Symposium on Photo-interpretation in Paris 1966. Nor was Eastern Europe unaffected by this enormous flourishing of research based on aerial photography, important studies were conducted in Russia (on the remains of centuriation), Poland (systematic territorial research) and Yugoslavia (on the layout of Greek colonies in Dalmatia). More recently, following the fall of the Berlin Wall, work has been conducted in Slovakia and Romania.

From this point onwards, even in Italy, which by then had largely caught up with the other nations, the method spread thanks to the work of proficient scholars: as well as the work of Schmiedt at the Istituto Geografico Militare in Florence, also worthy of mention are the activities of the Istituto di Topografia di Roma e dell'Italia antica of "La Sapienza" University of Rome headed by Castagnoli.

The comparison is useful, concerning the last few years, with foreign colleagues who have for a longer time been developing the activity of aerial recognition and who have promoted and fuelled discussion and comparison in a sector whose fields of action was certainly limited by restrictive norms, now fortunately abolished. Nevertheless, this scientific activity was always vital and dynamic, with deep roots, and it is historically testified in the boundless specialized bibliography.

It must be reaffirmed, however, that this line of research is valid only if founded upon solid cultural bases and connected to a well-rooted tradition of studies, with professionalism and

Fig. 2.6 Siracusa, the *Neapolis* area photographed from a biplane bomber, the Caproni Ca3. Under the wing of the biplane are the ruins of the amphitheatre and theatre (Ceraudo 2004)

competences tied up to the activity on the territory. We risk starting with inadequate phenomena: some abstractions are unfortunately too technical and, in line with much present-day thinking, are more interested in the projects than in the works themselves, or there may be confusion, due to the lack of formation of a basis, as a result of which the instruments used for the research (we allude, in this sphere, to surveys, aerial recognition and relatively oblique photographs) have sometimes been taken over by disciplines (Fig. 2.9).

Among these "tools", the use of aerial photography has increased notably in different directions: on the one hand the areas interested in the experiences of archaeological photo-interpretation have increased, and on the other hand there is a stronger interest in cartographic representations of the territory, both as basic cartography – an essential support for knowledge and for guardianship – and as photogrammetry adapted for archaeological use.

From the methodological point of view, I remain convinced that the use of aerial photography must be tightly tied to the primary demand of contextualization and the topographical position of the find – its trace – and to its precise survey. The design phase, which is the action to fix a defined object in space and in this case to position it on the map (cartographic positioning), even if as a trace, constitutes the essential

Fig. 2.7 *Palmyra* view from SW through the Valley of the Tombs in an oblique aerial photo of Poidebard in 1937 (Denise and Nordiguian 2004)

presupposition for the knowledge and protection of the cultural heritage (Fig. 2.10). In the specific case of archaeological traces, even if they are individuated, interpreted and described, but not georeferenced with aerial photogrammetric restitution, they will remain abstract elements, uprooted from their context, and only a passing moment in the research of a determined territory, on which it would thus be impossible to effect exhaustive studies or to practise any action of guardianship.

Even if the digital image is confidently set out to be the only tool to be exploited, the existence of an enormous quantity of traditional aerial images on film, a lot of them still "unpublished", preserved in the aerial photographic archives and still to be read and elaborated, makes it essential to maintain procedures and the "know-how" necessary to competently extract the data contained in them. It is worth remembering that a stereoscopic strip of vertical aerial photographs is readable (and therefore measurable) in three dimensions and that non-perceivable data, at times on a single frame, analogue or digital, can be extrapolated with traditional techniques that permit the employment of suitable instrumentation that can be used for the emphasized perception of the relief (stereoscopes) (Figs. 2.11 and 2.12). In my opinion, a superior refinement of archaeological photo-interpretation is possible that elaborates and will not neglect even the smallest signs that are potentially contained within the aerial images, in the attempt to recover data from indexes that are fragmentary or barely visible on the ground. This is undoubtedly less sensational than some amazing oblique photographs but equally important for an integrated and scientifically valid activity of research.

It is obvious that the data elaborated by the reading of the aerial photographs (vertical and oblique, historical and recent), in the specific case of archaeological traces, obligatorily requires a punctual check on the ground to be able to pass from the level of generic indication to that of archaeological evidence of all the effects: a presumed archaeological trace, seen on an aerial image, has necessarily to be connected to objective data, that can be checked only after

Fig. 2.8 R.A.F. aerial photograph of March 15, 1944. At the foot of Monte Cassino, with the Abbey already heavily damaged, and the area of the modern town have been bombed to devastating effect. The damage is clearly visible through the dense smoke and dust near the remains of the Roman city of *Casinum* (Ceraudo 2004)

Fig. 2.9 At the *top*, historical aerial photo of the town of Arpi (IGM 1954); in the *middle* and *below*, in comparison, vertical aerial photos (Aerofotogrammetrica Nistri 1997) and oblique (LabTAF 2005) of two sections of the old town (Ceraudo 2008)

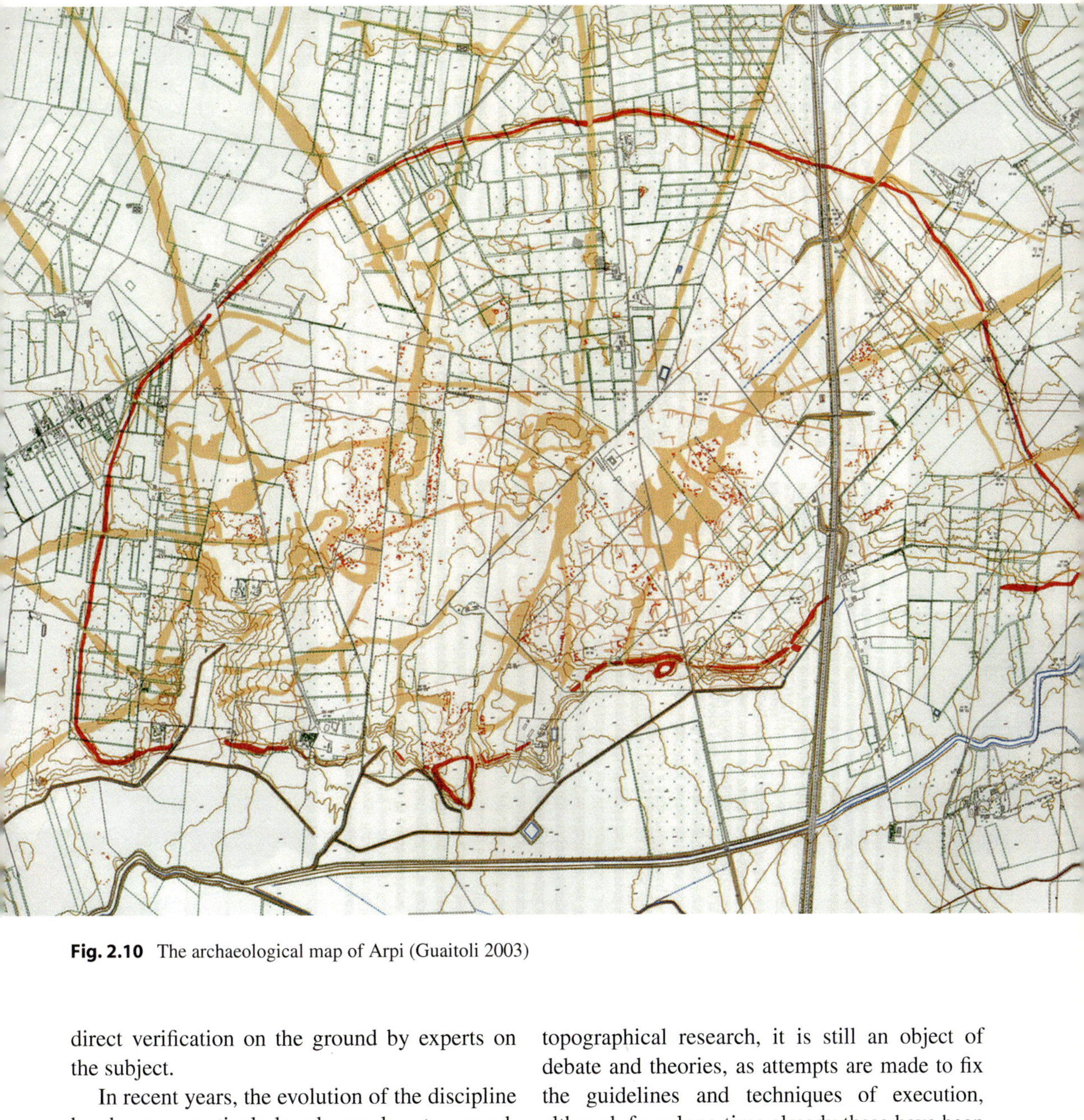

Fig. 2.10 The archaeological map of Arpi (Guaitoli 2003)

direct verification on the ground by experts on the subject.

In recent years, the evolution of the discipline has become particularly advanced, not so much as regards the basic methodology of the research, by now fixed exactly on the lines established at the end of 1800, but in terms of the availability of new instruments derived from technological progress and from close integration with other disciplines, in both the humanistic and natural sciences fields. Rediscovered in these last few years by sectors of study and research that were previously unconcerned with the problems of topographical research, it is still an object of debate and theories, as attempts are made to fix the guidelines and techniques of execution, although for a long time already these have been defined and routinely applied by employees.

A comparison is necessary, even in these different ways of working, so as to be able to direct our discussion towards the need for refinement and development, a need which is implicit in scientific research.

The limits and merits of this instrument of investigation have, in reality, been well known for a long time to all those people who regularly operate in the sector. "New" different terminologies are added to the old wording, all of which, among other things, are inherent in the concept and the methodology of the topographical investigation of the territory. To the specific subject of "ancient topography" are added landscape archaeology, field survey and total archaeology. These are unexceptionable terms in themselves, although perhaps more modern and attractive, but they are signs of the fact that there was the need to express a certain multiplicity of interventions on the territory; this multiplicity does not always works out as an enrichment or with a precise definition, but is sometimes a symptom of the introduction of elements of confusion that are unfortunately not always confined to the formal level, but at times risk infecting also the substance of the subject. From the terminology,

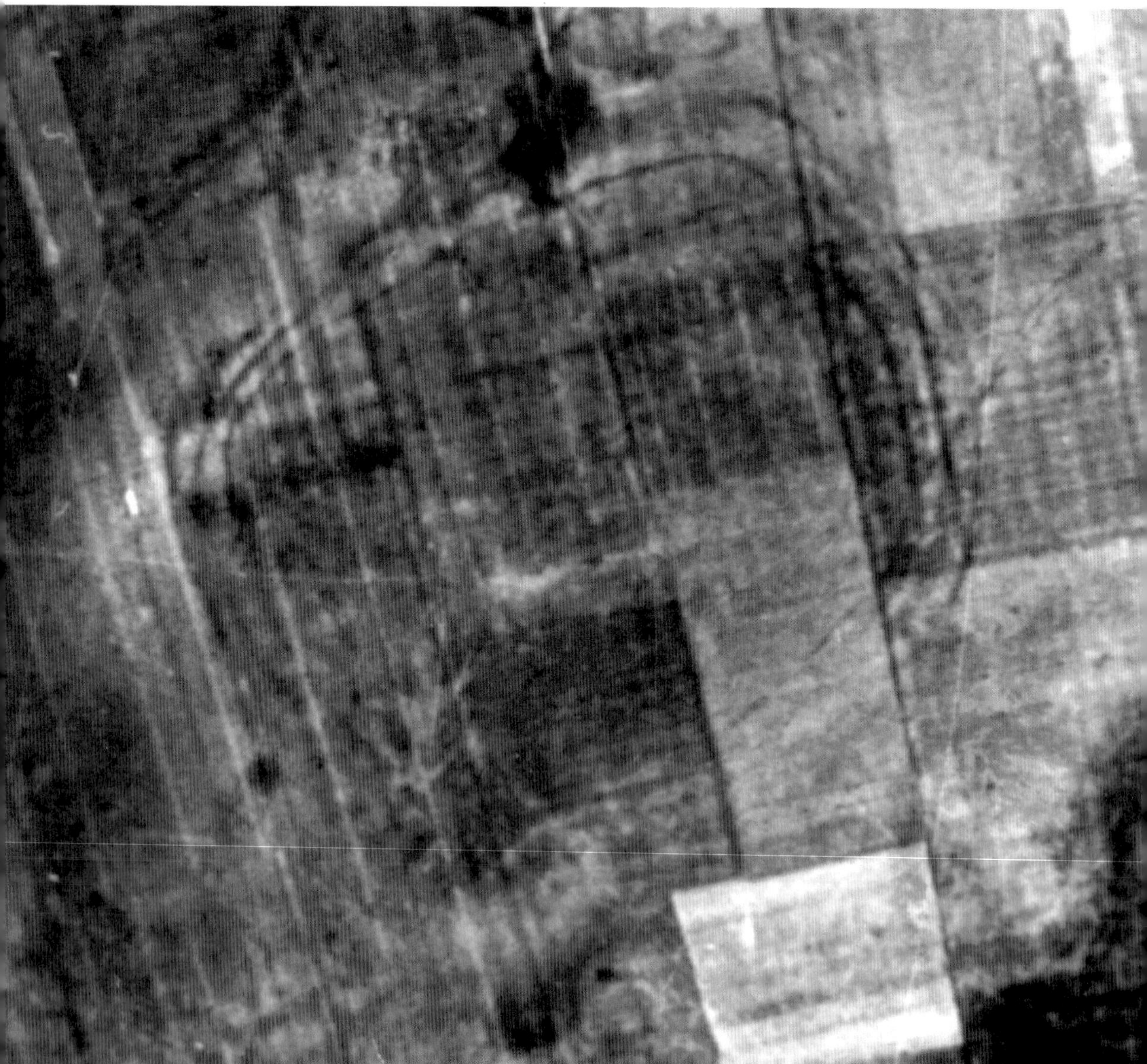

Fig. 2.11 Neolithic village near Masseria Fongo, S of Foggia. (**a**) Vertical photo, IGM May 1955; (**b**) oblique aerial photo of May 2005 (Archive LABTAF)

Fig. 2.11 (continued)

sometimes used in a provincial way, it is justifiable to deduce a certain confusion between the means and the goal or rather between the means of study and the instruments that are useful for the research and the scientific goals of the research itself, with an excess of evaluation or a contortion of the traditional instruments of investigation which we are now accustomed to using.

In the meantime, unfortunately, there has been an increase in the abandonment, looting and destruction of the territory, with frequent peaks of cancellation of a less developed morphology that itself constituted historical testimony. To arrest this folly, which is unfortunately very widespread, it would not be enough to rely on the increased availability of technologies whose effect currently remains, for the most part, confined within the limbo of good intentions. Agricultural and public activities, great infrastructural works, cementing over of the outskirts

Fig. 2.12 Veio. On the *left*, oblique aerial view of the central area of the ancient city (27/09/2010), on the *right* the same area in a vertical photo (29/09/2010) (Archive LABTAF)

and the coasts and building abuses are progressively and irreparably destroying our archaeological heritage.

The last few years have seen significant development in the use of aerial reconnaissance and aerial photography in studies of ancient topography, with archaeologists acquiring their own oblique images, which, together with new remote sensing systems and technologies, represent the greatest advance in the sector: reference can be made here to infrared (false colour and thermal) photographic images, multispectral and hyperspectral scanning sensors, radar and LiDAR (Fig. 2.13) systems and the continuous evolution of the use of satellite images (Fig. 2.14) (see Chaps. 4, 5, 6, this volume).

2.2 Aerial Photography Techniques

Aerial photography and aerial reconnaissance are tools with numerous applications in archaeology: in searching for and documenting new evidence, graphic restitution and the presentation and conservation of sites.

The use of aerial photography is thus not limited to the identification and discovery of new archaeological sites, but is a practice which over the years has acquired increasing importance in archaeology, and now plays a fundamental role in all phases of research, from interpretation to documentation, not to mention its potential in the safeguarding and monitoring of the sites themselves. Aerial photographs may be either vertical or oblique images, and their combined use makes it possible to resolve many of their respective limitations and exploit their individual characteristics to the full. The difference between vertical and oblique aerial photographs lies in the techniques by which they are acquired. Vertical photographs are taken with the axis of the camera lens perpendicular to the earth's surface, using sophisticated instrumentation mounted on aeroplanes precisely for that purpose. Initially, vertical photography had a purely military or cartographic function; today it is used above all for environmental monitoring and the planning of new communication networks and infrastructure. In the archaeological field, it has the advantage of providing a synoptic and objective view of the context in question at the moment of the shot, but

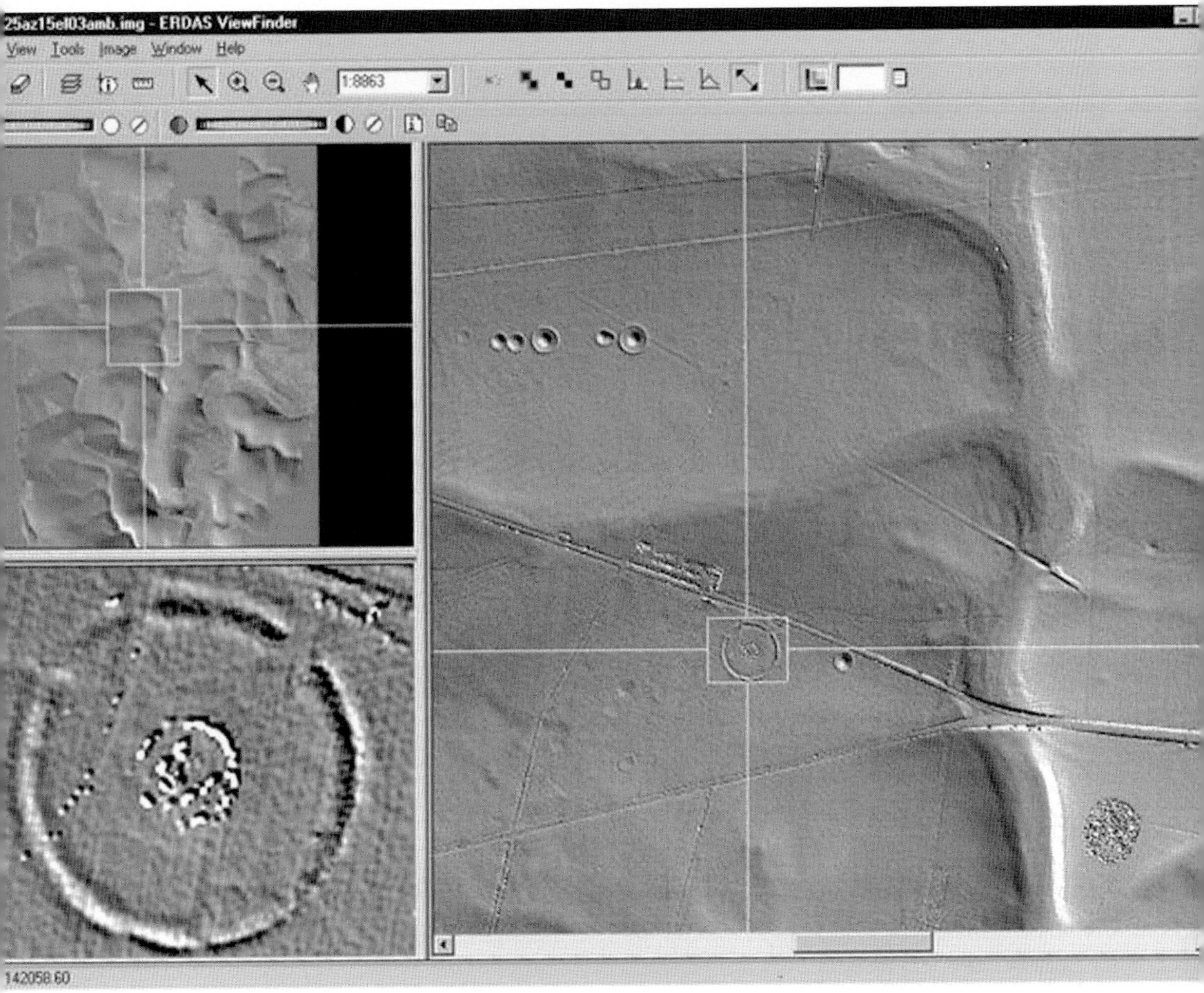

Fig. 2.13 LiDAR images of Stonehenge taken to test the potential of this new technique (Bewley 2004)

it can also illustrate its phases of development, as documented by successive images over the years. The main limits of vertical photography are its extremely high cost and the fact that in almost all cases the archaeological evidence appears by chance, since flights are only rarely undertaken especially for archaeological purposes. The aeroplane acquires the images by making a series of flights during which the photographs are taken automatically at regular intervals, so that each photograph partially overlaps the previous one and the subsequent one. The overlaps provide a three-dimensional view of the territory being photographed, thereby avoiding gaps in the documentation.

Oblique aerial photographs are taken at a sharp angle to the earth's surface and provide data that is more intuitive and easier to read. They are considered much more suitable for archaeological applications than vertical images, because they are special views selected during the flight by the archaeologist and because they can be acquired under the best conditions in terms of visibility, light and readability of the surface. Moreover, they can be produced at very reasonable cost, they do not require special photographic equipment and ordinary tourist aircraft can be used. However they have the disadvantage of not providing complete and exhaustive documentation of the

Fig. 2.14 Satellite image of *Hierapolis*, 25-3-2005 QuickBird 2 (Scardozzi 2007)

area being studied. In addition, any evidence that the archaeologist does not recognise, or any views which he or she feels are not worth recording, ends up not being photographed.

Used in combination, vertical and oblique images increase the amount and quality of the information considerably, exploiting on the one hand their ability to provide an overview and on the other their potential for identifying previously unreported archaeological sites and expanding our knowledge of elements that have only been partially described.

2.3 Principles of Archaeological Photo-Interpretation

The correct approach to the interpretation of aerial images must be comprehensive; reading an aerial photograph does not mean trying to identify just the elements that indicate past human activities, but must use "the modern" as an element of contrast that helps to bring out the residual components of the ancient landscape. In photo-interpretation the factors that determine the nature of the objects represented in the aerial photographs are shape, size, shadow, tone, texture and associated characteristics. While the first two factors are rather intuitive, the others repay further consideration. Indeed, some objects in some cases are barely comprehensible in the image while their shadow, larger than the object itself and more sharply contrasted, can be much simpler to understand (as is often the case with poles and pylons carrying electricity cables).

Concerning tone, this depends on the colour of the object, the angle of incidence of the light that strikes it and the nature of its surface: the smoother it is, the paler it appears in the photo. The texture arises from the combination of small details whose limited dimensions prevent them from being perceived individually, but which combine to form an image with identifiable characteristics: for example, the various types of crops such as vineyards, orchards and olive groves, where we do not distinguish the individual plants but rather their overall effect. Lastly, the associated characteristics are the result of the way in which the element in question is inserted in and associated with the context.

2.4 Genesis and Classification of Archaeological Traces

A very important aspect of aerial photography as an investigative tool in archaeology is archaeological traces. Archaeological traces are the result of a process by which an archaeological object makes an impression in a photographic image not by itself but by means of the effects it has on some of the elements surrounding it, covering it or hiding it. These elements include humidity, humus, vegetation and relief, to which may be added conceptual factors such as the topographical anomalies sometimes seen in the image of a landscape.

The identification of traces is one of the main objectives of aerial photography for archaeological purposes, and the choice of when to fly generally depends on this. Normally, favourable conditions in terms of light and visibility are preferred and the hours of the day when the sun is low on the horizon, so as to exploit the positive effects of the incident light and the resulting shadow.

Aerial photography sometimes highlights objects that are barely or not at all visible on the ground; their degree of visibility in the photographic image ranges from almost imperceptible to strikingly obvious. The photographic process detects the objects in question not in themselves, but indirectly via a series of effects that they have on the surrounding environment. This is why we speak of "traces". The different ways in which these objects reveal their presence depend on the quality of the elements involved in the procedure, which can be illustrated schematically in the following way:

Object→effects on adjacent elements
mediators→trace

The traces can be seen in the photographic restitution of particular nuances of colour (or greyscale in the case of black-and-white images), in distinctive aspects of the morphology of the landscape and in particular patterns of altimetric variation of the terrain, which is often minimal.

It is above all the *overview* provided by aerial photographs which enables the tonal shifts and nuances of colour to be recorded.

The appearance of the landscape depends on a whole series of factors connected with the natural aspects of the environment and the present and past human activities that have shaped it; the traces of those previous impacts are obviously less evident and more fragmentary.

The presence of hidden objects can alter the appearance of the terrain, influencing the shape of the surface, the degree of humidity and the characteristics of the plant cover.

The above considerations are valid for any type of hidden object, but our interest is obviously in objects of an archaeological nature. By identifying the factors that highlight the presence of the various categories of archaeological object, it is possible to draw up a classification of the traces. Archaeological traces may thus be subdivided into *damp-marks*, *grass-weed-crop-marks*, *soil-sites*, *shadow sites*, *topographical anomalies* and *legacy marks*.

2.4.1 Damp-Marks

Damp-marks are seen on terrain with no vegetation cover (generally ploughed fields) in the form of tonal shifts. The phenomenon arises from the fact that the terrain takes on different grades of colour depending on how wet it is. Indeed, after a rain shower, the ground tends to present a patchwork of different colours, reflecting variations in the water content and in the absorption of the soil. In soil that has been "disturbed", either by an irregular settling of the geological layers or by buried elements, after a period of heavy rain, at a certain moment during the drying out process, the soil is characterised by patches with different water content, which essentially depends on the different local thickness of the humus. For example, ancient-walled structures buried at shallow depths below the surface form a sort of upward extension of the underlying bedrock, with a consequent significant thinning of the layer of humus, which will thus hold less water than the area surrounding it and will tend to dry out more rapidly, taking on a paler colour. In contrast, overlying a negative archaeological element such as a pit or trench, there will be a thicker layer of humus, which holds more water and takes longer to dry out, with the consequent appearance of darker patches. Damp-marks are visible for a short period of time, until the terrain dries out.

Another element that affects the visibility of damp-marks is the depth below ground of the archaeological element; if it is too deep, then the effect of the rain will not be visible and the remains may also be affected by rising damp from below. It is not possible to give a precise measure, since it is necessary to take account not only of the depth of the deposit but also of the size and nature of the artefact and the type of terrain, as well as the usual meteorological and climatic variables. Sometimes, there is an "inversion of tone" of the damp-mark, meaning that counterintuitively, a buried-walled structure is signalled by a dark trace and a filled pit by a pale trace. In the former case the phenomenon is generally caused by near-permanent masses of water resulting from the presence of rubble or buried material from collapsed ancient buildings that is able to hold moisture. In the second case the inversion is due to the presence in the pit of clayey soils or very fine sand that accumulate when the negative archaeological elements are filled in very slowly by waters drained from the surrounding land.

Not just rain but all kinds of precipitation are able to trigger indicators of remains, if conditions permit: in some cases the thermal conditions of the terrain, influenced by the presence of structures near the surface, cause tiny anomalies in the melting of snow or winter frost, clearly highlighting the layout of the buried remains.

2.4.2 Grass-Weed-Crop-Marks

The mechanisms behind this category of trace are the same as those of the class described above. The main difference lies in the presence of plant coverage, which acts as a mediator for the appearance of the hidden objects. In the vast majority of cases, the vegetation involved in this process is made up of grasses, usually crops but sometimes weeds, in fields left fallow or used for grazing. In rare cases it might be shrub vegetation or even trees. Indeed, the health of the plants depends on the right quantities of water and nutrients being available; thus, where the vegetation has a greater quantity of moisture and humus, it germinates earlier and grows faster, greener and more densely. Local variations in the "fertility" of the soil are therefore chromatic indicators: dark in the case of negative archaeological elements that have been filled in, pale in the case of buried structures. The deeper the deposit of archaeological material, the larger the archaeological element in question and the plants which mediate its appearance need to be. For example, ancient walls buried in the terrain at a depth of a few decimetres normally disturb the root systems of cereals and grazing plants; structures lying at considerably greater depths generally do not directly affect the roots of grasses and cereals, which do not reach that far down. However, the presence of particularly thick walls or fortifications may be felt indirectly by herbaceous vegetation, due to a local decrease in the quantity of moisture in the soil, and directly by shrub vegetation, whose roots extend to greater depths. In the case of truly imposing structures buried very deep, early leaf senescence in deciduous trees has been reported.

The state of conservation of an artefact also conditions the photographic restitution of the trace: a structure whose walls have been razed to the level of the ground can take the form of a pale quadrangle, while if the walls are conserved to a certain height, it can create a "bath" effect, leading to accumulation of moisture and consequently a dark quadrangular trace. Another type of trace produced by vegetation is the effect generated by local concentrations of organic material, which can give rise to areas of more intense plant growth even when moisture levels are no greater than the surrounding soil, as in the case of hut floors and shaft tombs.

Some underwater archaeological structures that are not directly visible in themselves (since they are similar in colour to the sand of the sea bed) can only be seen due to the seaweed that grows on them, which makes them darker.

2.4.3 Soil-Marks

These are seen on terrain that has no vegetation cover and take the form of areas of different coloration from that of the context; the tonal shift is more easily detected if the terrain is moist and has been deeply ploughed and harrowed. They are formed due to the presence in the soil of materials that alter its surface texture, causing changes in its reflectivity and thus its photographic colour, or of materials that directly influence the colour of the terrain itself. Usually these materials have originated from the disintegration of ancient structures that were subjected to ploughing. They are visible in photographs in the form of pale patches as a result of the pulverisation of the mortar. Dark areas are due to the presence of much coarser materials that make the surface of the soil much "rougher" (and thus less reflective) or are due to high concentrations of organic material which is generally darker in colour.

2.4.4 Shadow Sites

The surface of the terrain reflects the geological bedrock below it, replicating its forms albeit in a softer and attenuated way. By the same principle, buried archaeological elements sometimes reveal themselves in altimetric patterns that are so subtle and gradual as to be invisible to direct observation. Using aerial photography, however, an expert eye can detect them via a three-dimensional reading or even using individual photographs, if they are taken with the sun low on the horizon (long shadows highlight even small changes in elevation). We are dealing here with micro-relief traces. This indicator can be used for the identification of practically any type of archaeological object, unless the terrain has been levelled mechanically. The relationship between trace and object is direct: a rise corresponds to the wall, a slight depression to the pit or trench. Even the shape, though greatly "softened", is maintained; in the case of macrostructures, such as buildings used for public spectacles, it is possible to detect a difference between the outer perimeter and the inside of the building, while in other cases we have only a generic rise, roughly corresponding to the volume of the construction.

This category of traces includes underwater structures that have the same colour as the sand of the seabed, from which they protrude only very slightly, and thus they can be detected only with reference to their shadow or their stereoscopic volume.

2.4.5 Topographical Anomalies

All the traces described fall within the category of anomaly but there are cases in which the archaeological object is perceived via the mediation of conceptual rather than physical anomalies. This category includes evidence that is foregrounded because it clashes with the general context.

2.4.6 Legacy Marks

This category includes indicators generated by archaeological elements that have remained above ground but, due to their extremely fragmentary nature, have little indicative value in themselves. Rather, their importance stems from the possibility they provide of a philological reading aimed at the reconstruction of the ancient situation. Alternatively, they may be archaeological objects that have been handed down to our times not in themselves but thanks to the survival, partial or total, of their function.

The classic example is the remains of the centuriation; when a piece of archaeological evidence of this type has been handed down to us in an almost complete state, the analysis can proceed without difficulties. In this case we are not dealing with traces in the narrow sense, since the boundaries are not physically those of ancient times but rather elements of the modern landscape that replicate them. A quite different case is when the remains of centuriation are now in such a fragmentary condition that their identification requires a broader study based on the detection of anomalous elements that seem to have some logical criterion in common. When subjected to careful analysis, discontinuous, scattered fragments of the ancient division of farmland, which have survived in the form of short stretches of walls and hedges, ditches, field boundaries and rural lanes, diluted and camouflaged in the more modern rural fabric, are found to have a common orientation and are located at regular intervals. On the terrain they can be physically verified, while the overview provided by the aerial image facilitates the task of recognising their original layout.

References

Bewley R (2004) Understanding the past. Aerial survey, remote sensing, interpretation and management. Archeologia Aerea Studi di Aerotopografia Archeologica 1:37–45

Ceraudo G (2004) Un secolo e un lustro di fotografia aerea archeologica in Italia (1899–2004). Archeologia Aerea Studi di Aerotopografia Archeologica 1:47–68

Ceraudo G (2008) Aerial archaeology in Italy. Recent research and future prospects. The contribution of LabTAF at the University of Salento. Archeologia Aerea Studi di Aerotopografia Archeologica 3:219–232

Denise F, Nordiguian L (2004) Une aventure archéologique: Antoine Poidebard, photographe et aviateur. Parenthèses, Marseille

Guaitoli M (ed) (2003) Lo sguardo di Icaro. Le collezioni dell'Aerofoteca Nazionale per la conoscenza del territorio. Campisano, Roma

Piccarreta F, Ceraudo G (2000) Manuale di Aerofotografia Archeologica. Metodologia, tecniche e applicazioni. Edipuglia, Bari

Scardozzi G (2007) L'urbanistica di Hierapolis di Frigia: ricerche topografiche, immagini satellitari e fotografie aeree. Archeologia Aerea Studi di Aerotopografia Archeologica 2:83–134

Shepherd EJ (2007) Il Rilievo topofotografico di Ostia dal pallone (1911). Archeologia Aerea Studi di Aerotopografia Archeologica 2:15–38

Undistorting the Past: New Techniques for Orthorectification of Archaeological Aerial Frame Imagery

3

Geert Verhoeven, Christopher Sevara, Wilfried Karel, Camillo Ressl, Michael Doneus, and Christian Briese

Contents

3.1 Aerial Archaeological Frame Footage: An Introduction and Overview

3.1.1 One Hundred Years of Status Quo

Since Joseph Nicéphore Nièpce (1765–1833) invented 'drawing with light' in the 1820s, photography can almost celebrate its second centenary. Archaeological aerial photography covers approximately one half of that time

G. Verhoeven (✉)
Department of Archaeology, Ghent University, Ghent, Belgium

VIAS – Vienna Institute for Archaeological Science, University of Vienna, Vienna, Austria

LBI for Archaeological Prospection and Virtual Archaeology, Vienna, Austria
e-mail: geert.verhoeven@ugent.be

C. Sevara
Initiative College for Archaeological Prospection, VIAS – Vienna Institute for Archaeological Science, University of Vienna, Vienna, Austria

W. Karel
VIAS – Vienna Institute for Archaeological Science, University of Vienna, Vienna, Austria

Department of Geodesy and Geoinformation, Vienna University of Technology, Vienna, Austria

C. Ressl
Department of Geodesy and Geoinformation, Vienna University of Technology, Vienna, Austria

M. Doneus
VIAS – Vienna Institute for Archaeological Science, University of Vienna, Vienna, Austria

LBI for Archaeological Prospection and Virtual Archaeology, Vienna, Austria

C. Briese
LBI for Archaeological Prospection and Virtual Archaeology, Vienna, Austria

Department of Geodesy and Geoinformation, Vienna University of Technology, Vienna, Austria

C. Corsi et al. (eds.), *Good Practice in Archaeological Diagnostics*, Natural Science in Archaeology,
DOI 10.1007/978-3-319-01784-6_3, © Springer International Publishing Switzerland 2013

span. The first aerial image was taken in 1858 from a tethered hot-air balloon by Gaspard-Félix Tournachon – also known as Nadar – from the village of Petit Bicêtre (Colwell 1997; Newhall 2006). It was not, however, until June 1899 that the first (European) archaeological photograph, of the forum in Rome, was taken from a balloon by Giacomo Boni (Castrianni 2008). Despite the first flight of a manned, motor-driven machine built by Orville and Wilbur Wright in 1903, archaeologically significant pictures were not captured from an aeroplane until World War I (Barber 2011). In this first phase of archaeological aerial reconnaissance, much credit must be given to O.G.S. Crawford (1886–1957). This Englishman is considered to be the inventor of scientific aerial reconnaissance, and his work in the 1920s and beyond was the basis for the future development of aerial archaeology (e.g. Crawford 1924, 1929, 1933; Crawford and Keiller 1928).

Since Crawford and other pioneers of aerial archaeology such as Antoine Poidebard (1878–1955) and Theodor Wiegand (1864–1936), it has been recognised that archaeological remains can show up on the earth's surface in a number of ways. Aside from standing structures (e.g. bridges, theatres, fortifications) which are directly visible from the ground as well as the air, most archaeological remains are partly eroded or only exist as sub-surface archaeological features, showing up on the surface under certain conditions as *visibility marks*: i.e. indirect indicators of archaeological residues due to the changed properties of the soil matrix or the local topography (Crawford 1924; Scollar et al. 1990; Wilson 2000; Bewley and Rączkowski 2002; Brophy and Cowley 2005; see Chap. 2 in this volume). Apart from the less frequent flood and wind marks, archaeologists generally differentiate between four main types of marks:

- Soil marks – due to varying chemical and physical properties affecting the soil colour on the surface
- Crop/vegetation marks – due to variable growth and vigour of the vegetation
- Shadow marks – when earthworks are thrown into relief by low slanting sunlight
- Snow/frost marks – due to differential snow accumulations and differential melting of snow or frost

To date, the common practice of active archaeological aerial photographic reconnaissance is quite straightforward and seems not to have changed over the past century (Verhoeven 2009a). In general, images are acquired from the cabin of a low-flying aircraft using a small- or medium-format hand-held photographic/still frame camera equipped with a lens that is commonly uncalibrated (Wilson 1975; Crawshaw 1995). Once airborne, the archaeologist flies over targeted areas, trying to detect possible archaeologically induced anomalies in the landscape. Once an archaeological feature is detected, it is orbited and documented from various positions (generally from an oblique point of view) on the digital camera sensor or a specific panchromatic, true colour, monochromatic infrared or (false-) colour infrared film. This type of aerial photographic reconnaissance has been the workhorse of all archaeological remote-sensing techniques since it is one of the most cost-effective methods for site discovery and the non-invasive approach yields easily interpretable imagery with abundant spatial detail (Wilson 2000; Palmer 2005).

However, no matter how efficient this reconnaissance approach can be in certain areas and periods, its main disadvantage is the fact that the whole flying strategy is *observer directed* (Palmer 2005) and generates extremely selective (i.e. biased) data that are totally dependent on an airborne observer recognising archaeological phenomena. Thus sub-surface soil disturbances that are visually imperceptible at the time of flying (e.g. Verhoeven 2009a), or those that are simply overlooked, will not make it into a photograph. To counteract this, several authors have already questioned this strategy of observer-directed survey and pointed out the advantage of a so-called unbiased, vertical approach (Palmer 1996, 2007; Doneus 1997, 2000; Mills 2005; Coleman 2007). Although the observer-directed flying method might yield vertical photographs as well, the vast majority of the photographs will be oblique in nature. This means that the optical axis of the imager intentionally deviates more than 3° from the vertical to the earth's surface (Schneider 1974). Depending on the visibility of the horizon, the image is then further classified as low oblique (i.e. horizon is not included) or high oblique (Harman et al. 1966).

3.1.2 The Vertical Debate

In a strictly vertical sortie, every parameter is set to make sure that all photographs are nadir/vertical images. In effect, this means that photographs will be acquired with expensive, accurately calibrated, built-in (versus hand-held), gyro-stabilised and low distortion mapping frame cameras (often referred to as metric or cartographic cameras – Slater et al. 1983). These cameras are solidly housed and operated in bigger and higher-flying aeroplanes. Images are acquired in parallel strips at regular intervals, generally with a large frame overlap: in one flight strip, each photograph has a generally accepted degree of overlap of circa 60 % ± 5 % (figures to 90 % can be found as well, see Schneider 1974) with the following and preceding image (longitudinal overlap). Adjacent strips have on average an overlap of 25–40 % (lateral overlap) (Read and Graham 2002). The camera is pointing directly down to the earth to acquire (near) nadir photographs. Because a perfect vertical is almost never achieved, an image with an angle of less than or equal to 3° is called vertical (Estes et al. 1983).

Archaeological resources often appear on verticals through what has been termed the *serendipity effect* (Brugioni 1989): a circumstance in which photosets yield unanticipated or 'bonus' material which was not the primary objective during original data collection. Unlike oblique aerial photography for archaeological purposes, those vertical surveys are generally executed to acquire basic material for (orthophoto) map generation (Falkner and Morgan 2002). Although this approach generates geographically unbiased photographs of large areas in a very fast manner, its adversaries remark that several issues militate against the effective use of those vertical photographs for archaeological purposes. Of those, the fact that imagery is not captured at the perfect oblique angle to maximise the visibility of archaeological information (Crawshaw 1997) is often seen as the strongest argument to not fly (or even use) verticals. On the other hand, vertical footage offers an advantage in mapping, as the induced geometrical distortions are much less than those embedded in oblique footage (Imhof and Doolittle 1966; see part 2). Since the data are by default captured in stereo pairs, they are also perfectly suited to create analogue or digital 3D stereo models. Additionally, the high spatial resolution and comparatively broad coverage of standard vertical mapping images make them valuable for a holistic view of the landscape as well as for the primary discovery of individual archaeological features.

As a result, many aerial archaeologists have extracted much valuable information from verticals (e.g. Moscatelli 1987; Kennedy 1996; Doneus 1997), and a few studies have proven the undeniable and often complementary value of verticals after a thorough comparison with obliques from the same area (e.g. Zantopp 1995; Doneus 2000; Palmer 2007). In reality, even those archaeologists that favour obliques over a blanket vertical coverage will incorporate verticals into their research, simply because many valuable historic aerial photographs were acquired with a (near-)vertical approach (Stichelbaut et al. 2009; Hanson and Oltean 2013). Since these photograph series are able to illustrate change through time, they provide valuable data regarding landscape change and indirect land use impact on archaeological resources (Cowley and Stichelbaut 2012).

3.1.3 The Rise of the Unmanned Machines

Finally, it needs to be mentioned that both oblique and vertical frame images can also be acquired from low-altitude unmanned platforms. Since the beginning of aerial photography, researchers have used all kinds of devices (from pigeons, kites, poles and balloons to rockets) to take still cameras aloft and remotely gather aerial imagery (see Verhoeven 2009b for an archaeological overview). To date, many of these unmanned devices are still used for what has been referred to as low-altitude aerial photography or LAAP (Schlitz 2004). In addition to these more traditional camera platforms, radio-controlled (multi-) copter platforms have recently added a new aspect to LAAP (Fig. 3.1). The overwhelming amount of

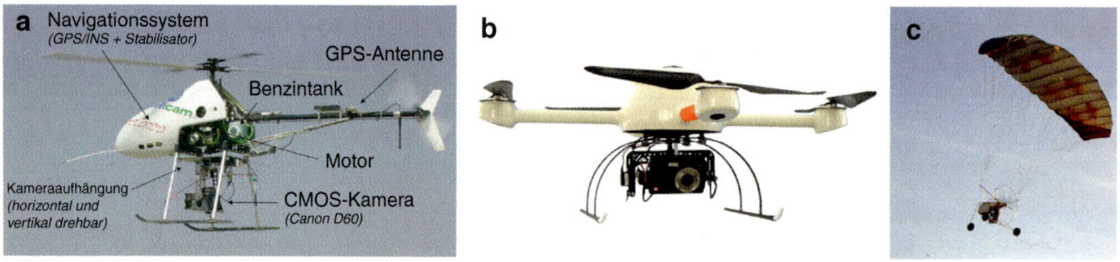

Fig. 3.1 (**a**) Example of a remotely controlled helicopter to acquire digital aerial imagery (Reproduced from Eisenbeiss et al. 2005) (**b**) The Microdrone MD4-200 quadcopter (Microdrones GmbH 2008) (**c**) Remotely controlled paraglider (Krijnen 2008)

brands and types (heli-, dual-, tri-, quad-, hexa- and octocopters), together with the wide variety of navigation options (e.g. altitude and position hold, waypoint flight – Eisenbeiss 2009; Eisenbeiss and Sauerbier 2011) and camera mounts, indicate that these platforms are here to stay for some time. Given the multitude of still camera types and the image quality they are currently capable of, endless combinations of low- and high-cost LAAP solutions are available. In addition, LAAP allows for the exploitation of new imaging techniques, as it is often only a matter of lifting the appropriate device. In this way several archaeological studies have utilised close-range near-infrared photography (e.g. Whittlesey 1973; Aber et al. 2001; Verhoeven et al. 2009a; Wells and Wells 2012) or even less straightforward near-ultraviolet imaging (e.g. Verhoeven 2008a; Verhoeven and Schmitt 2010; Wells and Wells 2012).

3.1.4 The Mapping Paradigm

Despite this large variety of still frame images and means to acquire them (actively or passively), their archaeological information cannot (or will not) be exploited efficiently as long as the images are not thoroughly interpreted (i.e. interpretatively mapped – cf. Doneus et al. 2001) and integrated with other data sources. The lack of this interpretative mapping is often encountered and may have multiple causes. Availability of resources may be one cause, but one of the most important ones is likely the time-consuming (and often difficult) georeferencing process (Verhoeven et al. 2012a). As a result, millions of aerial photographs are just stored in archives, waiting for their archaeological potential to be explored. Obviously, aerial archaeology is in need of fast, straightforward and accurate georeferencing approaches that allow orthophoto production of a wide variety of images: old or new, acquired in a vertical or oblique manner from low or high altitudes.

This chapter elaborates on such an approach and presents a method to automate the important but recurring task of orthophoto generation. The approach proposed here attempts to overcome the conventional georeferencing problems related to archaeological aerial frame images, which means that in this chapter imagery resulting from panoramic and line cameras is not included. To this end, the methodology exploits some of the technological improvements in hardware configurations as well as state-of-the-art algorithms mainly developed in the fields of computer vision and photogrammetry: two disciplines that research the recovery of 3D content from 2D imagery using – to a certain extent – their own specific approaches (Hartley and Mundy 1993). Before outlining the method (Sect. 3.3), the concept of georeferencing and all the sources of geometrical image deformations that have to be taken into account will be outlined in Sect. 3.2. Section 3.4 will illustrate these concepts with several case studies. In addition to this illustrative purpose, these case studies will also provide some more in-depth knowledge about specific aspects of particular aerial image types. A conclusion, presenting some future aims and remarks, will then finalise this chapter.

3.2 Aerial Frames Offer Deformed Views

3.2.1 (Digital) Aerial Images

Aerial imaging is facilitated by the use of an airborne remote-sensing instrument that gathers the earth's spatially, temporally, radiometrically and spectrally varying upwelling electromagnetic radiation and uses this to generate (digital) images (see Schott 2007 for a good treatise of this subject). In past decades, this detection of radiation was usually accomplished by a photographic emulsion sensitised into one or more spectral regions of the visible and near-infrared electromagnetic spectrum. Although geometrical processing of these film frames was performed for decades in an analogue – and later analytical – way, they are normally scanned now to enable a digital processing of the aerial image.

To date, most airborne imaging devices provide digital products directly since the detection is usually accomplished by the conversion of incoming electromagnetic radiation (expressed as *at-sensor radiance*) into an electrical output signal which is subsequently digitised into digital numbers (*DNs*). Most digital image capture systems comprise optical elements such as lenses, mirrors, prisms, gratings and filters that gather the radiation and focus it onto an imaging sensor. This imaging sensor itself consists of several (often millions) of individual optical detectors (also called *photodetectors* – Norton 2010) that can detect the incoming radiation and generate a signal in response to it (Verhoeven 2012a). In this chapter, all imaging sensors are considered to be frame sensors, since they consist of an array of individual photodetectors arranged in a rectangular frame. Moreover, they are assumed to work in the optical radiation spectrum, commonly accepted to reach from the ultraviolet to the infrared (Ohno 2006; Palmer and Grant 2010). Additionally, for the remainder of this chapter, image and photograph are assumed to mean digital image.

Whether they are generated by scanning the analogue film frame or directly produced by the digital imaging sensor, the fundamental building blocks of any digital image are called pixels or pels, coined terms for picture elements (see Billingsley (1965) and Schreiber (1967), respectively, for the first use of these terms). In the case of a digital imaging sensor, each photodetector commonly produces one pixel. An array of pixels is called a digital image, which can be mathematically represented as an $M \times N$ matrix of numbers, M and N indicating the image dimensions in pixels. Pixels are thus determined by a pair of pixel coordinates (r, c indicating row and column) and a certain value or grouping of values that contains information about its measured physical quantity (Smith 1997). Just as a pixel of a common digital colour photograph contains three samples or DNs at the same location to represent the amount of radiation captured in three individual spectral bands, a greyscale image consists of one DN per pixel. Images can thus be represented by O matrices of $M \times N$ elements, in which O equals the amount of spectral bands that are sampled (Bernstein 1983). Every image is also characterised by a certain bit depth, which determines the resolution by which the amplitudes of the continuous analogue radiation signal can be mapped onto a discrete set of digital values. Consider an 8 megapixel digital image, 4,000 pixels in width and 2,000 pixels in height. If the image is an 8-bit greyscale image, every pixel has an integer DN between 0 and 255. 16-bit integer pixels could contain values between 0 and 65 535. Digital images are thus said to be sampled (spatially, spectrally and temporally) and quantised (radiometrically, defined by the number of bits) representations of a scene, defined by a multidimensional matrix of numbers.

However, the analogue real-world signal (in the form of electromagnetic radiation arriving at the imaging sensor) is degraded in various ways. As a result, the final digital image is never a faithful reproduction of the real-world scene. Aside from the spectral and radiometric transformations that occur, the geometric three-dimensional (3D) properties are mapped to a two-dimensional (2D) plane (Fig. 3.2). This mapping result (i.e. the final image) is influenced by a wide variety of factors such as earth curvature, film and paper shrinkage, nonplanar image film plane, atmospheric refraction effects, optical distortions, tilt and relief displacements (Imhof and Doolittle 1966). Not only

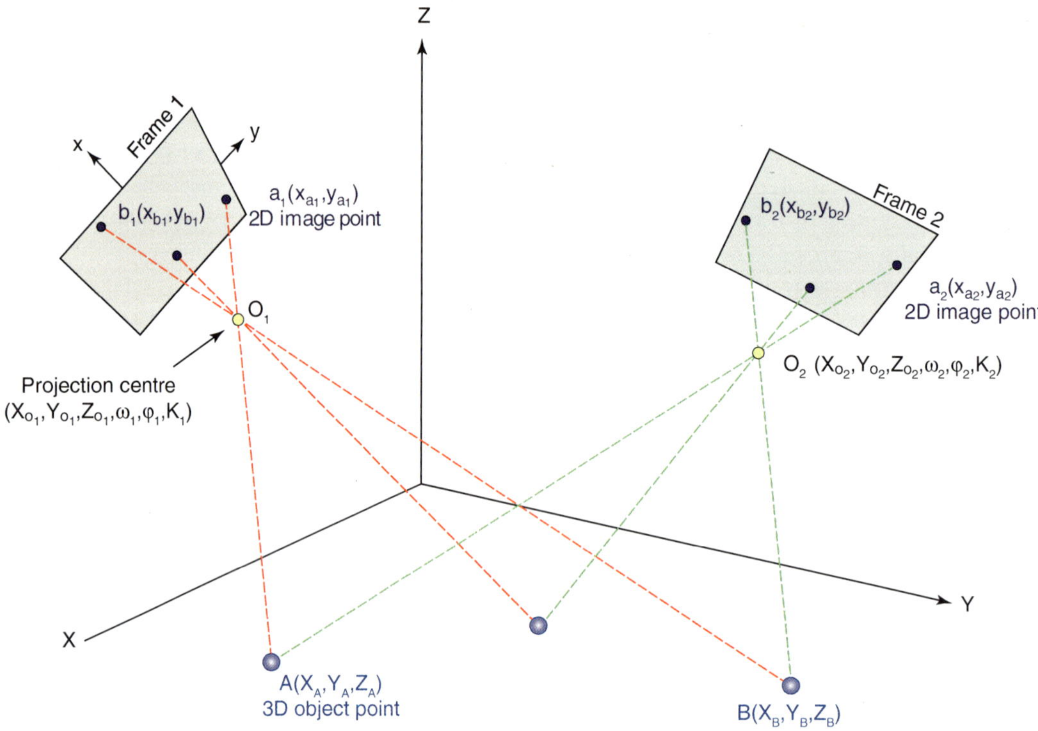

Fig. 3.2 Mapping of 3D object points onto 2D points in two aerial frame images

does every individual aerial image suffer from these geometrical deformations, but they also vary from frame to frame due to variations in the flying height and camera tilt. Compensating for them through some kind of geometric correction is essential for accurate mapping and information extraction. Since the geometric errors induced by the optics, the topographical relief and the tilt of the camera axis contribute most to image deformations; they will be shortly reviewed below.

3.2.2 Optical Distortions

In photogrammetry and computer vision, the geometry of central perspective projection is used to model the formation of an image mathematically (Mundy and Zisserman 1992; Buchanan 1993). In the field of photogrammetry, this is expressed by the *collinearity equation* which states that the object point, the camera's projection centre and the image point are located on a straight line and the image is formed on an exact plane (Fig. 3.2). Lens distortions (radial and decentring), atmospheric effects (mainly refraction) and a nonplanar image sensor are factors which prevent this. Since digital image sensors are by default treated as perfectly planar surfaces (Wolf and Dewitt 2000) and refraction is a very specific topic that is only of major importance when imaging from rather high altitudes and off-nadir angles (Hallert 1960; Gyer 1996), only lens distortions will be considered here.

In the case of an ideal camera, which would be a perfect central projection system in which projection implies a transformation of a higher-dimensional 3D object space into a lower-dimensional 2D image space (Mikhail et al. 2001), the lens imaging system would be geometrically distortionless (Billingsley et al. 1983). The mathematical parameters describing this ideal situation are the principal distance and the principal point (forming the so-called *interior/inner orientation*; see below). However, since optical distortions are always present in real cameras, the image points are imaged slightly off of the location they should be at according to the central projection. To metrically work with airborne images, every image point must be reconstructed to its location according to this ideal projective camera (Gruner et al. 1966). Therefore, the deviations from the perfect situation are modelled by suitable distortion parameters, which complete the interior orientation. All the parameters of the interior orientation (also called *camera intrinsics*) are determined by a *geometric camera calibration* procedure (Sewell et al. 1966). After this geometric camera calibration, all parameters that allow for the building of a model that can reconstruct all image points at their ideal position are obtained, thereby fulfilling the basic assumption used in the collinearity condition. More specifically, the main elements of interior orientation which camera calibration should determine are the following:

- *Principal distance* (*PD*): the distance measured along the optical axis from the perspective centre of the lens (more exactly the rear nodal point of the optical system) to the image plane (more exactly the principal point of the image) (Mikhail et al. 2001). When the camera is focused at infinity, this value equals the focal length f of the lens (Wolf and Dewitt 2000). For close-range focusing this is no longer the case and the principal distance will increase. This means that any change in focus or zoom produces a new calibration state. In aerial mapping cameras applied for vertical surveys, the calibrated focal length f_c is often given, which equals the principal distance that produces an overall mean distribution of lens distortion (Slater et al. 1983).
- *The location of the principal point* (x_p, y_p): this is the second essential quantity to adequately define the internal camera geometry. It can be defined as the intersection of the optical axis of the lens system with the focal plane (Mikhail et al. 2001). This means that the location of the principal point can change with different zoom settings, but it will always be close to the image centre. In an ideal camera the principal point location would coincide with the origin of the image coordinate system.
- *Radial lens distortion parameters* (k_1, k_2, k_3, k_4): in optics, distortion is a particular lens aberration, but one that does not reduce the resolution of an image (Gruner et al. 1966; Slater et al. 1983). Radial lens distortion is the central symmetrical component of lens distortion and occurs along radial lines from the principal point. Although the amount may be very small in aerial mapping cameras, this type of distortion is unavoidable (Brown 1956). In consumer lenses, radial distortions are usually quite significant. Generally, one to four k parameters are provided to describe this type of distortion. Radial distortion can have both positive (outward, away from the principal point) and negative (inward) values. Negative radial distortion is denoted as pincushion distortion (since an imaged square will appear to have its sides bow inward), while positive distortion is termed barrel distortion (because straight lines bow outward) (Gruner et al. 1966). Either positive or negative radial distortion may change with image height (Fig. 3.3), and its amount is also affected by the magnification at which the lens is used. It can also occur that one lens system suffers from both negative and positive distortion (Kraus 2007). Figure 3.3 depicts a typical distortion curve. On the left, the distortion scale is indicated in micrometres. In the graph, the distortion is plotted as a function of the radial distance r from the principal point.
- *Decentring lens distortion parameters* (p_1, p_2): this distortion can be broken down into asymmetric radial distortion and tangential lens distortion. Both distortions are caused by imperfections in the manufacture and alignment of individual lens elements during the construction of the lens (Brown 1966). Their magnitude is typically much smaller than that of radial lens distortion (Fig. 3.3) and conventionally described by two parameters p_1 and p_2 (Burnside 1985). Although it is generally not significant in aerial mapping lenses, decentring distortion is common in commercial lenses with variable focus or zoom.

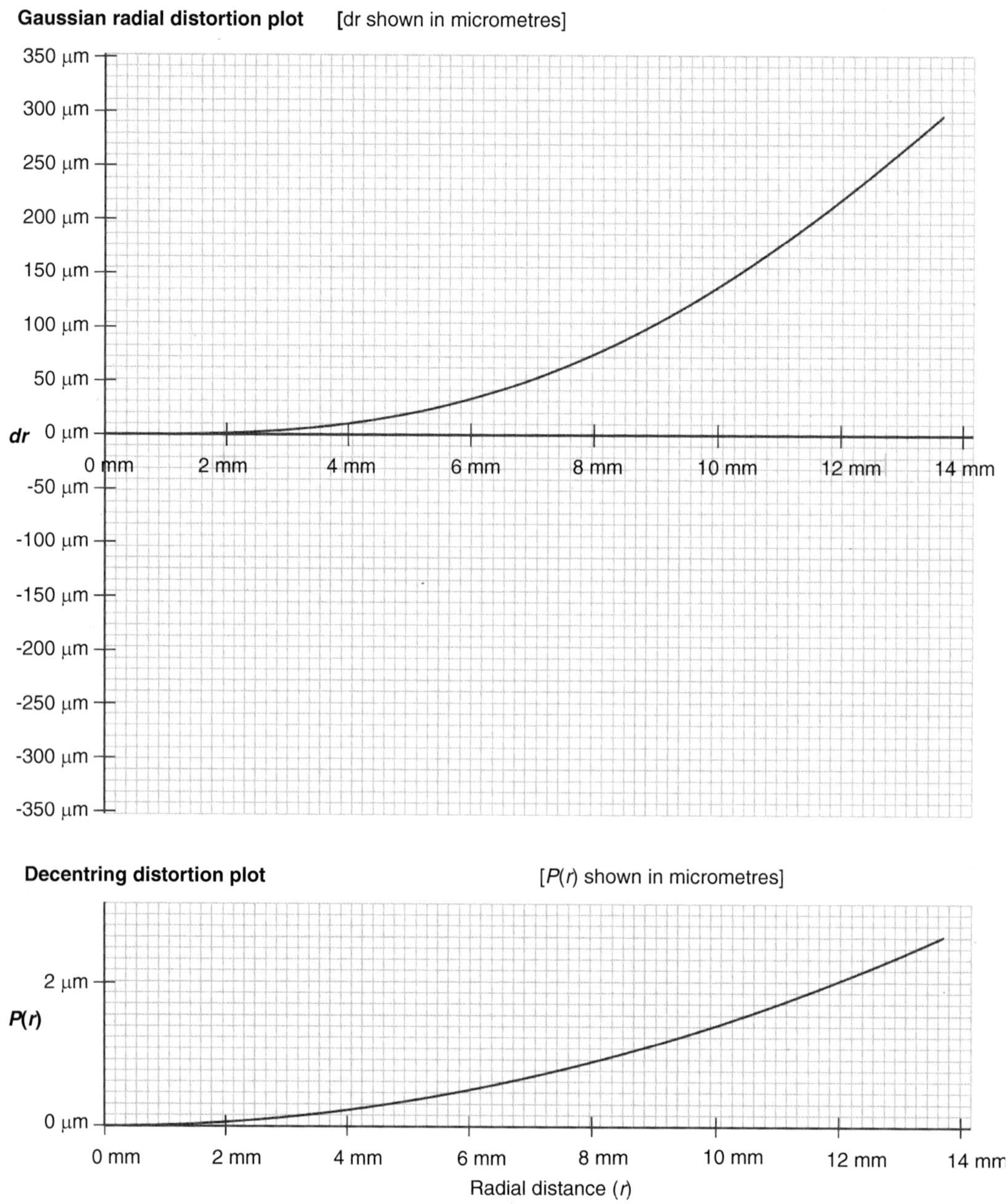

Fig. 3.3 Radial and decentring distortion plots of the AF Nikkor 24 mm f/2.8D (infinity focus). The radial distortion *dr* (expressed in micrometres) and decentring distortion *P*(*r*) are given as a function of radial distance *r* (mm)

In addition to the abovementioned parameters, several other camera characteristics can be calibrated: affinity in the image plane (consisting of aspect ratio (or squeeze) and skew (or shear)), unflatness of the film plane and the coordinates of the fiducial marks. The latter are used in analogue systems and provide a coordinate reference for the principal point and all image points, while also allowing for the correction of film distortion (Kraus 2007). Calibrating a digital frame camera is in many ways more straightforward than calibrating film cameras, since the individual sensor photodetectors are essentially fixed in position, which practically eliminates film distortion considerations (Wolf and Dewitt 2000). Fiducials are therefore not needed in digital cameras (Graham and Koh 2002). Moreover, zero skew (i.e. perpendicular axis) and a unit aspect ratio (i.e. photodetector width to height equals 1) can be assumed for digital frame cameras as well (Remondino and Fraser 2006; Xu et al. 2000; Szeliski 2011).

From the previous paragraphs, it should now be obvious that the nonmetric cameras conventionally used in archaeological oblique aerial reconnaissance are characterised by an adjustable principal distance, varying principal point and high-distortion lenses, while lacking film flattening and fiducial marks (in the case of analogue devices). Finally, it can be mentioned that there exists a wide variety of digital camera (auto-) calibration methods (see Remondino and Fraser (2006) for an overview). Although exceptions exist, the calibration methods applied in photogrammetry are tailored towards high accuracy and try to recover at least ten interior orientation parameters. Current computer vision methods (see Sect. 3.3) generally use camera models described by only four to five interior orientation parameters.

3.2.3 Tilt Displacement

A camera is placed at a certain location in space (in the air or on the ground) and is pointed in a certain direction. The location defines the projection centre O with three coordinates (X_O, Y_O, Z_O), and the direction is defined by three rotation angles roll, pitch and yaw (ω, φ, κ). Together, these six parameters establish the so-called *exterior/outer orientation* (Fig. 3.2) (Kraus 2007). Other terms for that are *camera extrinsics* or simply *pose*. Together with the interior orientation the position of the image is unequivocally defined. During a vertical photography flight, φ and ω are near to zero. When they equal zero, the result is a perfect nadir/vertical photo that does not need any correction for tilt displacement. The more tilted the photographic axis with respect to the ground surface, the more corrections need to be dialled in (Tewinkel et al. 1966).

These effects may be illustrated most clearly by considering the appearance of a regular grid and a circle on a completely flat terrain in both a vertical and a tilted photograph (Fig. 3.4, lens distortions are excluded for the sake of illustration). A vertical optical axis images the circle as a circle, while the net of squares remains unaltered as well. The same features photographed with a non-zero angle of tilt result in a distorted square net as well as an ellipse-like feature. The difficulty inherent to tilt displacements is the fact that it is often hard to detect while it yields constantly varying scale changes across the image (Dickinson 1969). When dealing with vertical photographs, there is just one nominal scale S that can be calculated by $S = PD/H$ (i.e. the ratio of the principal distance to the flying height H above the terrain) (Tewinkel et al. 1966). In this case, the scale is completely independent of the measurement direction. For tilted images, the scale will vary with direction (Estes et al. 1983). In the background of a tilted photograph, the scale is smaller than the scale in the foreground. The projective transformation of a tilted aerial image to a horizontal plane to remove these tilt displacements (and thus scale differences) is called *(planar) rectification* (Spurr 1960; Altenhofen and Hedden 1966; Dickinson 1969).

For convenience, the tilt in Fig. 3.4 is considered to be acting only along the direction of flight (φ). In practice, tilt will act in random directions due to a combination of non-zero φ and ω angles and rectification will be needed to correct for these displacements. That is why rectification is also said to transform an oblique aerial photograph to an equivalent vertical image (Wolf and Dewitt 2000). However, the rectified image will only be completely identical to the vertical image geometry in absence of lens distortions and perfect flatness of the imaged scene, since any terrain undulation will cause so-called relief (or topographic/elevation) displacements and those even affect perfect nadir images.

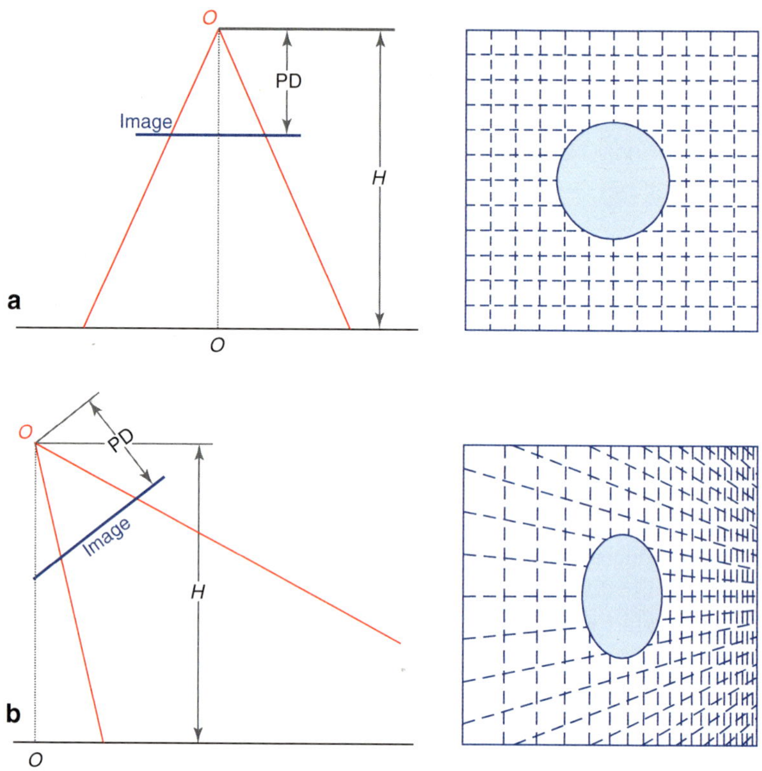

Fig. 3.4 (**a**) Vertical image, (**b**) Oblique image with resulting tilt displacement (*O* denotes the projection centre, *o* the nadir or plumb point, *PD* the principal distance and *H* the flying height above the terrain. The camera's field of view is indicated by the *red lines*.)

3.2.4 Relief Displacement

Image displacements are not only caused by tilt. Any (even tilt-free) aerial photograph will contain displacements due to topographic relief and other height differences (Tewinkel et al. 1966). Thus any feature lying below or above the horizontal reference surface will be misplaced in a planar rectification (Estes et al. 1983) due to the central perspective of the air photo (Hallert 1960).

In Fig. 3.5, the acquisition of a perfect vertical photograph is depicted. *KK′* indicates the average terrain height but can also be seen as any reference horizontal plane (called a datum surface). On the right, a green tower is shown. If the left top of this tower was to be depicted in a map, the orthogonal projection used to create maps would make it fall in point *z*, the same point which indicates the foot of the tower. In the aerial image, one also notices point *z*. Nevertheless, due to the central projection, the top is depicted in *z′* instead of point *z*. Consequently, the top of the tower has undergone a displacement of magnitude *p″*, resulting in a tower whose side is visible in the aerial image.

Although it may not be as visually obvious as in the case of buildings, imaged relief also suffers from this (Falkner and Morgan 2002). Consider the hill in the middle of Fig. 3.5. The top *y′* should normally be projected in point *y*, like on a map. However, in this case the projection also causes a displacement *p′* and instead of being depicted in *y*, the top is projected onto *y′* in the image. Following the same principle, the valley on the left also suffers from relief displacement (of magnitude *p*). In this case, it is not a displacement away from the centre, but towards it. Without regard to direction, this distance of such displacement is called *parallax*. In this respect, parallax gives a numeric value for the relief or topographic displacement.

Although this phenomenon complicates the mapping and interpretation of aerial imagery, it also enables humans to perceive three dimensions and calculate the height of objects from images (Spurr 1960). As the location of the nadir point does not suffer from this displacement

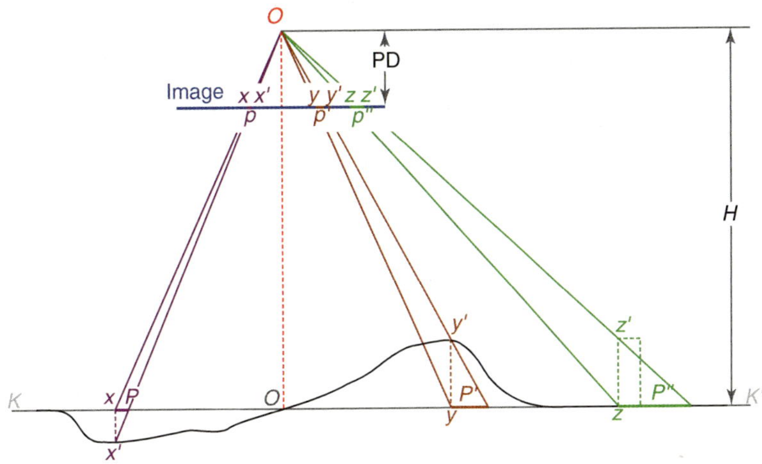

Fig. 3.5 The phenomenon of relief displacement and how it influences the geometry of a vertical image (*symbols* are explained in the main text)

(because its projection is a perfect orthogonal projection), relief displacement is always radial from the nadir or plumb point *o*. This is determined by the intersection of a vertical, constructed from the optical centre *O* towards the ground, and the image plane; this vertical axis is equal to the optical axis of the whole system in the case of a perfect vertical photograph – such as Fig. 3.5 (Tewinkel et al. 1966).

Geometric correction aims to compensate for most of these deformations. The result of such a correction must be an image with a geometric integrity like a map, i.e. an orthogonal projection to the horizontal reference plane. Just as rectification denotes the process of removing tilt from a photograph, relief displacements and other geometrical deformations (such as optical distortions) can be corrected through the process of *orthorectification* or *differential rectification* (Hassett et al. 1966; Turpin et al. 1966; Wolf and Dewitt 2000).

3.2.5 Georeferencing and Geometric Correction

Aerial photography provides a basis for gathering spatial data. Before archaeological information can be extracted from these sources in a way that is useful for mapping and further analysis, the aerial images must be *georeferenced* in an absolute manner. This process, which is also known as *ground registration*, assigns spatial information to any kind of spatial data (raster data such as imagery as well as vector data) to explicitly define their location and rotation in respect to a specific Earth-related coordinate frame.

Often, the geometry of these data is already corrected for any possible deformation. However, the process of georeferencing is often applied to geometrically distorted data as well. Although it is *sensu stricto* not covered by its definition, georeferencing can thus also involve the necessary steps to remove the optical distortions as well as tilt and relief displacements of the aerial image in order to place each image pixel on its true location on the Earth's surface. To do this, a wide variety of approaches and software solutions exist. In many cases, archaeologists fit tilted images to a flat surface by means of a projective transformation, a process introduced in the previous sections and denoted (planar) rectification (Hallert 1960; Altenhofen and Hedden 1966; Wolf and Dewitt 2000). Although these rectified images no longer suffer from tilt displacements, they still contain scale variations and displacements due to topographic relief (hills, buildings etc.). Consequently, projective transformations can only be considered 'archaeologically sufficient' when dealing with completely flat areas. If the aerial view suffers from relief displacements, georeferencing often employs polynomial corrections, spline algorithms or piecewise affine warpings embedded in archaeologically dedicated tools such as AirPhoto SE (Scollar 2002) and AERIAL (Haigh 2005). Although these approaches are very popular and might deliver fairly good metrical information when the terrain variations are quite moderate, the methods are often suboptimal because they do not (or only partly) eliminate all the image displacements, the distortion of the optics and – to a lesser extent – the atmospheric refraction. Consequently, this image georeferencing is well suited for rather small-scale mapping but inadequate for a detailed multi-temporal and multi-method analysis.

When one needs to mosaic several multi-temporal aerial observations into an extensive overall view of an archaeological region – hence serving as a basic information layer for further prospection and excavation, protection measures and heritage management – the aforementioned issues need to be dealt with. Therefore, planimetrically correct true orthophotographs are of the utmost value. However, these can only be achieved when more advanced ortho-correction approaches embedded in programs such as Leica Photogrammetry Suite or Trimble INPHO photogrammetric system are utilised. Although these more expensive packages offer rigorous orthorectification algorithms to produce superior geometric quality, they are limited by the fact that photogrammetric skills, interior orientation parameters and an accurate, high-resolution digital surface model (DSM) are essential, three conditions that are generally not met in aerial archaeology.

Irrespective of the method applied, the georeferencing of (individual) images is commonly determined with ground control points (GCPs), whose manual measurement and identification is a time-consuming operation that requires experience while being bound to certain prerequisites. As a result of all these issues, many archaeologically valuable aerial images never get properly georeferenced and stay hidden on local hard drives or in image archives.

3.3 A New Workflow

Since a variety of factors contribute to image deformation, imagery needs to be geometrically corrected in order to correspond as closely as possible to a map. At the same time, the workflow should be as straightforward and generally applicable as possible. Currently, cost-effective means are available for orthorectification of a wide variety of (archaeological) aerial frame imagery. These became possible due to the ever increasing technological improvements in computer hardware and the serious advances made the past 15 years in the scientific field of computer vision, which is often defined as the science that develops mathematical techniques to recover information from images. This image data can take many forms, such as multidimensional imagery from medical scanners, stereo photographs, video sequences or views from multiple still cameras. Initially, many computer vision applications were focused on robotic vision and inspection. As a result, the methods were characterised by few constraints and focused on a high degree of automation rather than the accuracy and reliability characteristic of photogrammetry (Remondino et al. 2012). However, the last decade has witnessed a shift of focus to more accurate 3D visualisations and virtual reality, along with many new insights in the geometry of multiple images (see Faugeras et al. 2001 or Hartley and Zisserman 2003 for a good overview).

Using techniques such as triangulation, an image point occurring in at least two views can be reconstructed in 3D (Fig. 3.2). However, this requires the knowledge of the interior and exterior orientations of the images. In computer vision, these orientation parameters are usually combined in the so-called *projection matrices* of the images (Robertson and Cipolla 2009), which can be determined by an approach called *structure from motion* (*SfM*; Ullman 1979). During this approach the relative projection geometry of the images is computed along with a set of 3D points that represent the scene's structure. SfM only requires corresponding image features occurring in a series of overlapping photographs captured by a camera moving around the scene (Fisher et al. 2005; Quan 2010; Szeliski 2011). Sometimes, this approach is also referred to as *structure and motion* (*SaM*), since both the structure of the scene and the motion of the camera (i.e. the different camera positions during image acquisition) are recovered.

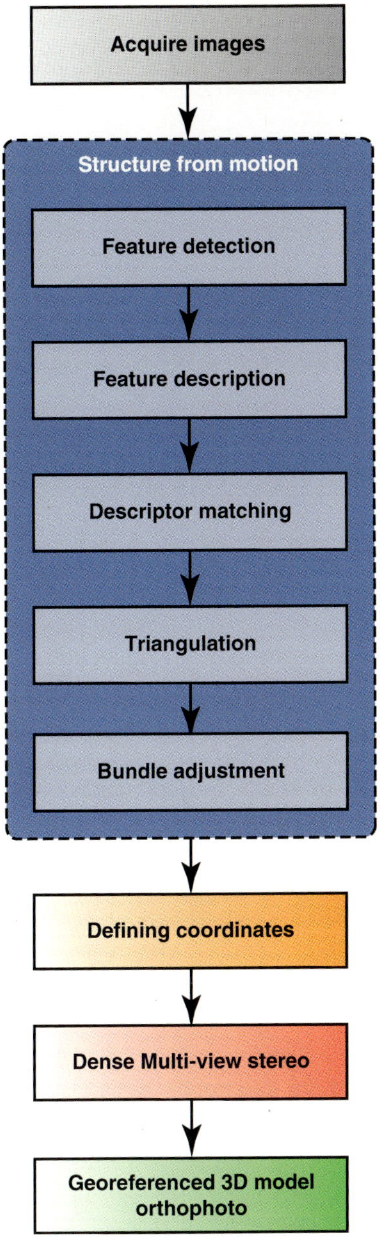

Fig. 3.6 The individual steps of the SfM + MVS processing pipeline (terminology is explained in the text)

In order to achieve this, SfM relies on algorithms that detect and describe local features for each image and then match those 2D points throughout the multiple images. Using this set of correspondences as input, SfM computes the locations of those interest points in a local coordinate frame (also called model space) and produces a sparse 3D point cloud that represents the geometry/structure of the scene. As mentioned previously, the camera pose and internal camera parameters are also retrieved (Hartley and Zisserman 2003; Szeliski 2011). Below, the SfM approach and the individual steps (Fig. 3.6) essential for its execution are outlined in greater detail. Afterward, some details are given about the subsequent process, *multi-view stereo* (*MVS*), as this last stage uses the SfM output to generate a dense 3D model needed for accurate image orthorectification.

3.3.1 SfM + MVS Pipeline

3.3.1.1 Image Acquisition

For an SfM + MVS approach, it does not matter if the images are acquired with a metric camera or not, or whether they are shot in a vertical or oblique pose. Attention should, however, be paid to the angular separation of images in order to ensure that it is not too large. This will maximise the likelihood that a stable image network can be achieved. Although several feature point extraction algorithms (see the next part) with particular strengths and weaknesses have since been developed, Moreels-Perona found out that no detector/descriptor combination performs well with viewpoint changes of more than 25–30° (Moreels and Perona 2007). Therefore, a sufficient image overlap is advised (around 60–80 % for vertical images), and it is preferable for every image to be captured from a unique location. Panning from the same location should thus be avoided (Tingdahl et al. 2012). Moreover, the objects being photographed need to possess sufficient unique texture. In general, all these assumptions can be met in aerial archaeological imaging.

Once the images are acquired, the second stage of the pipeline can be executed. This stage is denoted the SfM algorithm and consists of several individual processing steps (some authors consider only the last two steps in this stage as the SfM algorithm, but Fig. 3.6 groups all these individual computing steps into one SfM stage). For the sake of clarity, all the individual steps will be defined below.

3.3.1.2 Feature Detection

Feature detection is the first step of many computer vision and photogrammetry-related applications, such as panorama stitching, object recognition, camera calibration, robot localisation and SfM. In past decades, a wide variety of feature detectors have been developed. Aside from their effectiveness, they vary widely in computational complexity and the type of features they detect. Although approaches exist that detect edges, ridges and regions of interest (e.g. Kadir and Brady 2001; Jurie and Schmid 2004; Matas et al. 2004; Deng et al. 2007), the image features used in most SfM approaches comprise interest points (IPs).

IPs represent image locations that are in a certain way exceptional and are locally surrounded by distinctive texture. Additionally, they should be stably defined in the image and scale spaces and reproducible under different imaging conditions. In technical jargon, it is said that IPs should have a high *repeatability*, which means that they should be invariant to any change in illumination, image noise and basic geometric transformations such as scaling, translation, shearing and rotation. In the last 10 years, several new algorithms have been proposed to compute such IPs (e.g. Features from Accelerated Segment Test or FAST (Rosten and Drummond 2005)). However, most detector techniques are based on:

- Hessian-based detectors (Lindeberg 1998)
- Harris-based detectors (Harris and Stephens 1988)

This means that frequently mentioned algorithms such as SIFT (Scale Invariant Feature Transform (Lowe 2004)), SURF (Speeded-Up Robust Features (Bay et al. 2006, 2008)) and ASIFT (Affine-SIFT (Morel and Yu 2009; Yu and Morel 2011)) use variants of the abovementioned detectors (the popular SIFT and SURF detectors both rely on Hessian-based detectors). Figure 3.7a shows IPs computed with SURF. The airborne image in the figure was acquired on the 4th of September 2012 at around 11.00 h using an Olympus PEN E-P2 (a 12.3 megapixel mirrorless Micro Four Thirds camera) equipped with an Olympus M. Zuiko Digital 17 mm f/2.8 lens, mounted on a radio-controlled Microdrone MD4-1000 quadcopter. The aerial frame depicts a part of the excavated Roman city wall of *Carnuntum* (Austria).

3.3.1.3 Feature Description

Since the aim is to find correspondences between these IPs – which means that an algorithm has to find out which IPs are a 2D representation of the same physical 3D point – the IPs have to be described. This task is fulfilled by so-called *feature descriptors* or *feature vectors*. Such a descriptor computes a feature vector with local characteristics to describe a local patch (whose size can vary – Fig. 3.7b) of pixels around each IP (Schmid and Mohr 1996). Just as the IP, this vector should be invariant (i.e. robust to detection displacements, image noise and photometric plus geometric deformations). Various methods also exist to describe the patch around each IP:

- Gradient Location and Orientation Histogram (GLOH) (Mikolajczyk and Schmid 2005)
- Speeded-Up Robust Features (SURF) (Bay et al. 2006, 2008)
- Scale Invariant Feature Transform (SIFT) (Lowe 2004)
- Local Energy based Shape Histogram (LESH) (Sarfraz and Hellwich 2008)
- ASIFT (Affine-SIFT) (Morel and Yu 2009; Yu and Morel 2011)
- Histogram of Oriented Gradients (HOG) (Dalal and Triggs 2005)

In the end, an image feature can be defined as an IP and its descriptor. Note that several IP detectors also define their descriptor (e.g. SIFT, SURF, ASIFT). As can be expected, several authors have tried to compare the performance of various detector and descriptor combinations (e.g. Mikolajczyk and Schmid 2003, 2005; Mikolajczyk et al. 2005; Moreels and Perona 2007; Tuytelaars and Mikolajczyk 2007; Juan and Gwon 2009).

3.3.1.4 Descriptor Matching and Pairwise Image Orientation (Fundamental Matrices)

Finally, all descriptor vectors are matched between different images by associating each IP from one image to the other IPs of the remaining

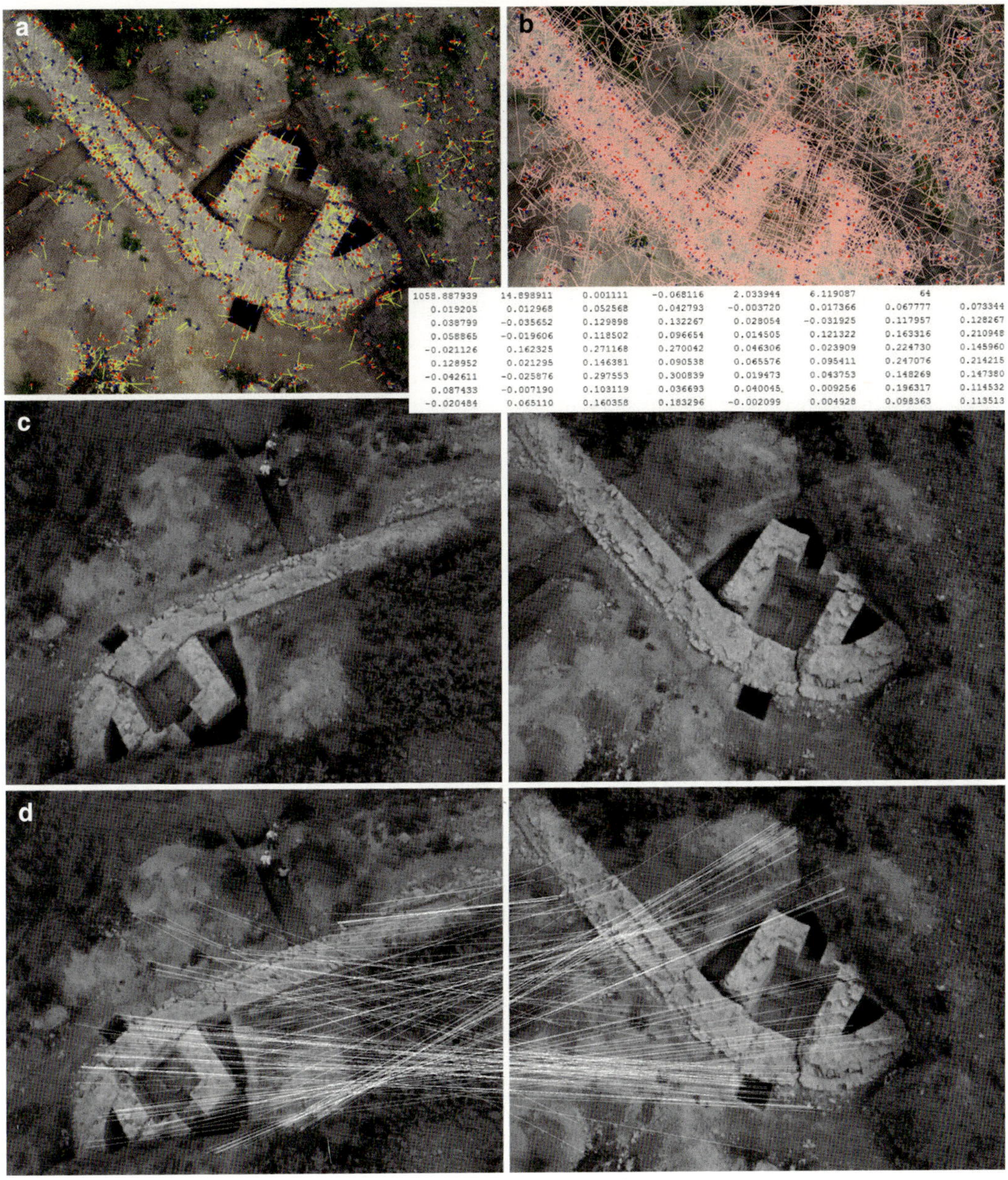

Fig. 3.7 (**a**) SURF IPs computed from an airborne image using ImageJ SURF (Labun 2009). The 1,852 IPs are accompanied by their orientation vectors whose lengths indicate the strength of the computed IPs. (b) The 1,852 SURF IPs with their descriptor windows. The inset shows the vector describing one of these IPs. Computations were performed with ImageJ SURF (Labun 2009). (**c, d**) The difference between two image matching routines. While (**c**) used the SIFT detector and was unable to find any matching points, ASIFT was applied for (**d**). This test was performed using the ASIFT online demo application at http://demo.ipol.im/demo/my_affine_sift/ (Yu and Morel 2011)

images. To compute a match, a distance between the descriptors is generally used (e.g. the Euclidean distance). The dimension of the descriptor has a direct impact on the time this takes, and fewer dimensions are desirable for fast IP matching. However, lower-dimensional descriptor vectors are generally less distinctive than their high-dimensional counterparts. Besides, approximate but fast methods exist (e.g. approximate nearest neighbour searches in kd-trees), while slow but rigorous matching procedures such as quadratic matching can also be applied.

A robust outlier detection algorithm such as RANSAC (RANdom SAmple Consensus (Fischler and Bolles 1981)), ORSA (Optimized Random Sampling Algorithm (Moisan and Stival 2004)), LMedS (Least Median of Squares (Rousseeuw 1984)) or MAPSAC (Maximum A Posteriori SAmple Consensus (Torr 2002)) will ensure the rejection of probable false matches by testing them for consistency. This is done for all possible image pairs by checking if their putative matches fulfil the so-called *epipolar geometry constraint*: i.e. that the displacements of IPs are a possible result solely of the motion of the camera between both images. At the end of this process, the *fundamental matrices F* of the image pairs are obtained: each of them is a 3×3 matrix depending on seven parameters that describes the motion (i.e. *relative orientation*) from the first to the second image. When dealing with calibrated cameras or pinhole camera models, the *essential matrix E* is used; in case of an image triplet, the trifocal tensor *T* can be applied (Robertson and Cipolla 2009). Because the fundamental matrix describes the correspondences in more general terms, it is used with uncalibrated cameras. This has very important implications as the matching can be performed without initially calibrating the cameras. Finally, the complete set of image correspondences (called tie points in photogrammetry) for the whole image sequence is obtained after considering all meaningful image pairs. The set of corresponding IPs thus obtained functions, together with the fundamental matrices, as input for the last steps of the SfM computation.

Figure 3.7c–d shows, however, that this input varies widely according to the algorithms applied to obtain this set of image correspondences. The differences between two image matching routines are illustrated, both of them trying to reliably identify and match two aerial images. In Fig. 3.7c, the SIFT detector is used while Fig. 3.7b uses the ASIFT approach. All IPs are then coded with the SIFT descriptor. The matching process first computes the Euclidean distance between an IP descriptor in the first image with all the descriptors found in the second image and uses its values to define whether IPs are considered as matched. Afterward, the ORSA algorithm is applied to filter out the false matches. The example shows that ASIFT retrieves the matches – indicated by the white lines – even under large changes of viewing angle, while there is a total failure in finding image correspondences using SIFT IPs. This is due to the nature of the algorithms used. While SIFT can only deal with a similarity invariance (i.e. invariant to four parameters describing translation, rotation and zoom) and less viewpoint change from one image to another, ASIFT is fully affine invariant. This means that ASIFT possesses invariance for the four similarity degrees of freedom as well as for the two angles defining the camera axis orientation. To achieve this, it simulates rotation and tilt on the images and can therefore deal with frames whose viewing angle is very different (Morel and Yu 2009; Yu and Morel 2011).

3.3.1.5 Triangulation

Relying on the algorithms that detect, describe and match local feature points throughout the multiple images, SfM computes the locations of those feature points in a local coordinate frame, creating a sparse 3D point cloud that represents the geometry/structure of the scene. This determination of a point's 3D position when observed from two or more cameras (Fig. 3.2) is called *image triangulation* (Szeliski 2011). However, image triangulation requires the knowledge of the images' interior and exterior orientation. These are obtained after combining all the relative orientations of the image pairs in form of their fundamental matrices.

SfM can accomplish this as it is based on the *projective reconstruction theorem*, which states: given a set of point correspondences in two views defined by the fundamental matrix, the 3D scene geometry and images' projection matrices (which comprise all the orientation parameters) may be reconstructed from these correspondences alone, and any two such reconstructions from these correspondences are projectively equivalent (Hartley 1994; Szeliski 2011). However, rather than a projective reconstruction, a metric reconstruction is wanted: i.e. one in which orthogonal planes are at right angles, parallel lines stay parallel and the reconstructed 3D model is a scaled version of reality. This can be accomplished by running a simultaneous *self-calibration/auto-calibration* to define the camera's interior orientation. The latter is stored for each image in the *intrinsic parameter matrix K* (Hartley and Zisserman 2003; Moons et al. 2008).

3.3.1.6 Bundle Adjustment

Up to now all images were dealt with in pairs, for each of which a fundamental matrix was computed (in a linear way by minimising a physically non-meaningful quantity – the so-called *algebraic error*). Afterwards the oriented image pairs were combined to form the complete block of images and to yield the structure of the scene. The results obtained this way are, however, suboptimal because not all overlapping images are used at the same time and the discrepancies in the structure (caused by small errors during the feature measurement phase) are not optimally minimised. To overcome these problems, the final stage of most SfM algorithms is bundle adjustment. Bundle adjustment iteratively optimises the 3D structure and the projection matrices of all images simultaneously by performing a robust non-linear minimisation of the actual measurement errors, also known as *re-projection errors* (Triggs et al. 2000). The technique was developed half a century ago in the field of photogrammetry but is now also largely applied in the computer vision community. The term bundle adjustment comes from the fact that the bundles of rays connecting camera/projection centres to 3D scene points are adjusted to minimise the sum of squared differences between the observed and re-projected image points (Szeliski 2011).

This means that an SfM approach can recover the scene structure and camera projection matrices from image correspondences alone without prior knowledge about camera poses or interior orientation (Hartley and Zisserman 2003; Szeliski 2011). There is thus no real need to use calibrated cameras and optics during the image acquisition stage (Quan 2010), which makes the procedure very flexible and well suited for almost any kind of imagery, particularly for completely unordered photo collections such as those that can often be found in aerial archives. It needs to be noted, though, that it is still more accurate to recover the significant interior orientation parameters in a separate calibration routine using a dedicated image network geometry (Remondino and Fraser 2006).

3.3.1.7 Defining a Coordinate Reference System

It is essential to understand that the SfM output is characterised by a scale ambiguity. This means that if the entire scene is scaled by some factor and the distance between the camera positions is simultaneously scaled by the same scale factor, the projections of the scene points in the image will remain exactly the same. The reconstructed 3D scene obtained after a standard SfM approach is thus expressed in a local coordinate framework and equivalent to the real-world scene up to a global scaling, rotation and translation. These parameters can only be recovered via the use of additional data, which in turn define a coordinate reference system (CRS). According to Barazzetti et al. (2011), this can be achieved in two ways:

- Import at least three spatially well-distributed GCPs with known altitude values and transform the complete model into an absolute CRS with a Helmert similarity transformation. Although more GCPs are advisable, three is the minimum since seven parameters (three translations, one scale and three rotations) must be determined for this spatial transformation. Since this operation is performed after the SfM computation and does not introduce any external constraint, it will not improve the initially obtained SfM result.
- Import highly accurate camera positions or a minimum of three GCPs and use them as constraints in the bundle adjustment. This rigorous approach is a better solution as it can correct for errors such as drift in the recovered camera and point locations (Snavely et al. 2006), avoids instability of the bundle solution (Remondino et al. 2012) while the SfM output is directly georeferenced (Verhoeven et al. 2012a).

3.3.1.8 Dense Multi-view Stereo (MVS)

At this stage a georeferenced sparse 3D reconstruction of the scene is available. 'Sparse' because it is only based on the reconstructed set of IPs. However, with the now known orientation of the images, it becomes possible to create a texture-mapped dense 3D model and compute orthophotographs. The essential step in this process is the computation of this denser 3D model. Alternatively, one could interpolate the sparse set of 3D points, but this would yield a far from optimal result. Therefore, it is better to run a multi-view stereo (MVS) algorithm to compute a dense estimate of the surface geometry of the observed scene. Because these solutions operate on pixel values instead of on feature points (Scharstein and Szeliski 2002; Seitz et al. 2006), this additional step enables the generation of detailed 3D meshed models (or dense point clouds) from the initially calculated sparse point clouds, hence reproducing fine details present in the scene.

Just as in all previous stages, MVS comes in many variants and a comparison of several approaches can be found in Seitz et al. (2006). However, since the publication of this paper by Seitz and his colleagues, many new algorithms have been developed. Although elaborating on them is outside the scope of this text, it might be worthwhile to notice that the most common algorithms can be divided into region growing patch-based approaches (e.g. Lhuillier and Quan 2005; Habbecke and Kobbelt 2006; Furukawa and Ponce 2010) and depth-map fusion pipelines (e.g. Mellor et al. 1996; Pollefeys et al. 2004; Goesele et al. 2006; Strecha et al. 2006; Bradley et al. 2008; Hirschmüller 2008). Obviously, each of those has its own specific pros and cons, generally striking a balance between accuracy and consistency (region growing approaches) versus a fast and elegant pipeline (depth-map fusion).

3.3.1.9 Georeferenced 3D Model and Orthophoto

The final georeferenced dense 3D model generated from these aerial images can be considered a DSM: a numerical representation of the topography and all its imposed structures such as trees and houses. As is known from conventional orthorectification (Manzer 1996), such a dense DSM is elementary when one wants to generate a so-called true orthophoto in which all objects with a certain height (such as houses, towers and trees) are also accurately positioned (Kraus 2002; Braun 2003). When combined with the previously calculated camera poses and interior orientation parameters, this dense DSM thus enables the generation of true orthophotos. Because the whole process takes most relevant geometrical degradations into account, the orthographic image is perfectly suited for archaeological purposes. For visualisation purposes, one could also export a textured 3D mesh which could be created by a texture mapping using a particular selection of the initial images.

3.3.2 Tools

3.3.2.1 Software

In recent years, SfM has received a great deal of attention due to Bundler (Snavely 2010) and Microsoft's Photosynth (Microsoft Corporation 2010): two SfM implementations that are freely available on the Web. To date, several SfM-based packages can be applied to obtain a (semi-) automated processing pipeline for image-based 3D visualisation. Often, these packages are complemented by an MVS approach (see Table 3.1). An overview of the accuracies that can be obtained in automated image orientation and camera calibration parameters with some of these packages is detailed in Remondino et al. (2012).

3.3.2.2 Hardware

Besides novel algorithms, the routine outlined above exploits some of the technological improvements in hardware configurations. Obviously, high-quality reconstructions with large image files are very resource intensive. All processing should therefore be undertaken on a multicore computer (or computing grid) with a 64-bit operating system and a large amount of RAM. Additionally, the graphics processing unit (GPU) can be considered one of the crucial hardware elements, as a high-performance GPU can greatly shorten processing times. Many

Table 3.1 Some commercial and freely available SfM and MVS packages

Company	Software	Free	SfM	MVS	Web	Orthophoto
Agisoft LLC	PhotoScan standard		X	X		
Agisoft LLC	PhotoScan professional		X	X		X
Matis laboratory (I.G.N.)	Apero	X	X			
Matis laboratory (I.G.N.)	MicMac	X		X		X
University of Washington and Microsoft Corporation	Bundler	X	X			
Microsoft Corporation	PhotoSynth	X	X		X	
University of Washington	VisualSFM	X	X	X		
AutoDesk	123D Catch	X	X	X	X	
KU Leuven	Arc3D	X	X	X	X	
Eos Systems Inc.	PhotoModeler Scanner		X	X		X
University of Illinois and University of Washington	PMVS2	X		X		
3Dflow SRL	3DF Samantha	X	X			
Henri Astre and Microsoft Corporation	PhotoSynth Toolkit	X	X	X		
CTU Prague	CMPMVS	X	X	X	X	X
Acute3D	Smart3DCapture		X	X		

SfM+MVS applications support the OpenCL (Open Computing Language) programming platform and can therefore access the GPU for executing very intensive computing during specific steps in the pipeline, although the steps that can be accelerated depend on the software. Still, better and more optimised algorithms are needed before time-efficient processing of large image sets on standard computers can take place (Verhoeven et al. 2012a).

3.4 Case Studies

SfM-based applications started to find their way into archaeological research about 10–15 years ago (e.g. Pollefeys et al. 1998, 2000, 2001, 2003, 2004; Pollefeys and van Gool 2002; El-Hakim et al. 2003). During the decade that followed, the SfM concept and dense matching techniques made great improvements and became capable of orienting very large datasets and delivering satisfactorily accurate dense 3D models (Barazzetti et al. 2011). Nowadays, an SfM and MVS pipeline can almost be considered a standard tool in many aspects of archaeological research (e.g. Ludvigsen et al. 2006; Lerma et al. 2011; Appetecchia et al. 2012; Bezzi 2012; Forte et al. 2012; Kersten and Lindstaedt 2012; Lo Brutto and Meli 2012; Opitz and Nowlin 2012). Although most of these studies use terrestrial images, there are some papers in which archaeological aerial frame images have also been used (e.g. Doneus et al. 2011; Verhoeven 2011, 2012c; Lo Brutto et al. 2012; Reinhard 2012; Remondino et al. 2011; Scollar and Giradeau-Montaut 2012; Verhoeven et al. 2012a).

The three case studies described below show the potential of this combined SfM+MVS method using diverse imagery (oblique and vertical, old and new, acquired in the visible and near-infrared spectral domain from manned and unmanned platforms) covering a variety of topographic settings. As these image sets predate the development of SfM-based approaches, they provide a perfect opportunity to evaluate the applicability of the method to older datasets. The case studies are presented in a common format: first, a short introduction to the site and the acquisition of the photographs are presented; secondly, the building of the orthophoto and possible drawbacks are addressed; and thirdly, each case study will also highlight some very specific advantages of this approach.

All 3D models and orthophotographs were computed using PhotoScan Professional edition (v. 0.8.1, build 877 and later) from Agisoft LLC. The choice for this software was based on its features, cost and completeness: it is currently the only commercial, frequently updated package that combines both SfM and MVS algorithms while additionally offering tools for generating orthophotographs, texture mapping and post-processing 3D models (Agisoft LLC 2012). Concerning the MVS stage, PhotoScan uses a pairwise binocular stereo approach to compute a depth estimate for almost every image pixel of each view. Afterward, several dense 3D reconstruction methods are provided, each differing in the way these individual depth maps are merged.

3.4.1 *Trea* (Italy)

Generally, the advised strategy when using PhotoScan is to solve the complex SfM math of as large as possible a set of images, without having to rely on virtual memory. Later, one can 'disable photos' and perform the subsequent dense reconstruction in parts (Verhoeven 2011). Although this approach is meant to tackle limited hardware resources, it opens up a completely new application field for aerial archaeologists. To illustrate this, a time series covering 6 years of aerial research on the Roman town of *Trea* (central Adriatic Italy, 43°19′ 06″ N, 13° 17'' 31″ E – WGS84) will be used.

In January 2000, Ghent University initiated the Potenza Valley Survey (PVS) project in the central Adriatic Region of Marche. This interdisciplinary geoarchaeological project has mainly been aimed at reconstructing the changing physical and human landscape along the Potenza River, one of Marche's major rivers. Aerial archaeological reconnaissance was identified from the start as one of the main survey techniques to be used due to its cost-effectiveness (Vermeulen 2002, 2004). Along the Potenza Valley lies the former Roman town of *Trea*, located on a hill surrounded by the heavily undulating landscape of the middle Potenza valley. The scene can thus be considered quite complex and the relief displacement in the aerial images very substantial. Although there have been a series of investigations into the character and extent of this city, almost nothing was known about its general layout and organisation before the systematic aerial campaigns of the PVS (see Moscatelli 1985). The survey results now allow for a near complete mapping of the main urban structures of this abandoned Roman city, such as the town defences, the internal street network and the main public and private buildings.

From the 208 images initially selected, 203 were aligned correctly in PhotoScan (Fig. 3.8a). This number is extremely high given the circumstances: a wide variety of cameras and lenses were used during the reconnaissance flights; the land cover varied from bare soil to crops in various phenological states; 39 images only recorded the radiance in the near-infrared (NIR) spectral band (see Verhoeven 2008b, 2012b; Verhoeven et al. 2009b for details on this). Unquestionably, this alignment result was facilitated by the fact that all images still had information about the focal length embedded in the Exif (exchangeable image file format) metadata tags, so that these values could be used to initialise the SfM step. To execute the dense reconstruction stage, a subset of 143 suitable images was used as input. The selection criteria for this were largely based on image scale, scene coverage and sharpness. This does not render the remaining images unusable, however. Once an accurate 3D model of the terrain is generated (Fig. 3.8b), every image or combination of images in the project can be transformed into an orthophoto through the use of the DSM for correction.

This way, it is possible to use only the NIR images (Fig. 3.8c-3) or those that best illustrate the crop marks (Fig. 3.8c-1) or soil mark state (Fig. 3.8c-2) or to generate a bespoke coverage. Not only does this approach speed up the processing of individual images (or related photo sets) considerably, but the final interpretation is more trustworthy as well: due to the heavy undulating nature of the terrain and the very steep slopes bordering the central plateau, most GIS packages and tools specifically developed for archaeological research (such as AERIAL or AirPhoto SE) will typically fail to accurately georeference these images. Although this might not seem to be a big issue when dealing with vague soil marks, the nature of the crop marks (faint and small) as well as the type of site (a complex Roman town with different phases) makes the accurate mapping of

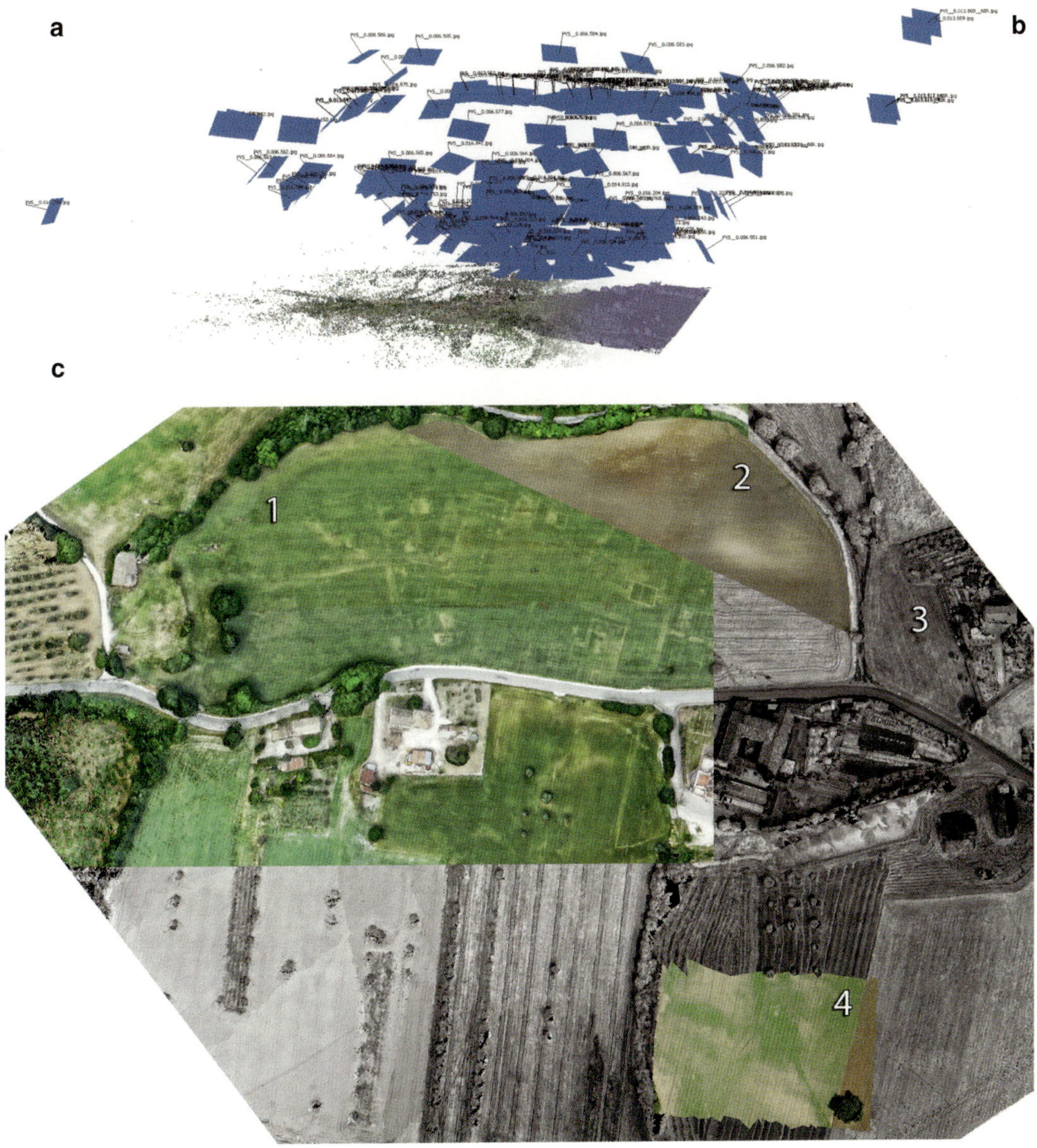

Fig. 3.8 (**a**) The relative position of all 203 camera stations. (**b**) The extracted DSM of *Trea*. (**c**) The integration of several orthophotos, showing crop marks (*1*) and soil marks (*2*) in the visible domain, the NIR terrain reflectance (*3*) and the orthorectification of an image (*4*) without any useable GCP

the features of the utmost importance for comparison of aerial footage from different years or to interpret the data with respect to a geophysical survey (for this case study, the georeferencing delivered a planar RMSE of 6.2 cm and an RMSE of 4.6 cm for the altitude component). Additionally, the whole process of orthophoto production is straightforward, fast and can deal with a variety of frame imaging sensors from which no calibration parameters need to be supplied. Moreover, as Fig. 3.8c-4 indicates, even individual images without any GCP can be transformed into orthophotos. The combination of these advantages largely overcomes the current drawbacks that archaeologists encounter in most (ortho)rectification approaches, certainly when dealing with larger areas (features of a palaeolandscape, extensive sites) or terrain undulation.

However, it should be noted that such an integrated approach only works when no major scene changes have taken place during the years of image acquisition. In the case study of *Trea*, the biggest surface difference was related to the phenological state of the vegetation: sometimes the fields were just harvested, while at other times the camera recorded the full canopy. Although it did not hamper the SfM stage, the DSM will obviously be influenced by this. Therefore, one can best use a set of images displaying the most common surface condition, after which a numerical form of the latter can be used to compute the orthophotos of more or less all images. This approach was used in this case study and did not result in archaeologically relevant positional differences of the computed orthophotos. In case the difference between different topographical conditions is too big, a multitude of DSMs should be computed to cover all possible surface states. In the worst case scenario, the landscape can have changed so drastically over time that image alignment will fail.

3.4.2 Kreuttal Region (Austria)

The acquisition of oblique aerial photographs is well suited for a computer vision approach. However, very ordered collections of vertical imagery can also be successfully processed into true orthophotos. Their high longitudinal and lateral overlap makes them very useful for 3D data extraction via photogrammetric means, but this also translates to high usability, automation and accuracy in an SfM-driven environment. This is not limited strictly to modern air photos, but can be used on high-quality historical air photo datasets as well. Furthermore, due to the high overlap of imagery, SfM-based data processing methodologies are able to extend the usability of these types of datasets into the 3D realm, allowing for the creation of not only 2D orthomosaics but 3D historical digital elevation models (hDEMs). Therefore, historic land use and land change can be evaluated from a topographic perspective, bringing a new dimension to archaeological landscape analysis (cf. Pérez Álvarez et al. 2013).

Of the many archives of vertical historical aerial images that exist, perhaps some of the most well known are The Aerial Reconnaissance Archives (TARA) and the National Archives and Records Administration (NARA) holdings. Located in Edinburg and Washington D.C., respectively, the total number of photographs in these archives is ca. 21 million (Cowley and Stichelbaut 2012; Cowley et al. 2013) dating from as early as 1918. Numerous national and regional archives also exist, of which a number are further detailed in Wilson (2000), Cowley et al. (2010) and Hanson and Oltean (2013). While the condition of materials in these archives can be highly variable, they are nevertheless vast and largely unique sources of information, and lack of proper camera and lens data for many of the photos contained therein is not necessarily an obstacle to successful reconstruction with SfM-based approaches.

The case study presented here examines the use of historical vertical datasets in the Kreuttal region of Lower Austria (48° 26′ 40″ N, 16° 27′ 01″ E – WGS84). Situated roughly 25 km north of Vienna, the Kreuttal contains traces of past land use from the Neolithic to the Modern Historic eras. Archaeological sites in this topographically varied region manifest themselves on aerial photographs in the form of vegetation marks, soil marks and shadow marks, with a number of upstanding and particularly well-preserved hill forts from the Bronze and Iron Age visible in the forest during off-leaf seasons. Two vertical datasets, acquired in March of 1945 and 2010, have been chosen from among the large archive of air photographs of the region to showcase the uses and issues involved in the processing of historic vertical datasets with SfM applications.

Sortie 15SG-1374, acquired on 23 March 1945, consists of 20 images acquired as part of an allied sortie over Lower Austria at the end of World War II. Images were acquired stripwise, west–east then east–west, at a scale of ca. 1:10 500 (Fig. 3.9a). Acquired from TARA through a local Austrian partner, the images came with no other camera or mission information. All images were 1,200 spi (samples per inch) scans of prints, many of which contain significant localised error due to warping and other degradation as a result of age and possibly improper storage before being acquired by TARA (Fig. 3.9b). Images were not

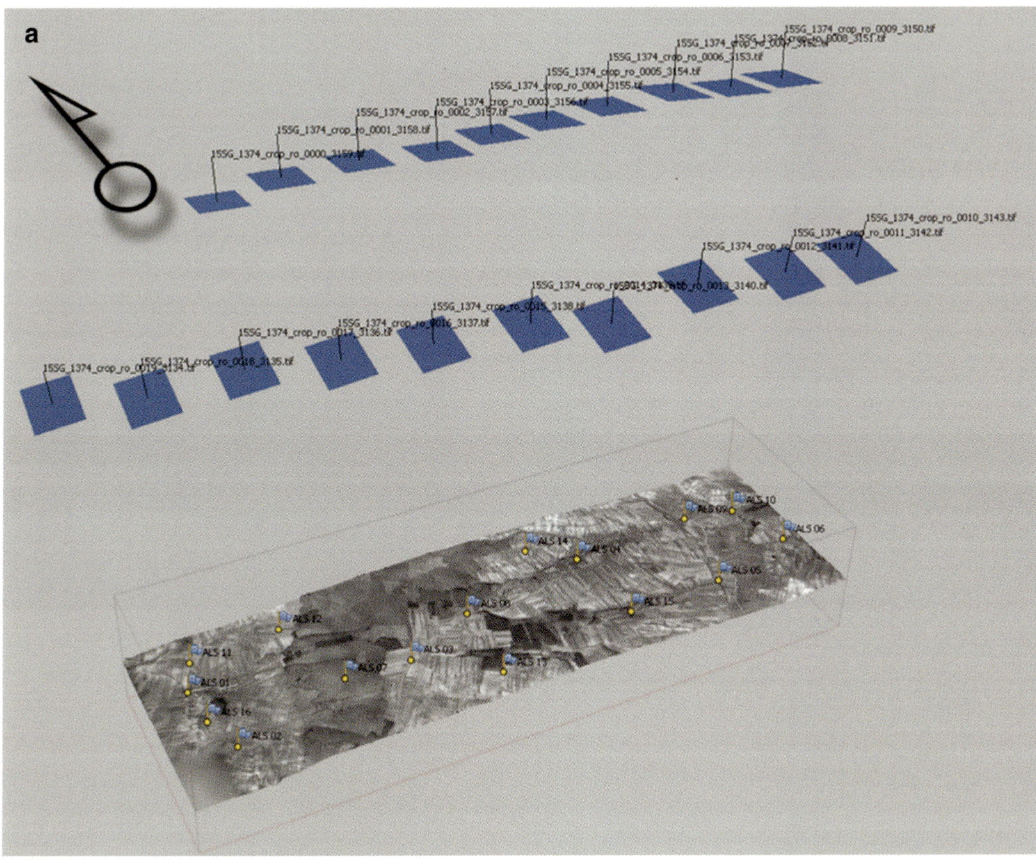

Fig. 3.9 (**a**) Reconstruction of flight path for sortie 15SG-1374. (**b**) Sample image from sortie 15SG-1374. (**c**) Reconstruction of flight path for flight 02100301. (**d**) Sample image from flight 02100301

Fig. 3.9 (continued)

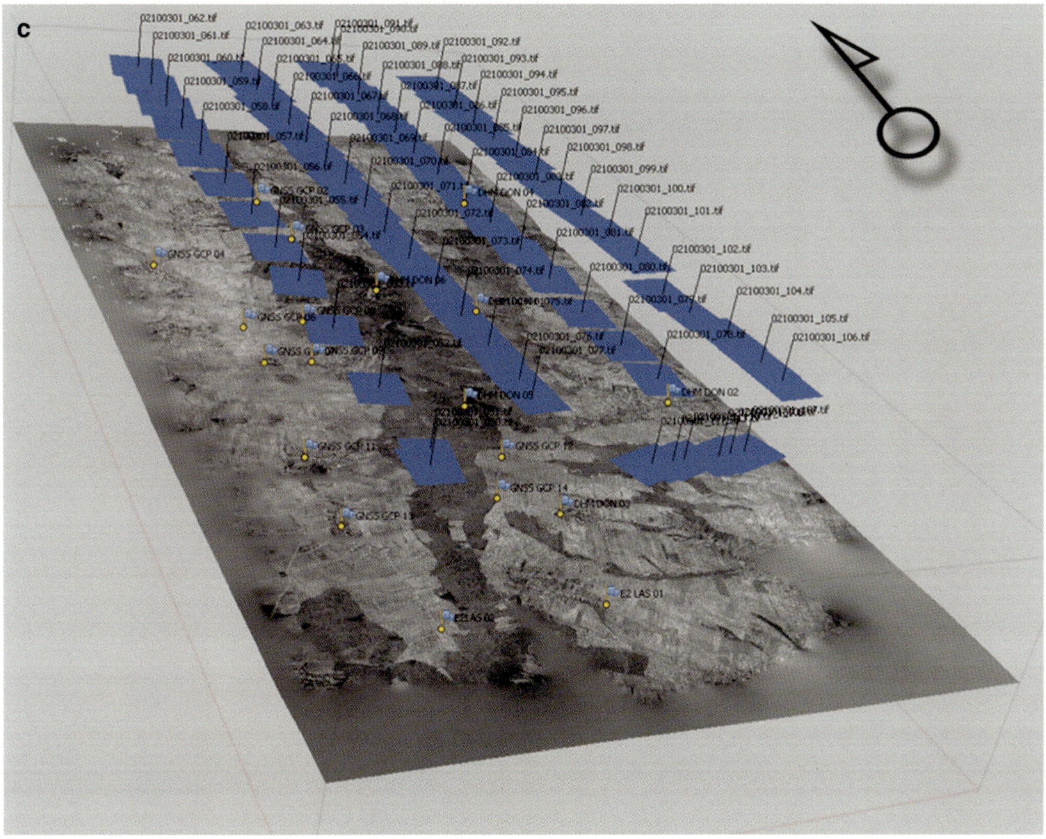

Fig. 3.9 (continued)

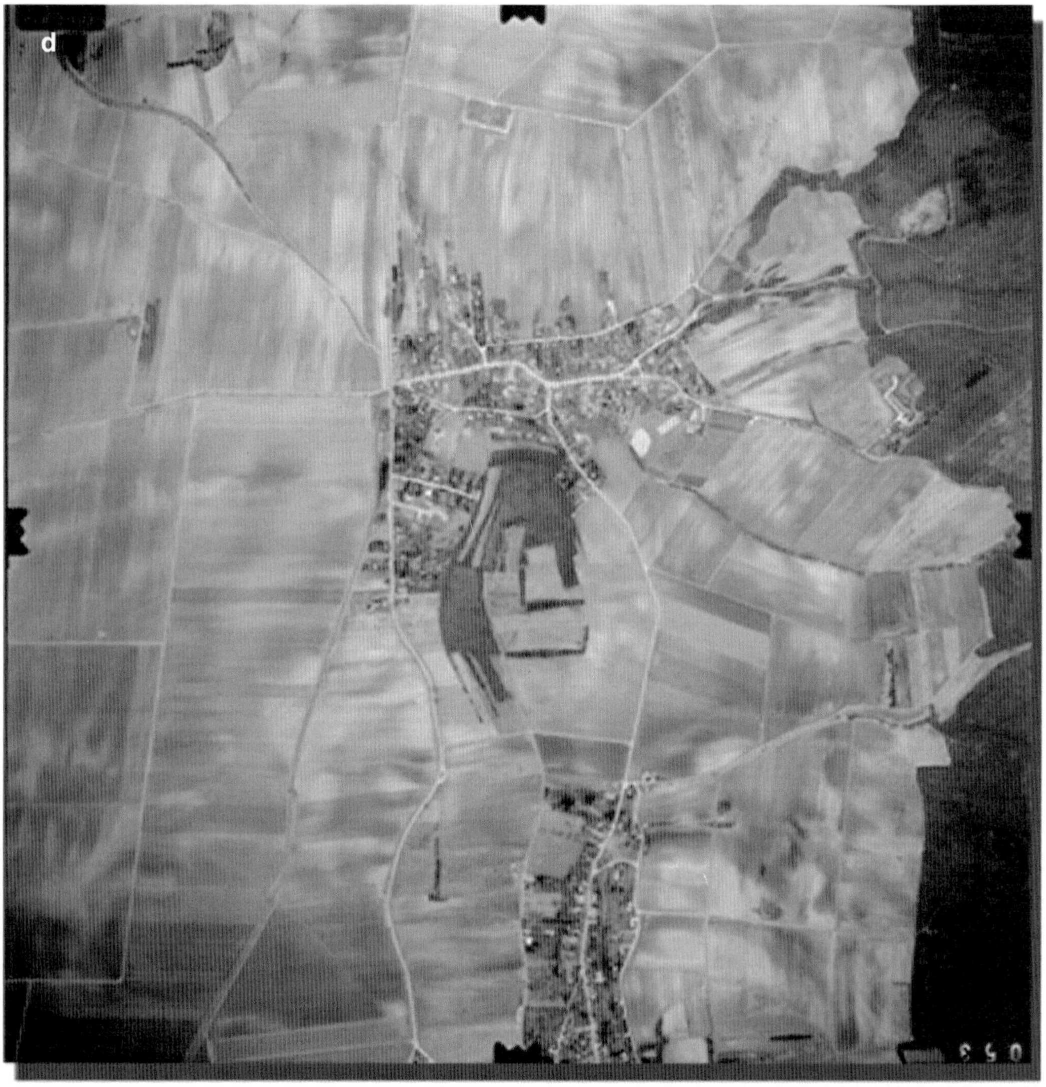

Fig. 3.9 (continued)

uniformly sharp, and many areas, including borders and fiducial marks, had to be masked so as not to interfere with reconstruction. Furthermore, as the images are scans of 'predigital' photographs, they contain no Exif data or calibration data which the software could use in the SfM phase.

Despite all this, PhotoScan was able to align and match all 20 images as delivered by the archive. However, there were significant issues with camera pose estimation. This was due to the fact that, as a by-product of the scanning process, all images had different pixel dimensions. This issue was resolved by loading all of them into a photo editor, aligning them via their fiducial marks and cropping them to identical dimensions. Once this was completed, camera pose estimation improved significantly. GCPs were then placed in order to georeference the dataset while further refining camera calibration and pose by treating the GCPs as constraints in a subsequent bundle adjustment. This presented its own obstacles as landscape change was significant enough over the intervening 58 years as to make it extremely difficult to locate unchanged reference points. Through extensive comparison with other datasets a number of GCPs were eventually identified, with a 50 cm spatial resolution DSM generated from airborne laser scanning (ALS) data used to acquire GCP coordinates.

After masking, GCP placement and several bundle adjustments, the final model was able to achieve a total distributed georeferencing error of 3.02 m, the majority of that being in the RMSE(Z) (*X* error 0.562 m, *Y* error 0.854 m, *Z* error 2.842 m). This was largely due to the degraded quality of the prints causing excessive localised distortion in the 3D reconstruction. In this instance, 2D orthomosaics proved the most useful output as the 3D hDEM was extremely noisy and still contained significant local error. This could be corrected by further post-processing methods to reduce noise and correct for residual local distortion (Sevara 2013).

Flight 02100301 was acquired on the 1st of March 2010 by the Austrian Military at the request of the Aerial Archive at the University of Vienna (Doneus et al. 2001). This flight consisted of 63 images and was flown stripwise north–south to south–north at a scale of 1:10,000 (Fig. 3.9c). Unlike sortie 15SG-1374, all camera parameters for this flight are known and interior orientation data were readily available. Images were scanned from negatives using a Vexcel UltraScan 5000 photogrammetric scanner (Doneus et al. 2007) at a resolution of 5,080 spi. As a result, the images from flight 02100301 are of a significantly higher quality than those of sortie 15SG-1374 (Fig. 3.9d). Images still needed to be masked and the same issues were still present with regard to lack of Exif data as with 15SG-1374. However, since all camera parameters were known, these could be entered manually into PhotoScan.

With all of these factors significantly improving alignment and pose estimation, initial results were already far more accurate. Due to the high quality of the scan process, all images were the same dimensions, obviating the need to manually crop them. GCP placement was also significantly easier, due to the recent nature of the dataset. GCPs were acquired from the same DSM as for sortie 15SG-1374. Once GCPs were placed and the model was cleaned and optimised by an additional bundle adjustment, re-projection error dropped to below 1 pixel. The total distributed error for this dataset was 0.89 m utilising 17 of the 19 GCPs, the error being more evenly distributed this time (*X*, 0.49 m; *Y*, 0.59 m; *Z*, 0.44 m).

In this instance, both 2D and 3D products generated from flight 02100301 were of extremely high quality. The 2D orthomosaic corresponded in horizontal quality to that of orthomosaics generated in Leica Photogrammetry Suite (LPS) using the same dataset, with significant improvement over the LPS dataset in heavily wooded and variegated terrain due to the high accuracy of the hDEM used for orthorectification. The hDEM provided a correspondence of <50 cm when analysed against independently collected ground control using a Leica GPS 500 RTK receiver. Furthermore, accurate 3D data could also be acquired for upstanding prehistoric earthworks in the area.

As can be seen from this case study, SfM-based approaches to orthomosaic generation and terrain reconstruction also work with historic datasets in a way that far exceeds the original intended use of the data. However, results can be highly variable and depend heavily on both the quality and quantity of original photographs, much as the other case studies in this section illustrate. Further information regarding this case study can be found in Sevara (2013).

3.4.3 Pitaranha (Portugal-Spain)

Ancient quarry sites are a good example of the multifaceted nature of certain archaeological sites. The often complex morphological and topographical characteristics of quarry landscapes, as well as the severe modification of the terrain configuration by both intensive quarrying and the intricate logistical extraction infrastructure complicate their survey. Since an accurate digital representation of the topographical surface is elementary to the spatial analysis of quarry sites and the availability of an orthophoto map a necessary prerequisite for fast and effective site navigation, the acquisition of such information is a crucial component of efficient quarry research. To this end, a cost-effective technique was developed to map the Roman quarry of Pitaranha, located on the present-day border between Portugal and Spain, some 200 m northeast of the village of Pitaranha (Alentejo, Portugal; 39° 22′ 13″ N, 07° 18′ 49″ W – WGS84). Historically, the quarry mainly provisioned the nearby Roman town of *Ammaia* (Vermeulen and Taelman 2010). Several periods of intensive building in the Roman town suggest large-scale quarrying at Pitaranha during the first centuries AD (Taelman et al. 2009). A thorough mapping of the site was deemed necessary in order to fully comprehend the particular mechanisms of the quarry.

After establishing a dense network of well-distributed GCPs (Fig. 3.10a), an unmanned low-altitude Helikite-based aerial system was used (Fig. 3.10b) to acquire aerial still imagery (detailed information on the development and construction of the Helikite platform can be found in Verhoeven et al. 2009a). For this case study, the Helikite platform was equipped with a 10 megapixel Nikon D80 reflex camera fitted with a Nikkor 20 mm f/3.5 AI-S. Although this lens suffers from quite some optical distortions, its resolving power – certainly in the centre of the image – is great, while it also offers a large angular field of view (61° by 43°) and is very light (235 g).

As a result of unstable wind conditions (i.e. thermal airstreams alternated with windless areas) and strong electromagnetic interference during camera and platform control, an unstructured collection of about 1,400 digital photographs was necessary to cover almost the entire quarry site. The scales of these images varied enormously, while the camera orientations – and to a certain extent the flight path – were almost random and certainly not as structured as initially intended. Since the ground-sampling distance (GSD) varied between approximately 3 and 8 cm, this variation was expected to be challenging because high-resolution detail would be attenuated with low-resolution geometries extracted from the images taken at high altitudes. Obviously, all these factors are normally not encountered in the highly structured datasets acquired by conventional aerial survey, such as those of the previous example.

In a first step, the complete image dataset was reduced to a more manageable photo collection of 377 sharp and well-exposed images. Altering the parameters resulted in different SfM solutions of which only the most accurate one was retained for subsequent MVS processing. After the calculation of a detailed continuous 3D surface, the final orthophotograph (Fig. 3.10c) was computed and its positional accuracy determined. To incorporate all possible uncertainties in the computed dataset (including those introduced by the control coordinates), the 95 % confidence interval was calculated and expressed according to the NSSDA standard (Federal Geographic Data Committee – Subcommittee for Base Cartographic Data 1998). In the end, the horizontal accuracy turned out to be 13.7 cm, while the overall absolute vertical accuracy value was 31 cm. Given that the source material consisted of an extremely unordered image collection of vertical, low and high oblique aerial photographs

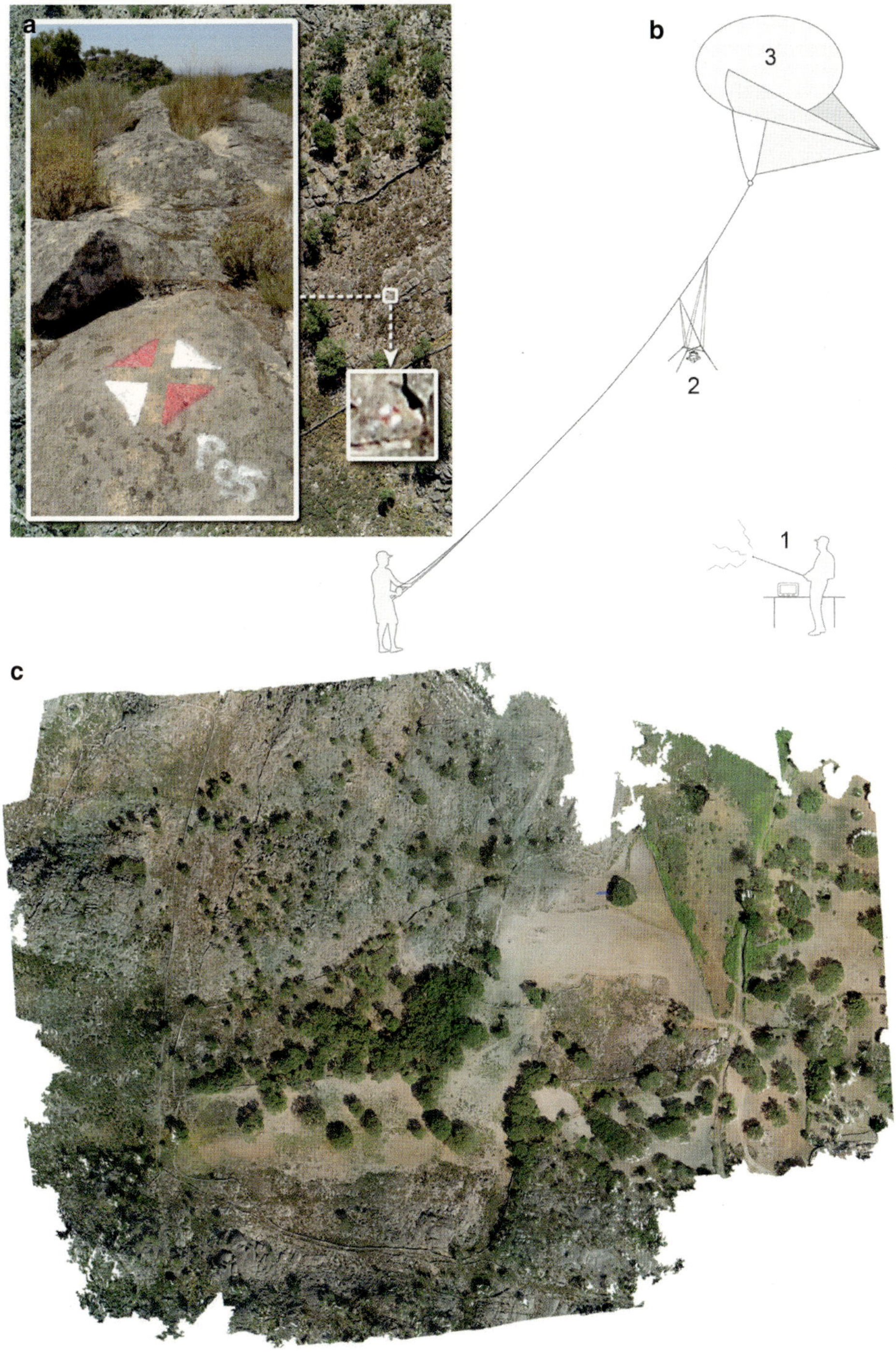

Fig. 3.10 (**a**) One of the oblique aerial images taken with the Helikite platform. The insets show one of the applied ground targets and how it is rendered in the final aerial photograph (**b**). A schematic overview of the Helikite aerial photography system, consisting of a Helikite (*1*), a digital still camera (*2*) and a camera operator with live video (*3*). (**c**) The final orthophotograph of the quarry

that are characterised by a GSD of minimum 8 cm, all acquired with non-metrical lens that suffered from a good deal of distortion, the reported positional accuracy of these datasets is considered very good (planimetric) to good (altimetric) and certainly better than initially expected. Moreover, at the moment of orthophoto production, the version of PhotoScan used did not allow to run a bundle adjustment which included the GCPs. As a result, the GCPs could not be applied to further optimise the SfM output but only to transform the complete model into an absolute CRS with a Helmert similarity transformation. Following the accuracy guidelines of the American Society for Photogrammetry and Remote Sensing (ASPRS), the RMSE values mean that the orthophoto can be used at a class 1 hard copy scale of 1:200 and contour lines with 50 cm intervals can be derived from the DSM (American Society for Photogrammetry and Remote Sensing 1990). More details on the rigorous assessment of the positional accuracy of this orthophoto and DSM can be found in Verhoeven et al. (2012b).

Conclusion

Straightforward orthophoto production is very important in the discipline of aerial archaeology. In this article, computer vision algorithms (structure from motion and multi-view stereo) complemented by proven photogrammetric principles (such as bundle adjustment) were exploited to present an integrated, cost-effective, semi-automated orthophoto production of archaeological aerial (uncalibrated) frame images. This approach is straightforward and requires no assumptions with regard to the camera projection matrix, extensive photogrammetric and computer vision knowledge of the user or the topography of the scenes. Moreover, simplicity is combined with geometrical quality due to the fact that the inner camera calibration parameters are automatically computed and a dense DSM is extracted and applied in a final phase to generate true orthophotos. As a result, this method largely accounts for most relevant kinds of geometrical degradations and is capable of generating 3D models and orthophotos that are perfectly suited for archaeological purposes. Further, only minimal technical knowledge and user interaction are required. Finally, this approach can also work in the total absence of any information about the instrument the imagery was acquired with, although it is still advised to have at least information on the focal length of the imaging system applied. The extra investments needed for software and computing hardware are recovered easily when taking the time and cost savings of map production into account.

This option of fast and accurate orthophoto production is very welcome for aerial archaeologists, given their current approaches which are not tailored to deal either with individual aerial frame images lacking sufficient ground control or with large amounts of photographs from different cameras shot in different seasons. This newly available method offers the enormous advantage that, besides a handful of GCPs, there are only standard photographic recording prerequisites. One simply needs to make sure that enough overlapping and sharp aerial images are acquired. Even though this might involve flying one or more orbits of the scene of interest (for the oblique approach) or vertical strips with up to 80 % overlap, this method will afterwards prove itself in terms of orthophoto quality and – in most occasions – processing speed, certainly when a larger area must be mapped or uneven terrain is involved. Furthermore, the case studies have shown that a large variety of old and new images can be processed into orthophotos whose accuracy is sufficient for large-scale archaeological photo mapping, as well as being visually appealing.

Of course, it is not all roses. First of all, it was indicated that the processing is very computer resource intensive, while the method is not applicable for the individual image. At least two – but preferably more images – are needed for accurate DSM computation. In addition, erroneous alignment of the imagery can occur when dealing with very large photo collections, images that suffer from excessive noise or blur, highly oblique photographs or photographs that have a very dissimilar appearance (e.g. due to major underexposure or changing topographic terrain parameters). Additionally, several authors have already noted that the accuracy of the final products and the recovered camera parameters is often less than results yielded by the expensive and rigorous photogrammetric approaches (Remondino et al. 2012). However, differences are often small, while the approach presented here is superior in versatility and flexibility. The latter point cannot be overestimated, as many archived images do not fulfil the constraints (e.g. camera parameters) that are essential for accurate and straightforward georeferencing using any of the more standard georeferencing approaches by non-photogrammetrists. Currently, the biggest disadvantage of most available SfM-based software packages is the lack of computed metrics and tools in order to inspect the image orientation and matching reliability and accuracy.

Finally, the approach presented here is currently semi-automatic and automation only makes sense when it seriously reduces or completely eliminates steps in a process. In the case of archaeological orthophoto generation, these are the recurring steps of visualising and selection of the images, selecting the essential geodata (GCPs) and setting all the parameters for the subsequent execution of the algorithms. Since this is currently considered to be the bottleneck in large-scale archaeological projects with thousands of images, a project which aims at the creation of completely automatic solutions for orthophoto generation (including the GCP selection) of archaeological aerial photographs was initiated in 2012 (funded by the Austrian Science fund, P 24116-N23). This would offer possibilities for the consistent creation and updating of archaeologically relevant cartographic data in our rapidly changing landscapes.

Acknowledgements This article has been written within the framework of the Austrian Science Fund (FWF): P 24116-N23. The case study from the Potenza Valley Survey project was made possible thanks to support from Belgian Science Policy (Interuniversity Attraction Poles, project P6/22). The Ludwig Boltzmann Institute for Archaeological Prospection and Virtual Archaeology (archpro.lbg.ac.at) is based on an international cooperation of the Ludwig Boltzmann Gesellschaft (A), the University of Vienna (A), the Vienna University of Technology (A), the Austrian Central Institute for Meteorology and Geodynamic (A), the office of the Provincial Government of Lower Austria (A), Airborne Technologies GmbH (A), RGZM (Roman-Germanic Central Museum) Mainz (D), RAÄ (Swedish National Heritage Board) (S), IBM VISTA (University of Birmingham) (GB) and NIKU (Norwegian Institute for Cultural Heritage Research) (N).

References

Aber JS, Aber SW, Leffler B (2001) Challenge of infrared kite aerial photography. Trans Kansas Acad Sci 104:18–27. doi:10.1660/0022-8443(2001)104[0018:COIKAP]2.0.CO;2

Agisoft LLC (2012) Agisoft PhotoScan user manual. Professional edition, version 0.9.0. http://downloads.agisoft.ru/pdf/photoscan-pro_0_9_0_en.pdf. Accessed 13 Feb 2013

Altenhofen RE, Hedden RT (1966) Transformation and rectification. In: Thompson MM, Eller RC, Radlinski WA, Speert JL (eds) Manual of photogrammetry, vol II, 3rd edn. American Society of Photogrammetry, Falls Church, pp 803–849

Álvarez P, Antonio J, Herrera VM, Martínez del Pozo JÁ, de Tena MT (2013) Multi-temporal archaeological analyses of alluvial landscapes using the photogrammetric restitution of historical flights: a case study of Medellin (Badajoz, Spain). J Archaeol Sci 40:349–364. doi:10.1016/j.jas.2012.08.025

American Society for Photogrammetry and Remote Sensing, Specifications and Standards Committee (1990) ASPRS accuracy standards for large-scale maps. Photogramm Eng Remote Sens 56:1068–1070

Appetecchia A, Brandt O, Menander H, Thorén H (2012) New methods for documentation and analysis in building archaeology: prestudy. a project funded by the Swedish National Heritage Board, R & D funds, Lund. http://www.arkeologiuv.se/cms/showdocument/documents/extern_webbplats/arkeologiuv/publikationer_uv/rapporter/uv_rapport/2012/uvr2012_001.pdf. Accessed 4 Feb 2013

Barazzetti L, Remondino F, Scaioni M (2011) Automated and accurate orientation of complex image sequences. In: 3D-ARCH 2011: 3D virtual reconstruction and visualization of complex architectures, Proceedings of the 4th ISPRS international workshop, Trento, Italy, 2–4 Mar 2011. ISPRS

Barber M (2011) A history of aerial photography and archaeology. Mata Hari's glass eye and other stories. English Heritage, Swindon

Bay H, Tuytelaars T, Gool L (2006) SURF: speeded up robust features. In: Aleš L, Horst B, Axel P (eds) Computer vision, 9th European conference on computer vision (ECCV 2006, Graz, Austria, May 7–13, 2006), Proceedings, part I, vol 3951, Lecture notes in computer science. Springer, Berlin, pp 404–417

Bay H, Ess A, Tuytelaars T, van Gool L (2008) SURF: speeded up robust features. Comput Vis Image Underst 110:346–359

Bernstein R (1983) Image geometry and rectification. In: Colwell RN, Simonett DS, Ulaby FT (eds) Manual of remote sensing, vol. 1: Theory, instruments and techniques, 2nd edn. American Society of Photogrammetry, Falls Church, pp 873–922

Bewley R, Rączkowski W (eds) (2002) Aerial archaeology. Developing future practice, vol 337, NATO science series I: life and behavioural sciences. IOS Press, Amsterdam

Bezzi L (2012) 3D documentation of small archaeological finds. http://arc-team-open-research.blogspot.com.br/2012/08/3d-documentation-of-small.html. Accessed 11 October 2012

Billingsley FC (1965) Digital video processing at JPL. In: Electronic Imaging Techniques I, vol 15. SPIE, Bellingham

Billingsley FC, Anuta PE, Carr JL, McGillem CD, Smith DM, Strand TC (1983) Data processing and reprocessing. In: Colwell RN, Simonett DS, Ulaby FT (eds) Manual of remote sensing, vol. 1: Theory, instruments and techniques, 2nd edn. American Society of Photogrammetry, Falls Church, pp 719–792

Bradley D, Boubekeur T, Heidrich W (2008) Accurate multi-view reconstruction using robust binocular stereo and surface meshing. In: CVPR 2008. IEEE conference on computer vision and pattern recognition, 23–28 June 2008. IEEE, Anchorage, pp 1–8. doi:10.1109/CVPR.2008.4587792

Braun J (2003) Aspects on true-orthophoto production. In: Fritsch D (ed) Photogrammetric week '03. Wichmann Verlag, Heidelberg, pp 205–214

Brophy K, Cowley D (eds) (2005) From the air. Understanding aerial archaeology. Tempus, Stroud

Brown DC (1966) Decentering distortion of lenses: the prism effect encountered in metric cameras can be overcome through analytic calibration. Photogramm Eng Remote Sens 32:444–462

Brown DC (1956) The simultaneous determination of the orientation and lens distortion of a photogrammetric camera. Air Force Missile Test Center Technical Report 56–20. Florida

Brugioni DA (1989) The serendipity effect of aerial reconnaissance. Interdiscip Sci Rev 14:16–28. doi:10.1179/030801889789798357

Buchanan T (1993) Photogrammetry and projective geometry: an historical survey. In: Integrating photogrammetric techniques with scene analysis and machine vision, Orlando, FL, USA, 11 Apr 1993. SPIE, Bellingham, pp 82–91. doi:10.1117/12.155817

Burnside CD (1985) Mapping from aerial photographs, 2nd edn. Collins, London

Castrianni L (2008) Giacomo Boni: a pioneer of the archaeological aerial photography. In: Remote sensing for archaeology and cultural heritage management: proceedings of the 1st international EARSeL workshop, CNR, Rome, Arracne, Rome, September 30–October 4, 2008, pp 55–58

Coleman S (2007) Taking advantage: vertical aerial photographs commissioned for local authorities. In: Mills J, Palmer R (eds) Populating clay landscapes. Tempus, Stroud, pp 28–33

Colwell RN (1997) History and place of photographic interpretation. In: Philipson WR (ed) Manual of photographic interpretation, 2nd edn. American Society of Photogrammetry and Remote Sensing, Bethesda, pp 3–47

Cowley DC, Stichelbaut BB (2012) Historic aerial photographic archives for European archaeology. Eur J Archaeol 15:217–236. doi:10.1179/1461957112Y.0000000010

Cowley D, Standring RA, Abicht MJ (eds) (2010) Landscapes through the lens. Aerial photographs and historic environment, vol 2, Occasional publication of the Aerial Archaeology Research Group. Oxbow Books, Oxford/Oakville

Cowley DC, Ferguson LM, Allan W (2013) The aerial reconnaissance archives: a global aerial photographic collection. In: Hanson WS, Oltean IA (eds) Archaeology from historical aerial and satellite archives. Springer, New York, pp 13–30

Crawford OGS (1924) Air survey and archaeology, vol 7, Ordnance survey professional papers, New series. Ordnance Survey, Southampton

Crawford OGS (1929) Air photographs of the Middle East: a paper read at the evening meeting of the Society on 18 March 1929. Geogr J 73:497–509

Crawford OGS (1933) Some recent air discoveries. Antiquity 7:290–296

Crawford OGS, Keiller A (1928) Wessex from the air. Oxford University Press, Oxford

Crawshaw A (1995) Oblique aerial photography: aircraft, cameras and films. In: Kunow J (ed) Luftbildarchäologie in Ost- und Mitteleuropa/Aerial archaeoloy in Eastern and Central Europe: internationales symposium, Kleinmachnow, Land Brandenburg, 26–30, September 1994, vol 3, Forschungen zur Archäologie im Land Brandenburg. Verlag Brandenburgisches Landesmuseum für Ur- und Frühgeschichte, Potsdam, pp 67–76

Crawshaw A (1997) Letter. AARGnews 14:59

Dalal N, Triggs B (2005) Histograms of oriented gradients for human detection. In: Proceedings of the IEEE Computer Society conference on computer vision and pattern recognition, San Diego, CA, USA, 20–25 June 2005. IEEE Computer Society, Los Alamitos, pp 886–893. doi:10.1109/CVPR.2005.177

Deng H, Wei Zhang, Mortensen E, Dietterich T, Shapiro L (2007) Principal curvature-based region detector for object recognition. In: Proceedings of the 2007 IEEE conference on computer vision and pattern recognition CVPR '07, Minneapolis, MN, USA, 18–23 June. IEEE, Piscataway, pp 1–8. doi:10.1109/CVPR.2007.382972

Dickinson GC (1969) Maps and air photographs. Edward Arnold, London

Doneus M (1997) On the archaeological use of vertical photographs. AARGnews 15:23–27
Doneus M (2000) Vertical and oblique photographs. AARGnews 20:33–39
Doneus M, Eder-Hinterleitner A, Neubauer W (2001) Archaeological prospection in Austria. In: Archaeological prospection: fourth international conference on archaeological prospection, Vienna, 19–23 Sept 2001. Austrian Academy of Sciences, Vienna, pp 11–33
Doneus M, Briese C, Fera M, Fornwagner U, Griebl M, Janner M, Zingerle M-C (2007) Documentation and analysis of archaeological sites using aerial reconnaissance and airborne laser scanning. In: Anticipating the future of the cultural past: proceedings of the XXI international CIPA symposium, Athens, Greece, 1–6 Oct 2007, The ISPRS international archives of the photogrammetry, remote sensing and spatial information sciences. CIPA, Athens, vol XXXVI-5/C53, pp 275–280. ISSN 1682–1750
Doneus M, Verhoeven G, Fera M, Briese C, Kucera M, Neubauer W (2011) From deposit to point cloud: a study of low-cost computer vision approaches for the straightforward documentation of archaeological excavations. In: Geoinformatics 6, XXIIIrd international CIPA Symposium, pp 81–88
Eisenbeiss H (2009) UAV photogrammetry. PhD thesis, ETH Zürich, Zürich. http://e-collection.library.ethz.ch/eserv/eth:498/eth-498-02.pdf#search=%22%28author:henri%20eisenbeiss%29%22. Accessed 11 Feb 2013
Eisenbeiss H, Sauerbier M (2011) Investigation of UAV systems and flight modes for photogrammetric applications. Photogramm Rec 26:400–421. doi:10.1111/j.1477-9730.2011.00657.x
Eisenbeiss H, Sauerbier M, Zhang L, Grün A (2005) Mit dem Modellhelikopter über Pinchango Alto. Geomat Schweiz 9:510–515
El-Hakim, SF, Beraldin J-A, Picard M (2003) Effective 3D modeling of heritage sites. In: Proceedings of the 4th international conference 3-D digital imaging and modeling, Banff, Canada, 6–10 October. IEEE Computer Society Press, Los Alamitos, pp 302–309
Estes JE, Hajic EJ, Tinney LR, Carver LG, Cosentino MJ, Mertz FC, Pazner MI, Ritter LR, Sailer CT, Stow DA, Streich TA, Woodcock CE (1983) Fundamentals of image analysis: analysis of visible and thermal infrared data. In: Colwell RN, Simonett DS, Ulaby FT (eds) Manual of remote sensing, vol. 1: Theory, instruments and techniques, 2nd edn. American Society of Photogrammetry, Falls Church, pp 987–1124
Falkner E, Morgan D (2002) Aerial mapping. Methods and applications, 2nd edn, Mapping sciences series. Lewis, Boca Raton
Faugeras O, Luong Q-T, Papadopoulo T (2001) The geometry of multiple images. The laws that govern the formation of multiple images of a scene and some of their applications. MIT Press, Cambridge
Federal Geographic Data Committee – Subcommittee for Base Cartographic Data (1998) Geospatial positioning accuracy standards. Part 3: National Standard for Spatial Data Accuracy (FGDC-STD-007.3-1998). Federal Geographic Data Committee, Reston
Fischler MA, Bolles RC (1981) Random sample consensus: a paradigm for model fitting with applications to image analysis and automated cartography. Commun ACM 24:381–395. doi:10.1145/358669.358692
Fisher RB, Dawson-Howe K, Fitzgibbon A, Robertson C, Trucco E (2005) Dictionary of computer vision and image processing. Wiley, Chichester
Forte M, Dell'unto N, Issavi J, Onsurez L, Lercari N (2012) 3D archaeology at Çatalhöyük. Int J Herit Digit Era 1:352–378. doi:10.1260/2047-4970.1.3.351
Furukawa Y, Ponce J (2010) Accurate, dense, and robust multiview stereopsis. IEEE Trans Pattern Anal Mach Intell 32:1362–1376. doi:10.1109/TPAMI.2009.161
Goesele M, Curless B, Seitz SM (2006) Multi-view stereo revisited. In: Proceedings of the 2006 IEEE Computer Society conference on computer vision and pattern recognition CVPR'06. IEEE Computer Society Press, Los Alamitos, 17–22 June 2006, vol. 2, pp 2402–2409. doi:10.1109/CVPR.2006.199.
Graham R, Koh A (2002) Digital aerial survey. Theory and practice. CRC Press/Whittles Publishing, Boca Raton
Gruner H, Pestrecov K, Norton CL, Tayman WP, Washer FE (1966) Elements of photogrammetric optics. In: Thompson MM, Eller RC, Radlinski WA, Speert JL (eds) Manual of photogrammetry, vol I, 3rd edn. American Society of Photogrammetry, Falls Church, pp 67–132
Gyer MS (1996) Methods for computing photogrammetric refraction corrections for vertical and oblique photographs. Photogramm Eng Remote Sens 62:301–310
Habbecke M, Kobbelt L (2006) Iterative multi-view plane fitting. In: Kobbelt L, Kuhlen T, Aach T, Westerman R (eds) Proceedings of the 11th international fall workshop vision, modeling, and visualization 2006, Aachen, Germany, 22–24 Nov 2006. Akademische Verlagsgesellschaft Aka GmbH, Berlin, pp 73–80
Hallert B (1960) Photogrammetry. Basic principles and general survey, McGraw-Hill civil engineering series. McGraw-Hill, New York
Hanson WS, Oltean IA (eds) (2013) Archaeology from historical aerial and satellite archives. Springer, New York
Harman WE Jr, Miller RH, Sidney Park W, Webb JP (1966) Aerial photography. In: Thompson MM, Eller RC, Radlinski WA, Speert JL (eds) Manual of photogrammetry, vol I, 3rd edn. American Society of Photogrammetry, Falls Church, pp 195–242
Harris C, Stephens M (1988) A combined corner and edge detector. In: Proceedings of the fourth Alvey Vision conference AVC88, University of Sheffield Printing Office; Sheffield, 31 August–2 September 1988. BMVA, pp 147–151
Hartley RI (1994) Projective reconstruction and invariants from multiple images. IEEE Trans Pattern Anal Mach Intell 16:1036–1041. doi:10.1109/34.329005
Hartley RI, Mundy JL (1993) Relationship between photogrammetry and computer vision. In: SPIE (ed) Integrating

photogrammetric techniques with scene analysis and machine vision, 11 Apr 1993, Orlando, FL, USA. SPIE, Bellingham, pp 92–105. doi:10.1117/12.155818

Hartley R, Zisserman A (2003) Multiple view geometry in computer vision, 2nd edn. Cambridge University Press, Cambridge

Hassett TJ, Mullen RR, Pilonero JT, Pugh HV, Freeman J, Speert JL (1966) Aerial mosaics and photomaps. In: Thompson MM, Eller RC, Radlinski WA, Speert JL (eds) Manual of photogrammetry, vol II, 3rd edn. American Society of Photogrammetry, Falls Church

Hirschmüller H (2008) Stereo processing by semiglobal matching and mutual information. IEEE Trans Pattern Anal Mach Intell 30:328–341. doi:10.1109/TPAMI.2007.1166

Imhof RK, Doolittle RC (1966) Mapping from oblique photographs. In: Thompson MM, Eller RC, Radlinski WA, Speert JL (eds) Manual of photogrammetry, vol II, 3rd edn. American Society of Photogrammetry, Falls Church, pp 875–917

Juan L, Gwon O (2009) A comparison of SIFT, PCA-SIFT and SURF. Int J Image Process 3:143–152

Jurie F, Schmid C (2004) Scale-invariant shape features for recognition of object categories. In: Proceedings of the 2004 IEEE Computer Society conference on computer vision and pattern recognition, CVPR 2004. IEEE Computer Society Press, Los Alamitos, 27 June–2 July, vol. 2, pp 90–96. doi:10.1109/CVPR.2004.1315149

Kadir T, Brady M (2001) Saliency, scale and image description. Int J Comput Vis 45:83–105. doi:10.1023/A:1012460413855

Kennedy D (1996) Aerial archaeology in the Middle East. AARGnews 12:11–15

Kersten TP, Lindstaedt M (2012) Potential of automatic 3D object reconstruction from multiple images for applications in architecture, cultural heritage and archaeology. Int J Herit Digit Era 1:399–420. doi:10.1260/2047-4970.1.3.399

Kraus K (2002) Zur Orthophoto-Terminologie. Photogramm Fernerkund Geoinf 6:451–452

Kraus K (2007) Photogrammetry. Geometry from images and laser scans, 2nd edn. Walter de Gruyter, Berlin-New York

Krijnen F (2008) A fresh look at aerial photography. http://www.aircatcher.com

Labun E (2009) ImageJ SURF. http://labun.com/imagej-surf/ImageJ_SURF_2009-12-01_08.19.jar

Lerma JL, Navarro S, Cabrelles M, Seguí AE, Haddad N, Akasheh T (2011) Integration of laser scanning and imagery for photorealistic 3D architectural documentation. In: Wang C-C (ed) Laser scanning, theory and applications. InTech, Shanghai, pp 413–430

Lhuillier M, Quan L (2005) A quasi-dense approach to surface reconstruction from uncalibrated images. IEEE Trans Pattern Anal Mach Intell 27:418–433. doi:10.1109/TPAMI.2005.44

Lindeberg T (1998) Feature detection with automatic scale selection. Int J Comput Vis 30:79–116. doi:10.1023/A:1008045108935

Lo Brutto M, Meli P (2012) Computer vision tools for 3D modelling in archaeology. Int J Herit Digit Era 1:1–6. doi:10.1260/2047-4970.1.0.1

Lo Brutto M, Borruso A, D'Argenio A (2012) UAV systems for photogrammetric data acquisition of archaeological sites. Int J Herit Digit Era 1:7–14. doi:10.1260/2047-4970.1.0.7

Lowe DG (2004) Distinctive image features from scale-invariant keypoints. Int J Comput Vis 60:91–110. doi:10.1023/B:VISI.0000029664.99615.94

Ludvigsen M, Eustice R, Singh H (2006) Photogrammetric models for marine archaeology. In: Proceedings of the IEEE/MTS OCEANS'06 conference and exhibition, Boston, MA, 18–21 Sept 2006. IEEE, Piscataway, pp 1–6. doi:10.1109/OCEANS.2006.306915

Manzer G (1996) Avoiding digital orthophoto problems. In: Greve C (ed) Digital photogrammetry: an addendum to the manual of photogrammetry. American Society of Photogrammetry and Remote Sensing, Falls Church, pp 158–162

Matas J, Chum O, Urban M, Pajdla T (2004) Robust wide-baseline stereo from maximally stable extremal regions. Image Vis Comput 22:761–767. doi:10.1016/j.imavis.2004.02.006

Mellor JP, Teller S, Lozano-Pérez T (1996) Dense depth maps from epipolar images, vol 1953, AI Lab technical memo. Massachusetts Institute of Technology/Artificial Intelligence Laboratory, Cambridge

Microdrones GmbH (2008) Key Information for md4-1000 http://www.microdrones.com/products/md4-1000/md4-1000-key-information.php. Accessed 21 April 2008

Microsoft Corporation (2010) Photosynth. Microsoft Corporation, Redmond, http://photosynth.net/

Mikhail EM, Bethel JS, Chris McGlone J (2001) Introduction to modern photogrammetry. Wiley, New York

Mikolajczyk K, Schmid C (2003) A performance evaluation of local descriptors. In: Proceedings of the 2003 IEEE Computer Society conference on computer vision and pattern recognition, CVPR 2003, Madison, WI, USA, 16–22 June 2003, vol. 2. IEEE Computer Society, Los Alamitos, pp 257–263. doi:10.1109/CVPR.2003.1211478

Mikolajczyk K, Schmid C (2005) A performance evaluation of local descriptors. IEEE Trans Pattern Anal Mach Intell 27:1615–1630. doi:10.1109/TPAMI.2005.188

Mikolajczyk K, Tuytelaars T, Schmid C, Zisserman A, Matas J, Schaffalitzky F, Kadir T, van Gool L (2005) A comparison of affine region detectors. Int J Comput Vis 65:43–72

Mills J (2005) Bias and the world of the vertical aerial photograph. In: Brophy K, Cowley D (eds) From the air: understanding aerial archaeology. Tempus, Stroud, pp 117–126

Moisan L, Stival B (2004) A probabilistic criterion to detect rigid point matches between two images and estimate the fundamental matrix. Int J Comput Vis 57:201–218. doi:10.1023/B:VISI.0000013094.38752.54

Moons T, van Gool L, Vergauwen M (2008) 3D Reconstruction from multiple images, part 1: Principles. Found Trends Comput Graph Vis 4:287–404. doi:10.1561/0600000007

Moreels P, Perona P (2007) Evaluation of features detectors and descriptors based on 3D objects. Int J Comput Vis 73:263–284. doi:10.1007/s11263-006-9967-1
Morel J-M, Guoshen Yu (2009) ASIFT: a new framework for fully affine invariant image comparison. SIAM J Imaging Sci 2:438–469. doi:10.1137/080732730
Moscatelli U (1985) Municipi romani della V regio Augustea: problemi storici ed urbanistici del Piceno centro-settentrionale (III – I sec. a.C.). PICUS Studi e ricerche sulle Marche nell'antichità 5:51–97
Moscatelli U (1987) Materiali per la topografia storica di *Potentia*. In: Paci G (ed) Miscellanea di studi marchigiani in onore di Febo Allevi. Facoltà di Lettere e Filosofia/Università di Macerata, Agugliano, pp 429–438
Mundy JL, Zisserman A (1992) Appendix – projective geometry for machine vision. In: Mundy JL, Zisserman A (eds) Geometric invariance in computer vision. MIT Press, Cambridge, pp 463–534
Newhall B (2006) The history of photography. From 1839 to the present, 5th edn. Museum of Modern Art, New York/Boston
Norton PR (2010) Photodetectors. In: Bass M, DeCusatis CM, Enoch JM, Lakshminarayanan V, Li G, MacDonald CA, Mahajan VN, van Stryland EW (eds) Handbook of optics, vol. II. Design, fabrication, and testing; sources and detectors; radiometry and photometry, 3rd edn. McGraw-Hill, New York, pp 24.3–24.102
Ohno Y (2006) Basic concepts in photometry, radiometry and colorimetry. In: Dakin JP, Brown RGW (eds) Handbook of optoelectronics. Taylor & Francis, Boca Raton, pp 287–305
Opitz R, Nowlin J (2012) Photogrammetric modeling + GIS: better methods for working with mesh data. ArcUser Spring:46–49
Palmer R (1996) Editorial. AARGnews 13:3
Palmer R (2005) If they used their own photographs they would not take them like that. In: Brophy K, Cowley D (eds) From the air: understanding aerial archaeology. Tempus, Stroud, pp 94–116
Palmer R (2007) Seventy-five years v. Ninety minutes: implications of the 1996 Bedfordshire vertical aerial survey on our perceptions of clayland archaeology. In: Mills J, Palmer R (eds) Populating clay landscapes. Tempus, Stroud, pp 88–103
Palmer JM, Grant BG (2010) The art of radiometry. SPIE, Bellingham
Pollefeys M, van Gool L, Vergauwen M, Cornelis K, Verbiest F, Tops J (2001) Image-based 3D acquisition of archaeological heritage and applications. In: Proceedings of the 2001 conference on virtual reality, archaeology, and cultural heritage, Glyfada, Greece, 28–30 Nov 2001. Association for Computing Machinery, New York, pp 255–262
Pollefeys M, van Gool L (2002) Visual modelling: from images to images. J Vis Comput Animat 13:199–209. doi:10.1002/vis.289
Pollefeys M, Koch R, Vergauwen M, van Gool L (1998) Virtualizing archaeological sites. In Proceedings of the 4th international conference on virtual systems and multimedia, VSMM 98, Gifu, Japan, 18–20 Nov 1998. IOS Press, Amsterdam
Pollefeys M, Koch R, Vergauwen M, van Gool L (2000) Automated reconstruction of 3D scenes from sequences of images. ISPRS J Photogramm Remote Sens 55: 251–267. doi:10.1016/S0924-2716(00)00023-X
Pollefeys M, van Gool L, Vergauwen M, Cornelis K, Verbiest F, Tops J (2003) 3D recording for archaeological fieldwork. IEEE Comput Graph Appl 23: 20–27. doi:10.1109/MCG.2003.1198259
Pollefeys M, van Gool L, Vergauwen M, Verbiest F, Cornelis K, Tops J, Koch R (2004) Visual modeling with a hand-held camera. Int J Comput Vis 59:207–232. doi:10.1023/B:VISI.0000025798.50602.3a
Quan L (2010) Image-based modeling. Springer, New York
Read RE, Graham R (2002) Manual of aerial survey. Primary data acquisition. CRC Press/Whittles Publishing, Boca Raton
Reinhard J (2012) Things on strings and complex computer algorithms: kite aerial photography and structure from motion photogrammetry at the Tulul adh-Dhahab, Jordan. AARGnews 45:37–41
Remondino F, Fraser C (2006) Digital camera calibration methods: considerations and comparisons. In ISPRS Commission V symposium 'image engineering and vision metrology', 25–27 Sept 2006. International Society for Photogrammetry and Remote Sensing, Dresden, pp 266–272
Remondino F, Barazzetti L, Nex F, Scaioni M, Sarazzi D (2011) UAV photogrammetry for mapping and 3d modelling: current status and future perspectives. In: Proceedings of the international conference on unmanned aerial vehicle in geomatics UAV-g, Zurich, Switzerland, 14–16 Sept 2011, vol 38(1/C22). International Archives of Photogrammetry, Remote Sensing and Spatial Information Sciences, Zürich
Remondino F, Del Pizzo S, Kersten TP, Troisi S (2012) Low-cost and open-source solutions for automated image orientation: a critical overview. In: Progress in cultural heritage preservation. In: Proceedings of the 4th international conference Euromed 2012, Lemessos, Cyprus. October 29–November 3, 2012. Springer, Berlin/Heidelberg, pp 40–54
Robertson DP, Cipolla R (2009) Structure from motion. In: Varga M (ed) Practical image processing and computer vision. Wiley, New York
Rosten E, Drummond T (2005) Fusing points and lines for high performance tracking. In: Proceedings of the tenth IEEE international conference on computer vision ICCV'05. IEEE Computer Society Press, Los Alamitos, 17–21 Oct 2005, vol 2, pp 1508–1515. doi:10.1109/ICCV.2005.104
Rousseeuw PJ (1984) Least median of squares regression. J Am Stat Assoc 79:871–880. doi:10.2307/2288718
Sarfraz MS, Hellwich O (2008) Head pose estimation in face recognition across pose scenarios. In: Proceedings of the third international conference on computer vision theory and applications VISAPP 2008, Funchal,

Portugal, 22–25 Jan 2008, vol 1. INSTICC, Setúbal, pp 235–242
Scharstein D, Szeliski R (2002) A taxonomy and evaluation of dense two-frame stereo correspondence algorithms. Int J Comput Vis 47:7–42
Schlitz M (2004) A review of low-level aerial archaeology and its application in Australia. Aust Archaeol 59:51–58
Schmid C, Mohr R (1996) Combining grey value invariants with local constraints for object recognition. In: Proceedings of the 1996 IEEE Computer Society conference on computer vision and pattern recognition CVPR '96, San Francisco, California, 18 June–20 June 1996. IEEE Computer Society Press, Los Alamitos, pp 872–877. doi:10.1109/CVPR.1996.517174
Schneider S (1974) Luftbild und Luftbildinterpretation, vol 11, Lehrbuch der allgemeinen Geographie. Walter de Gruyter, Berlin/New York
Schott JR (2007) Remote sensing. The image chain approach, 2nd edn. Oxford University Press, New York
Schreiber WF (1967) Picture coding. Proc IEEE 55:320–330. doi:10.1109/PROC.1967.5488
Scollar I, Giradeau-Montaut D (2012) Georeferenced orthophotos and DTMs from multiple oblique images. AARGnews 44:12–17
Scollar I, Tabbagh A, Hesse A, Herzog I (1990) Archaeological prospecting and remote sensing, vol 2, Topics in remote sensing. Cambridge University Press, Cambridge
Seitz SM, Curless B, Diebel J, Scharstein D, Szeliski R (2006) A comparison and evaluation of multi-view stereo reconstruction algorithms. In: 2006 IEEE Computer Society conference on computer vision and pattern recognition CVPR'06, vol. 1. IEEE, Washington, DC, pp 519–528
Sevara C (2013) Top Secret Topographies: Examining the potential for recovering two and three-dimensional archaeological information from historic reconnaissance datasets using image-based modelling techniques. Inl J of Heritage in the Digital Era 2:3
Sewell ED, Livingston RG, Quick JR, Norton CL, Case JB, Sanders RG, Goldhammer JS, Aschenbrenner B (1966) Aerial cameras. In: Thompson MM, Eller RC, Radlinski WA, Speert JL (eds) Manual of photogrammetry, vol I, 3rd edn. American Society of Photogrammetry, Falls Church, pp 133–194
Slater PN, Doyle FJ, Fritz NL, Welch R (1983) Photographic systems for remote sensing. In: Colwell RN, Simonett DS, Ulaby FT (eds) Manual of remote sensing, vol. 1: Theory, instruments and techniques, 2nd edn. American Society of Photogrammetry, Falls Church, pp 231–291
Smith SW (1997) The scientist and engineer's guide to digital signal processing, 1st edn. California Technical Publishing, San Diego
Snavely N (2010) Bundler: structure from motion for unordered image collections. Software
Snavely N, Seitz SM, Szeliski R (2006) Photo tourism: exploring photo collections in 3D. ACM Trans Graph 25:835–846
Spurr SH (1960) Photogrammetry and photo-interpretation. With a section on applications to forestry, 2nd edn. The Ronald Press Company, New York
Stichelbaut B, Bourgeois J, Saunders D, Chielens P (eds) (2009) Images of conflict. Military aerial photography and archaeology. Cambridge Scholars Publishing, Newcastle upon Tyne
Strecha C, Fransens R, van Gool L (2006) Combined depth and outlier estimation in multi-view stereo. In: Proceedings of the 2006 IEEE Computer Society conference on computer vision and pattern recognition, CVPR'06. IEEE Computer Society Press, Los Alamitos, 17–22 June 2006, vol. 2. pp 2394–2401. doi:10.1109/CVPR.2006.78
Szeliski R (2011) Computer vision. Algorithms and applications, Texts in computer science. Springer, New York
Taelman D, Deprez S, Vermeulen F, De Dapper M (2009) Granite and rock crystal quarrying in the *Civitas Ammaiensis* (north-eastern Alentejo, Portugal): a geo-archaeological case study. BABesch – Bulletin Antieke Beschaving 84:171–182
Tewinkel GC, Schmid HH, Hallert B, Rosenfield GH (1966) Basic mathematics of photogrammetry. In: Thompson MM, Eller RC, Radlinski WA, Speert JL (eds) Manual of photogrammetry, vol I, 3rd edn. American Society of Photogrammetry, Falls Church, pp 17–65
Tingdahl D, Maarten V, van Gool L (2012) ARC3D: a public web service that turns photos into 3D models. In: Stanco F, Battiato S, Gallo G (eds) Digital imaging for cultural heritage preservation: analysis, restoration, and reconstruction of ancient artworks, Digital imaging and computer vision series. CRC Press, Boca Raton, pp 101–125
Torr PHS (2002) Bayesian model estimation and selection for epipolar geometry and generic manifold fitting. Int J Comput Vis 50:35–61. doi:10.1023/A:1020224303087
Triggs B, Mclauchlan PF, Hartley RI, Andrew F (2000) Bundle adjustment – a modern synthesis. In: Triggs B, Zisserman A, Szeliski R (eds) Vision algorithms: theory and practice: proceedings of the international workshop on vision algorithms, Corfu, Greece, September 1999, vol 1883, Lecture notes in computer science. Springer, London, pp 298–372
Turpin RD, Ramey EH, Case JB, Coleman CG, Lynn WD, Michaelis OE (1966) Definitions of terms and symbols used in photogrammetry. In: Thompson MM, Eller RC, Radlinski WA, Speert JL (eds) Manual of photogrammetry, vol II, 3rd edn. American Society of Photogrammetry, Falls Church, pp 1125–1161
Tuytelaars T, Mikolajczyk K (2007) Local invariant feature detectors: a survey. Found Trends Comput Graph Vis 3:177–280. doi:10.1561/0600000017
Ullman S (1979) The interpretation of structure from motion. Proc R Soc B Biol Sci 203:405–426. doi:10.1098/rspb.1979.0006
Verhoeven G (2008a) Exploring the edges of the unseen: an attempt to digital aerial UV photography. In: Remote sensing for archaeology and cultural heritage management: proceedings of the 1st International EARSeL workshop CNR, Rome, September 30–October 4, 2008. Aracne, Rome, pp 79–83

Verhoeven G (2008b) Imaging the invisible using modified digital still cameras for straightforward and low-cost archaeological near-infrared photography. J Archaeol Sci 35:3087–3100. doi:10.1016/j.jas.2008.06.012

Verhoeven G (2009a) Beyond conventional boundaries. New technologies, methodologies, and procedures for the benefit of aerial archaeological data acquisition and analysis. PhD thesis, Nautilus Academic Books, Zelzate

Verhoeven G (2009b) Providing an archaeological bird's-eye view: an overall picture of ground-based means to execute low-altitude aerial photography (LAAP) in archaeology. Archaeol Prospect 16:233–249. doi:10.1002/arp.354

Verhoeven G (2011) Taking computer vision aloft: archaeological three-dimensional reconstructions from aerial photographs with PhotoScan. Archaeol Prospect 18:67–73. doi:10.1002/arp.399

Verhoeven G (2012a) Methods of visualisation. In: Edwards HGM, Vandenabeele PV (eds) Analytical archaeometry: selected topics. Royal Society of Chemistry, Cambridge, pp 3–48

Verhoeven G (2012b) Near-infrared aerial crop mark archaeology: from its historical use to current digital implementations. J Archaeol Method Theory 19:132–160. doi:10.1007/s10816-011-9104-5

Verhoeven G (2012c) Straightforward archeological orthophotos from oblique aerial images. SPIE Newsroom. doi:10.1117/2.1201210.004506

Verhoeven G, Schmitt KD (2010) An attempt to push back frontiers: digital near-ultraviolet aerial archaeology. J Archaeol Sci 37:833–845. doi:10.1016/j.jas.2009.11.013

Verhoeven G, Loenders J, Vermeulen F, Docter R (2009a) Helikite aerial photography: a versatile means of unmanned, radio controlled, low-altitude aerial archaeology. Archaeol Prospect 16:125–138. doi:10.1002/arp.353

Verhoeven G, Smet P, Poelman D, Vermeulen F (2009b) Spectral characterization of a digital still camera's NIR modification to enhance archaeological observation. IEEE Trans Geosci Remote Sens 47:3456–3468. doi:10.1109/TGRS.2009.2021431

Verhoeven G, Doneus M, Briese C, Vermeulen F (2012a) Mapping by matching: a computer vision-based approach to fast and accurate georeferencing of archaeological aerial photographs. J Archaeol Sci 39:2060–2070. doi:10.1016/j.jas.2012.02.022

Verhoeven G, Taelman D, Vermeulen F (2012b) Computer vision-based orthophoto mapping of complex archaeological sites: the ancient quarry of Pitaranha (Portugal-Spain). Archaeometry 54:1114–1129. doi:10.1111/j.1475-4754.2012.00667.x

Vermeulen F (2002) The potenza valley survey (Marche). In: New developments in italian landscape archaeology: theory and methodology of field survey, land evaluation and landscape perception, pottery production and distribution. Proceedings of a three-day conference held at the University of Groningen, 13–15 Apr 2000. Archaeopress, Oxford, pp 104–106

Vermeulen F (2004) Fotografia aerea finalizzata nelle Marche centrali: un progetto integrato. Archeologia Aerea Studi di Aerotopografia Archeologica 1:91–118

Vermeulen F, Taelman D (2010) From cityscape to landscape in Roman Lusitania: the Municipium of Ammaia. In: Changing landscapes: the impact of Roman towns in the Western Mediterranean. Proceedings of the International Colloquium, Castelo de Vide, Marvão. 15–17 May 2008. AnteQuem, Bologna, pp 311–324

Wells J, Wells R (2012). Kite aerial photography in the near infra-red and ultra-violet. http://www.armadale.org.uk/phototech05.htm. Accessed 11 February 2013

Whittlesey JH (1973) Balloons, 'flying mattresses' and photography. Expedition 15:30–39

Wilson DR (1975) Photographic techniques in the air. In: Wilson DR (ed) Aerial reconnaissance for archaeology, vol 12, Research report series. The Council for British Archaeology, London, pp 12–31

Wilson DR (2000) Air photo interpretation for archaeologists, 2nd edn. Tempus, Stroud

Wolf PR, Dewitt BA (2000) Elements of photogrammetry with applications in GIS, 3rd edn. McGraw-Hill, Boston

Xu G, Jun-ichi Terai, Heung-Yeung Shum (2000) A linear algorithm for camera self-calibration, motion and structure recovery for multi-planar scenes from two perspective images. In: Proceedings of the IEEE conference on computer vision and pattern recognition. IEEE Computer Society Press, Los Alamitos, 13–15 June 2000, pp 474–479. doi:10.1109/CVPR.2000.854886

Yu G, Morel J-M (2011) ASIFT: an algorithm for fully affine invariant comparison. IPOL. doi:10.5201/ipol.2011.my-asift

Zantopp R (1995) Methode und Möglichkeiten der Luftbildarchäologie im Rheinland. In: Jürgen K (ed) Luftbildarchäologie in Ost- und Mitteleuropa/Aerial archaeology in Eastern and Central Europe: Internationales Symposium, Kleinmachnow, Land Brandenburg, 26–30 Sept 1994, vol 3, Forschungen zur Archäologie im Land Brandenburg. Verlag Brandenburgisches Landesmuseum für Ur- und Frühgeschichte, Potsdam, pp 155–163

4 Roman Urban Survey: The Mapping and Monitoring of Complex Settlement Sites with Active Aerial Photography

Frank Vermeulen

Contents

F. Vermeulen
Department of Archaeology, Universiteit Gent, Ghent, Belgium
e-mail: frank.vermeulen@ugent.be

4.1 Introduction: Surveying Abandoned Roman Towns

Non-invasive field survey has been making a major contribution to our understanding of the rural landscapes of the Mediterranean for nearly 40 years (Keller and Rupp 1983; Macready and Thompson 1985; Bintliff et al. 2000; Francovich and Patterson 2000; Alcock and Cherry 2004). During that time the techniques used to map ancient settlement patterns have grown in sophistication from being a process of simply identifying sites in the landscape, to one which provided nuanced understandings of their layouts, chronologies and contexts. The 1980s and 1990s witnessed a surge in the number of regional survey projects, which developed increasingly intensive and refined methodologies. The introduction of desktop computers, GIS software and, more recently, mobile technologies (e.g. GPS receivers) contributed to increasingly efficient field procedures, and to enhanced possibilities for the storage and spatial analysis of large amounts of data concerning the rural landscape. One consequence of this was the growing realisation from the 1980s onwards that these same techniques also held out the promise of making a major contribution to our understanding of urban sites. Particularly the large towns, cities and ports of the Classical Mediterranean are a category of huge and complex, diachronic sites which until then were almost solely approached with archaeological excavations and traditional topographic work, typically centred on the more monumental

C. Corsi et al. (eds.), *Good Practice in Archaeological Diagnostics*, Natural Science in Archaeology,
DOI 10.1007/978-3-319-01784-6_4,

or visible structures. This new interest for the category of urban sites, which today are often partly or even fully abandoned, was spurred by seminal projects such as the study of Boeotian towns by Bintliff and Snodgrass (1985), as well as by the refinement of geophysical techniques and aerial photography that could be used for the fine-grained analysis required to bring out details of urban layout (Schmiedt and Castagnoli 1957; Scollar et al. 1990; Doneus 2004; Bourgeois and Meganck 2005; Barber 2011; Vermeulen et al. 2012a, b).

The consequence of all these developments, including the widespread use of GIS in archaeology, has been an upsurge in the non-destructive survey of urban sites, in the Mediterranean and beyond. Archaeologists of the classical periods have been quick to realise the potential offered by this technique (Christie and Augenti 2012; Vermeulen et al. 2012a, b). Large and complex urban sites which had hitherto been studied in a piecemeal approach that was largely predicated upon the monument-based interests of earlier scholars are the past decade increasingly being 'scanned' with survey techniques to rapidly generate plans of partial, or in some cases, complete townscapes. Parallel with some developments of large-scale integrated town survey in more northern regions of the classical world, such as at Wroxeter (Gaffney and Gaffney 2000; White et al. 2013) and *Carnuntum* (Doneus et al. 2001; Doneus and Neubauer 2005), these Mediterranean urban surveys were more and more dominating the large-scale survey efforts of whole teams of researchers, such as is well illustrated at the classical urban sites of *Falerii Novi* (Keay et al. 2000; Patterson 2004), *Portus* (Keay et al. 2005), *Potentia* (Vermeulen et al. 2006), *Italica* (Keay 2010), *Ephesos* (Groh 2012) and *Sagalassos* (Martens et al. 2012), to name but a few. This has also led to a revolution in how archaeologists approach urban sites, with survey techniques being used increasingly often to generate a plan of a town site prior to excavation as a way of ensuring that the excavation can be used to address site-specific questions in a way that had not been possible before. Cultural heritage management authorities have also benefited from this approach, with urban surveys providing them with a very effective tool for gauging the degree of archaeological survival on major urban sites in their care and choosing appropriate conservation strategies.

Most recently, research has begun to reveal the advantages of intensively integrating a range of different non-destructive techniques on urban sites, choosing those suites that are most appropriate for the nature of the town in question, such as is well demonstrated at the ancient urban sites of *Ammaia* (Corsi et al. 2012) and *Tanagra* (Slapšak 2012). The variety of techniques can be quite impressive, such as the application of different geophysical instruments (for georadar, magnetometer or earth resistance survey, etc.), different aerial photography approaches (such as flying with traditional airplanes, drones or balloons or using multispectral techniques of photography), geomorphological and geomatic approaches (coring, erosion modelling, DTM production …), etc. Therefore, the concept of integrated, non-invasive multi-method survey relates to a much wider range of techniques, and the overall methodology envisages a reasoned deployment of them all, or of a choice of them for systematic data acquisition at the site studied, by testing, sampling or total coverage (Slapšak 2012). Urban sites vary greatly between them in terms of applicability of non-invasive techniques to be deployed, depending on geological setting, nature of building materials and degree of preservation, and the whole site history, including duration and phasing, degree of urban change, restructuring and recycling in the lifetime, and degradation after abandonment of the site. Unlike regional surveys, urban prospections have no neutral background against which anomalies appear. The picture tends to be complex, and the tools to address this complexity are normally diverse, and adapted to each case under study. The approaches depend on a variety of factors specific to the local area: the natural setting and geomorphological changes through time, the scale of the site and its depth of monumentality and stratigraphy, the post-urban population presences and needs, the degree of medieval to modern efforts to rob or reclaim the terrain and its materials, sometimes earlier archaeological intrusions, etc. Furthermore, some sites are far more visible or more accessible than others, while some have landmarks which guide readings as to what is missing and some have been more damaged than others (Christie 2012, p. 285). So flexibility rather than uniformity will be the rule, and that requires a thorough understanding of the options at hand, and of the problems we can realistically explore by non-invasive archaeology.

The large sites of these former towns are in a way to be seen as open-air archaeological laboratories where a whole series of exciting advances in non-invasive archaeology are starting to make a very important understanding to urbanism in general and the Roman Empire in particular. In combination with astonishing new computer-based means of data visualisation, all of this work means that it is now possible to virtually reconstruct a buried town within a relatively short space of time, as opposed to the old and destructive excavation-centred approach that could take generations.

Among the wide array of approaches at hand today, active aerial photography of the sites and their surrounding landscape remains a very potent technique of recovery of the buried evidence. When not applied, as so often in the past, in a minimalistic or simple illustrative way, but in a well-considered, intensive, multi-method way, which allows good integration with other techniques, aerial photography has a major role to play in urban survey. In this chapter we will, therefore, focus on the specific contribution of this technique to the active mapping and further monitoring of such urban landscapes. This will be illustrated by some examples taken from our own experience with this technique, as well as from the experience of some colleagues in the very diversified field of aerial photography for the archaeological study of Roman urban sites. Although we will not deal here with the more 'passive' use of vertical photography taken for other purposes than archaeology (see contributions by Ceraudo (Chap. 2) and Verhoeven et al. (Chap. 3), this volume), the significance of this readily available aerial imagery for the study of former urban contexts can surely not be underestimated. This applies as well for certain historical photography as for the now widely available imagery from websites such as Google maps or Bing maps, particularly useful in regions where aerial archaeology flights have never been achieved.

4.2 Discovery and Monitoring

The discipline of aerial archaeology comprises the entire process from the acquisition and inventorying of aerial imagery, to the mapping of relevant features and their final interpretation as archaeological objects. It involves the study of all types of archaeological remains using data collected from an airborne platform: digital or film-based aerial photographs, airborne laser scanning, aerial imaging spectroscopy, etc. Of all archaeological remote-sensing techniques, active aerial photographic reconnaissance from a low-flying aircraft has been the standard since more than a century now (Bourgeois and Meganck 2005). Two main reasons make it one of the most effective methods for site discovery: the non-invasive approach yields easily interpretable imagery with abundant spatial detail, and the method is driven by the specific nature of the partly eroded or subsurface archaeological features. The latter typically show up on the surface under certain conditions as 'visibility marks': i.e. indirect indicators of archaeological residues due to the changing properties of the local topography or the soil matrix (Whimster 1989; Scollar et al. 1990; Wilson 2000; Bewley and Raczkowski 2002; Bourgeois and Meganck 2005; Brophy and Cowley 2005; Palmer 2005; Barber 2011).

The vast majority of the photographs taken from a low-flying aircraft will be oblique in nature, that is, taken at an angle to the earth's surface in order to take full advantage of the best visibility viewpoint of the surveyor. The views are directly selected during the flight by the archaeologist, under best conditions in terms of light, visibility and readability of the surface. The photographs are normally produced at very reasonable cost and with high flexibility. In general, the images are acquired using a small- or medium-format hand-held photographic camera. The archaeologist flies around in a certain area, according to a random or well-prepared strategic pattern, which simple GPS receivers can easily map afterwards. Once a possible archaeological anomaly is detected in the landscape, it is captured in an oblique photograph and used for subsequent study. Although aerial archaeologists have been experimenting with different filters and film emulsions (e.g. near-infrared sensitive media with a yellow filter to highlight certain crop marks), the majority of archaeological aerial photographs have been shot using media sensitive only to visible radiation. During the last decade, aerial archaeologists have started to use digital photographic cameras as their main working tool. Although the core functionality of still cameras has remained largely unaltered, the imaging sensors embedded in these digital devices are sensitive to both invisible near-ultraviolet (NUV) and near-infrared (NIR) radiation, which has been demonstrated to allow the production of imagery that can be sharper or often richer in archaeological information (Verhoeven 2008, 2011a; Verhoeven et al. 2009; Verhoeven and Schmitt 2010).

Specific to the study by way of active aerial photography of (partly or fully) abandoned Roman urban sites is that parts of these ancient towns are often already discovered in the past. The contribution of a systematic reconnaissance of these large sites lies, therefore, not so much in their initial finding, but more in their full comprehension as an urban landscape, including a first appreciation of their total size, their planned layout (wall circuits, street network, …), their relation to the general landscape (roads, field systems,…), their suburban areas, etc. Also recurrent is the first discovery, thanks to prolonged flying at different moments of the year, under ever-changing conditions of ground visibility, of monumental or other architectural features that populate such rich urban contexts, including theatres, amphitheatres, aqueducts, forums, large houses, and sanctuaries (Fig. 4.1). The specificity of Roman townscapes with their good architectural visibility, and often the homogeneity of the application of certain architectural models and plans, is of course a great help in this. This applies in particular to the phases of great expansion of the Roman world and of the heyday of its urban civilization, from mid-Republican times to Late Antiquity (circa fourth century BC to AD fifth century), when townscapes were created with an incredible good archaeological visibility. The sheer size of the former population centres and of certain architectural creations, the systematic choice of durable building materials and the sometimes quite impressive impact on the local topography of a place all contribute well to this present-day vision from the air, even on the hundreds of Roman town sites that have today left no or very few traces above ground. And even if the surface of such sites is still littered today with archaeological remains that have been (partly) studied and recorded from the ground, by way of ground-based photography, drawings or even traditional excavation, the aerial survey view and active recording from low or high altitude can often still add important information. This is particularly the case in landscapes of the Roman Empire, such as in North Africa or the East, which have not suffered so much the transformations, since later Medieval times and increasingly since the twentieth century, as is typical for large parts of western and southern Europe (Fig. 4.2).

As the object of our active aerial photography over these ancient town sites is so large, a specific strategy is needed, which differs considerably from the normal full landscape aerial surveys. In particular when the urban sites are located in active agricultural landscapes, it is seldom possible to photograph them totally and ideally during one flight only. Whichever the specific climatic conditions or season, and whichever the type of crops or vegetation covering the many different parts of the cityscape, regular and numerous flying will be necessary. Each flight will probably deliver some fragmentary information that was not picked up yet, even if many flights can be devoid of any new data, as is usual for general aerial surveys over whole landscapes. A crucial part of the strategy is that once a potential architectural structure or important new

Fig. 4.1 Exceptional aerial view taken in 2003 of the crop marks revealing buried structures of the Roman theatre of *Suasa* (Marche, Italy) (Courtesy of M. Destro and E. Giorgi)

feature is spotted, there is a regular follow-up of the area during different seasons and in different weather conditions. Important also is the idea that the town area will be controlled several times a year, which makes it possible to organise a real follow-up. We are convinced that this monitoring of the urban sites is an important element in a full comprehension of the many archaeological structures present in the soil and the only possible approach to its full complexity. Indeed, one encounters many examples of truly remarkable 'evolutions' of urban archaeological sites due to totally different detection opportunities over different moments, seasons or years. These repetitive observations conducted over a period of several years draw new details again and again from the soil. These can be joined, like pieces of a puzzle, into an extensive overall view of the urban landscape at one time in its development. With luck also diachronic evolutions of the town plan can be observed. The latter are particularly evident when features cross each other with different orientations or when the pattern of detected wall structures in an area is so dense that they can only be the result of regular rebuilding on the same spot.

My own experience from several urban surveys in Italy and southern France (Vermeulen 2004, 2011; Vermeulen and Verhoeven 2004; Vermeulen et al. 2005) where active aerial photography has been applied from a 'classic' manned aircraft (mostly bi- to four-sitters, but also ULMs) learns that very useful imagery of complex urban sites can also best be acquired from low-altitude unmanned platforms. Today all kinds of devices (kites, balloons, drones, …) are being used to take still cameras aloft and remotely gather aerial imagery (see Chap. 3 by Verhoeven et al., this volume). Especially radio-controlled multi-copter platforms are at present popularising aerial photography over large and complex sites. Their remarkable speed and image quality and their development into increasingly low-cost solutions make these platforms potent instruments

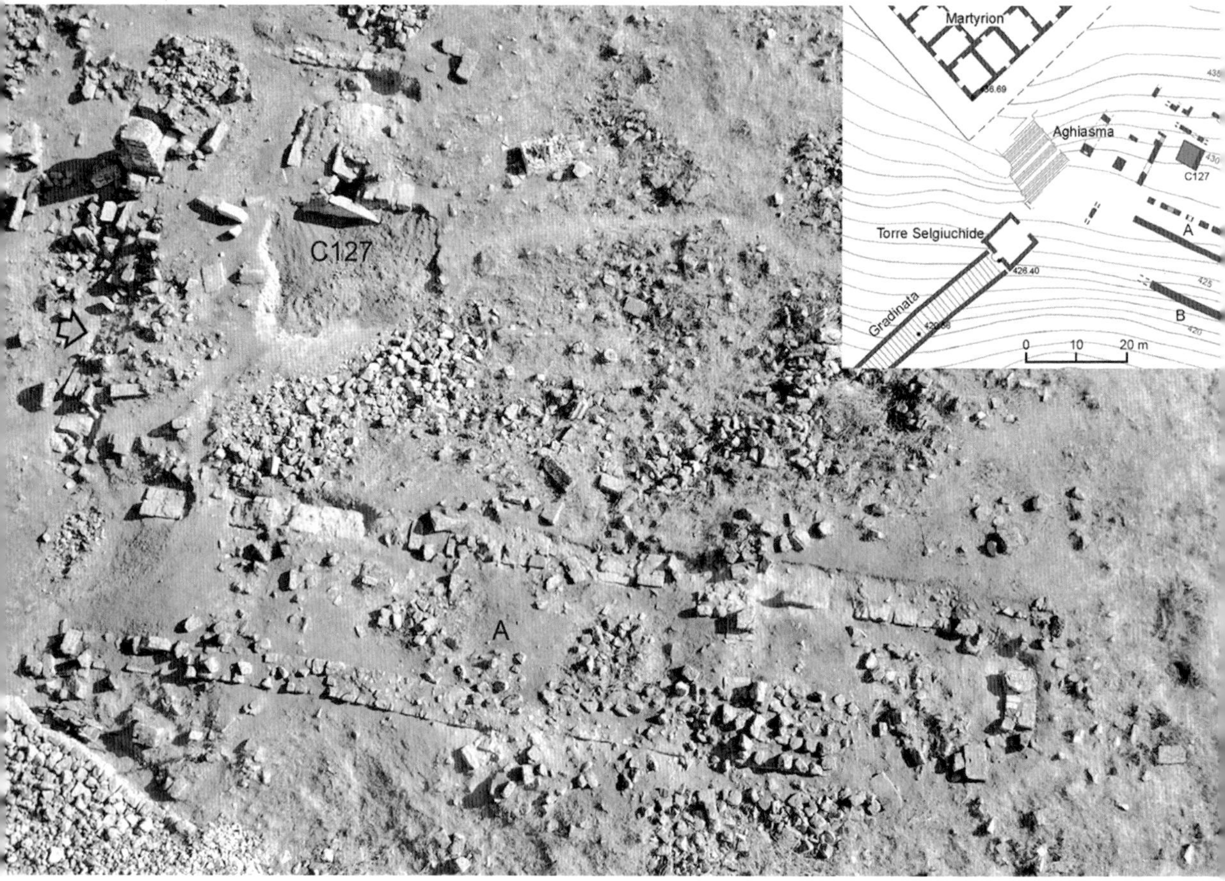

Fig. 4.2 Oblique aerial view from 2009 and mapping of archaeological structures visible above ground in an area south of the Martyrion of *St. Philip* in the ancient town of *Hierapolis* (Phrygia, Turkey) (Courtesy of G. Scardozzi)

for the aerial study and monitoring of (Roman) urban sites, wherever the local legal context allows archaeologists to use them. In addition, they stimulate the further exploitation of new imaging techniques, such as close-range near-infrared photography (Verhoeven 2008, 2011a) and near-ultraviolet imaging (Verhoeven and Schmitt 2010). These techniques, involving a wider use of the spectrum, seem particularly useful when surveying Roman urban landscapes, as the omnipresence of durable building materials in the subsoil guarantees good results when the moment of photography is well chosen. This is particularly well illustrated by a recent example from the *Portus* project, where parts of a dense economic neighbourhood along a Roman channel linking the Trajanic harbour of Rome with the River Tiber were visualised thanks to active near-infrared photography (Fig. 4.3).

The increasing use of such low-altitude unmanned platforms is not only of consequence for the further discovery of features. It will certainly enhance the possibilities for regular monitoring of the sites from the point of view of site management and curation. This is a main preoccupation when considering the huge amount of classical urban sites in the Mediterranean which can at best be preserved for the future and where non-invasive approaches to survey and small-scale stratigraphic operations are to be preferred to wide-scale excavation with all its consequences for site conservation.

4.3 Mapping, Interpretation and Integration

To fully understand such complex urban landscapes and to combine the evidence that may be recorded in sometimes hundreds or even thousands of photographs, the aerial photographs have to be mapped. Today, evidence from oblique photographs taken at different times, dates and altitudes (scales) can be quite easily compiled as orthophotos and maps, using GIS technology and a series of operations involving new software packages and suites (see Verhoeven 2011b; Verhoeven et al. 2012a, b and Chap. 3, this volume). The mapping process of the features detected on large and complex sites, such as abandoned urban settlements, where small and narrow fields with ever-changing vegetation cover prevails, can be very complicated. Many parts of such a site are visible only in certain moments of the year, for instance, during severe drought stress of the particular crop on a particular field. Consequently, mapping all of the photographs is like bringing pieces of a puzzle together and the only way to realise a coherent map of all archaeologically relevant features, which will never be seen on a single photograph. The mapping also needs often to be undertaken at different levels, i.e. at different scales, ranges of precision and with variety in the depiction of interpreted detail. It is clear that the

Fig. 4.3 Aerial image from the *Portus* project (Italy): (**a**) Conventional visible image acquired from a helicopter above the Roman town of *Portus* (N 41°46.497′, E 12°15.854′ – WGS84); (**b**) Near-infrared image of the same scene. Images were acquired with a Nikon D200 (**a**) and a modified Nikon D80+Hoya R72 filter (**b**) by Geert Verhoeven on September 21, 2009 at 15.20 h

Fig. 4.3 (continued)

use of GIS has much facilitated these operations (Fig. 4.4). High-precision mapping is particularly desirable when it is possible to link aerial archaeological evidence with results from excavations or geophysical measurements. The parallel use of (rectified) oblique photographs with larger scale vertical photographs, sometimes also used for producing a digital terrain model, allows to obtain higher precision.

Despite the present-day large variety of means to process and ortho-rectify the oblique imagery, their archaeological information will not be exploited efficiently as long as the image is not thoroughly interpreted, meaning interpretatively mapped and integrated with other data sources (Doneus 2001; Doneus et al. 2001; Bewley and Raczkowski 2002; Haigh 2005; Scollar 2002). A thorough understanding of ancient cityscapes is based on combining the interpreted evidence from various prospection methods, and the approach to the interpretation of aerial images must be comprehensive. As with all types of aerial archaeology, reading aerial photographs from an ancient townsite does not mean trying to identify only the elements that indicate past human activities related to the urban phase in Antiquity, but involves using all present-day landscape features as elements of contrast that helps to bring out the residual components of the ancient landscape. Furthermore, it is of great importance that the sometimes easily recognisable and regular Roman features are not the only ones to be filtered out, but also old structures and marks that might belong to other phases of the past. These might be of important interpretative value to the *longue durée* understanding of the site or help to explain the deterioration of Roman elements in the soils and, therefore, in surface visibility.

Fig. 4.4 Illustration of flight strategy and mapping in the survey project of the University of Bologna at the ancient port site of Classe (Ravenna, Italy) (Courtesy of F. Boschi)

Again GIS technology will play a crucial role in this interpretation process, together with all important expert knowledge and sometimes also with the availability of discussion and teamwork. Such interpretative mapping is the most time-consuming process, and the results are in a sense always provisional. It involves the use of image enhancement techniques, and most often interpretation is done image by image on screen in separate layers using different colours and attributes for different features. Again the uniformity of Roman building practice, such as in a coherent use of building materials for specific architectural contexts and a systematic use of similar depths and widths of buildings and infrastructures in a town, is particularly helpful for the aerial photography interpreter.

This approach of producing interpretation drawings of orthophotos one by one, and thereafter combining them in a layered sequence is particularly useful for several reasons. It allows to keep the interpretation process of the features as objective as possible and also to quickly estimate the geometrical accuracy of all images. The same GIS approach allows in a next stage to confront the interpretative map with mapped data from other sources, such as (old and new) excavation evidence and interpretative maps from intensive geophysical prospections over the same areas. Excellent work of this kind was produced over the past 15 years in projects involving the intensive study of Roman towns in the northern parts of the Empire, such as in the already mentioned towns of Wroxeter (Gaffney and Gaffney 2000; White et al. 2013) and *Carnuntum* (Doneus et al. 2001; Doneus and Neubauer 2005). More recently, since 2002, interpretative Roman town mapping based on the integration of active aerial photography, a series of other non-invasive survey techniques and legacy data from punctual excavations, was directed by the author in four towns of central Adriatic Italy.

4.4 The Potenza Valley Towns from the Air

In 2000 a central Adriatic valley lying south of Ancona was chosen by Ghent University as a case study area for intensive, regionally based, field research, and the Potenza Valley Survey (PVS) project was born. After a first phase of intensive fieldwork in order to map and study all retrievable occupation from later prehistory up to the early Middle Ages within the whole valley, the main survey efforts concentrated, from 2006 onwards, on intra-site city surveys of the four abandoned Roman town sites. Since 2009 the surveys were incorporated within the European funded programme Radio-Past, allowing for intensification and fine-tuning of some of the survey approaches.

One of the objectives of these intensive investigations on the coastal colony of *Potentia* and the inland municipal towns of *Ricina*, *Trea* and *Septempeda* was to map in as detailed a manner as possible the major town structures, without having to rely on new and expensive excavations. The survey methodology involved first the collection, analyses and reinterpretation of the existing evidence, such as from earlier fieldwork (essentially small-scale excavations on *Trea* and *Potentia*), from the existing vertical photography (such as RAF pictures of the 1940s and several Italian flights from the 1950s onwards) and from early and current maps of the area. The main new data capture was achieved via a series of active aerial photography operations, large-scale geophysical prospections and intensive artefact surveys, accompanied by a set of necessary topographic measurements and regular field observations (including coring) concerning the geomorphology of the sites and their wider environments. Within the wider programme of active aerial photography set up by the PVS team (Vermeulen 2004, 2011; Vermeulen and Verhoeven 2004), the four urban sites were continuously monitored from the air between 2001 and 2013, with a total of some 30 individual flights (circa 50 flight hours). The potential of the sites for such an approach is excellent: apart from the present-day presence of a series of modern buildings and roads, the major parts of the ancient intramural areas, as we can now delimit them, were and are still currently in use as arable land. Even if regular ploughing of the sites further contributes to their erosion, this activity has allowed fruitful aerial reconnaissance over more than a decade. Especially intense flying with a light two- or four-seat aircraft during certain dry spring seasons (e.g. 2003, 2009) has produced some remarkable images, locating and visualising many aspects of the hidden urban topographies (Figs. 4.5 and 4.6). The regular flights over the intramural areas and beyond resulted in excellent aerial views of crop marks as well as soil marks. The former, often sharp and linear, reveal mostly the better preserved buried structures of walls, road decks, sewers and floors. The soil marks are mostly the result of ploughed-up larger stone structures, such as the city streets, the circuit wall with towers, some public buildings and houses, but also of a combination of ploughed-up occupation layers, zones with different quantities of organic substance in the upper layers, and humidity traces caused by differential drying of the soil in some parts of the towns.

It is impossible to enumerate here the multitude of architectural structures brought to light in and around these cities. A synopsis of the major discoveries, which have now almost all been mapped carefully within the GIS system, would

Fig. 4.5 Aerial view in April 2002 on a set of crop marks from the regular street system of the abandoned Roman town of *Potentia* (Marche, Italy) (Photo by the author)

comprise the following (mainly Early Imperial) structures:

Septempeda: several new elements (gates, towers, etc.) of the already partly known city wall, the whole street network (which was quasi-totally unknown), many *domus* and other houses in the intramural part and a major extramural sanctuary.

Trea: parts of the circuit wall (with towers and gates) whose trace can now be fully mapped, the main pattern of city streets, the forum and most of its surrounding public buildings (several temples, shops, the basilica, a macellum) and a whole series of town houses and workshops.

Ricina: a large part of the formerly unsuspected wall circuit, an amphitheatre, a main temple complex, several *domus* and other private houses and probably the town's aqueduct.

Potentia: the full regular town grid with quasi-complete wall circuit, the three gates, all streets and building blocks, the forum with temple, other public buildings (e.g. a small theatre) and shops, several elements of private housing and finally three roads leaving the colonial coastal town flanked by photographically well-attested funerary monuments.

Important efforts have been undertaken to digitally enhance, rectify and map the most relevant archaeological features visible on this oblique imagery (Fig. 4.7). To enhance the possibilities for detailed and precise interpretative mapping of buried features, additional low-altitude high-precision

Fig. 4.6 Aerial view in May 2009 on a set of crop marks revealing the forum, streets, houses, etc. of the abandoned Roman town of *Trea* (Marche, Italy) (Photo by the author)

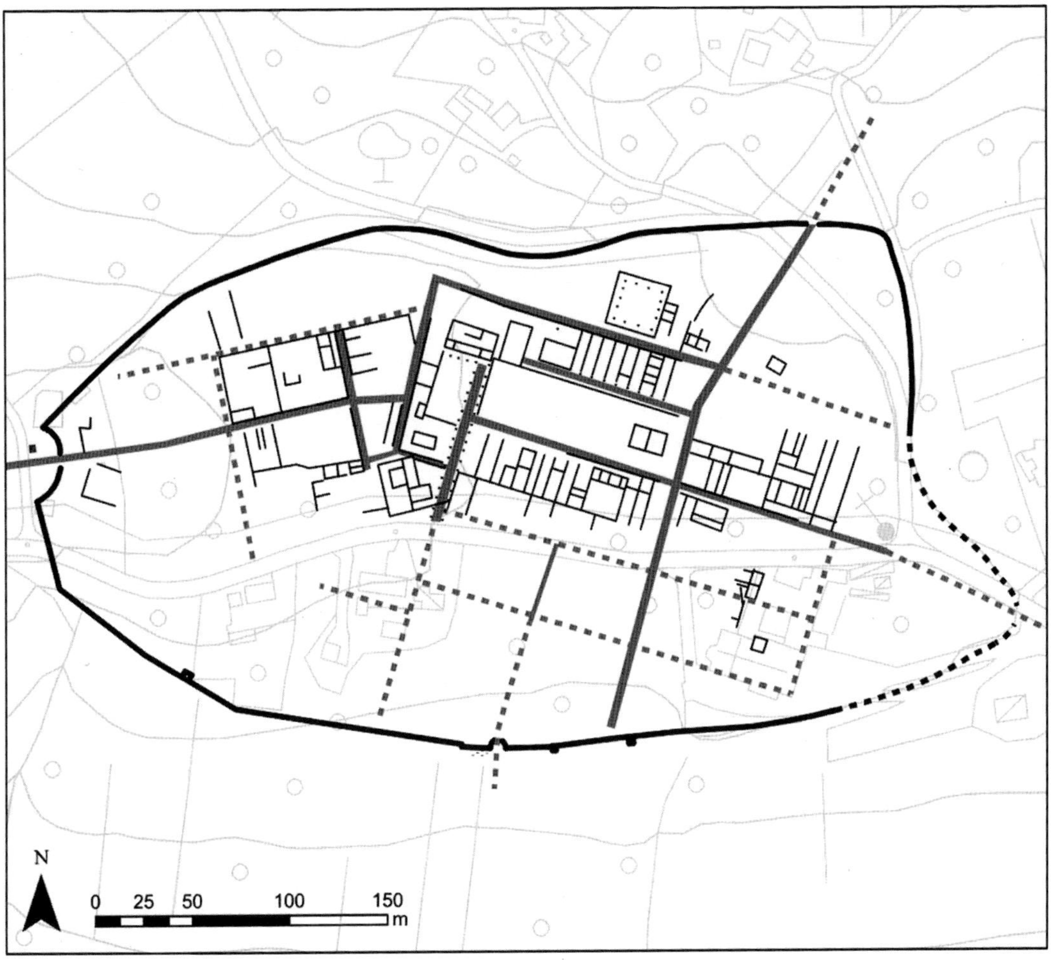

Fig. 4.7 Provisional mapping by the PVS team of the town plan of *Trea* (Marche, Italy) based on an integration of oblique aerial photo evidence, results from geophysical prospections and other topographic data

photography has been achieved. From 2006 to 2009 several flights were achieved over the four urban areas using a so-called Helikite. To deal with cloudy conditions (or other particular situations in which the shutter speed becomes too slow for conventional aerial photography) and to allow us to obtain high spatial resolution imagery of the large sites involved (in the visible, and also in the near-infrared and even ultraviolet, range), a stable, easily maintainable and remotely controllable construction was created. This combination of a helium balloon with a kite, linked to a set of additional devices (monitor, remote control, etc.), allowed certain details to be revealed which were not present in the more traditional aerial photography from a manned aircraft (Fig. 4.8). In 2013 also the first experiments started with a quadricopter allowing for even greater stability in low-altitude photography (Fig. 4.9).

These operations of aerial photography over Roman urban sites in the same region are only one (important) step for the systematic survey and mapping of the archaeological landscape of the abandoned Roman towns. The systematic full-coverage geophysical prospections of these sites, which started in 2004 and are still ongoing at the time of writing, will allow to create highly detailed maps of all recoverable but buried archaeological features. For several sites a fine integration is already achieved between the data from the aerial imagery and the results of

Fig. 4.8 Oblique NIR view taken from a Helikite of crop marks showing the presence of a gate in the late Republican circuit wall of *Septempeda* (Marche, Italy) (Photo by G. Verhoeven)

large-scale geophysical prospections and other field operations (e.g. Vermeulen et al. 2006, 2012b; Vermeulen 2012). The application side by side of a range of field methods allows also to achieve a certain ground-truthing of the aerial imagery and is also very helpful during the interpretative mapping process (Vermeulen et al. 2013). This integration work thus allows to make important contributions to the methodology of urban survey as well as to the interpretation of urbanism in this region of Italy.

Conclusion

It is particularly clear from recent research in many parts of the Roman world that the remote-sensing information, especially the one derived from intensive low-altitude aerial photography survey, is crucial in the process of revealing and studying ancient urbanism. It is most useful when this type of archaeological detection can be continued over a span of several years, making good use of the diversity of seasons and specifically adapted periods of 'archaeological flying'. Some examples prove that it can even be of use for the study of subperiods in the life of a Roman town, but here the lack of real stratigraphic data needs to be overcome by other approaches. However, when such active aerial research can be undertaken in close integration with intensive artefact surveys and field operations such as geophysical prospection, geomorphological observations and small-scale excavations, and when the full potential of

Fig. 4.9 Experiments with a quadricopter during aerial photography survey over the site of *Trea* (June 2013) (Photo by the author)

image processing and data capture (including NIR and UV photography) is included, it can revolutionise our comprehension of the occupation history of a well-delineated region and its population centres. The intensive aerial coverage by way of active oblique photography operations is not only essential for the study of urbanism, landscape evolution, town hinterlands and urban loss but has crucial potential when issues of heritage and management of large and complex sites are at stake. The further intensifying of the aerial approach, including the wider and more focused use of photography from new low-altitude platforms, also gives ammunition for dealing with recurrent phenomena such as robbing of sites for archaeological materials, the encroachment of modern communities, damage through ploughing or other agricultural works, land ownership issues and even threats from nature, such as erosion and river displacement. Together with the other non-invasive techniques, the highly visual method of urban survey using aerial photography should make landowners, local authorities, regional bodies, museums, etc. more aware and better informed of the value of this buried Roman urban heritage and the need to preserve and study it.

Acknowledgements The case studies from the Potenza Valley Survey project were made possible thanks to support from Belgian Science Policy (Interuniversity Attraction Poles, project P6/22) and the EC funded project 'Radiography of the Past' (Marie Curie action, Academia-Industry partnerships). I thank several colleagues (mentioned in the captions) for supplying me with illustrations.

References

Alcock S, Cherry J (eds) (2004) Side by side survey. Comparative regional studies in the Mediterranean world. Oxbow, Oxford

Barber M (2011) A history of aerial photography and archaeology. Mata Hari's glass eye and other stories. English Heritage, Swindon

Bewley R, Raczkowski W (eds) (2002) Aerial archaeology: developing future practice, vol 337, NATO

science series, series I: Life and behavioural sciences. Ios Press, Ohmsha
Bintliff J, Snodgrass A (1985) The Cambridge/Bradford Boeotian expedition: the first four years. J Field Archaeol 12:123–161
Bintliff J, Kuna M, Venclová N (eds) (2000) The future of surface artefact survey in Europe. Sheffield Academic Press, Sheffield
Bourgeois J, Meganck M (eds) (2005) Aerial photography and archaeology 2003. A century of information. Academia Press, Gent
Brophy K, Cowley D (2005) From the air. Understanding aerial archaeology. Stroud, Tempus
Christie N (2012) Urban landscape surveys: a view from the end. In: Vermeulen F, Burgers G-J, Keay S, Corsi C (eds) Urban landscape survey in Italy and the Mediterranean. Oxbow, Oxford, pp 285–289
Christie N, Augenti A (eds) (2012) Urbes extinctae. Archaeologies of abandoned classical towns. Ashgate, Farnham
Corsi C, Johnson PS, Vermeulen F (2012) A geomagnetic survey of the intra-mural area of Ammaia and its contribution to understanding Roman urbanism in Lusitania. J Rom Archaeol 25:121–145
Doneus M (2001) Precision mapping and interpretation of oblique aerial photographs. Archaeol Prospect 8:13–27
Doneus M (2004) Aerial archaeology in the landscape of Carnuntum. In: Archaeologia Aerea. Studi di Aerotopografia Archaeologica, I, 215–234. Libreria dello Stato/Istituto Poligrafico e Zecca dello Stato, Roma
Doneus M, Neubauer W (2005) Multiple survey techniques at Roman Carnuntum. Integrated prospection of the largest archaeological landscape in Austria. In: Chris M, Roger P, Stefano C (eds) In volo nel passato. Aerofotografia e cartografia archaeological. Biblioteca del dipartimento di archaeologia e storia delle arti e sezione archaeologica, Università di Siena, Siena, pp 272–279
Doneus M, Eder-Hinterleitner A, Neubauer W (2001) Roman Carnuntum: prospecting the largest archaeological landscape in Austria. In: Doneus M, Eder-Hinterleitner A, Neubauer W (eds) Archaeological prospection: fourth international conference on archaeological prospection, Austrian Academy of Sciences, Vienna, 19–23 Sept 2001). Austrian Academy of Sciences, Vienna, pp 47–59
Francovich R, Patterson H (eds) (2000) Extracting meaning from plough soil assemblages. Oxbow, Oxford
Gaffney C, Gaffney V (2000) Non-invasive investigations at Wroxeter at the end of the twentieth century. Archaeol Prospect 7(2):63–146
Groh S (2012) Strategies and results of the urban survey in the Upper City of Ephesus. In: Vermeulen F, Burgers G-J, Keay S, Corsi C (eds) Urban landscape survey in Italy and the Mediterranean. Oxbow, Oxford, pp 85–96
Haigh JGB (2005) From photographs to maps. A collaborative development. In: Bourgeois J, Meganck M (eds) Aerial photography and archaeology 2003. A century of information. Academia Press, Gent, pp 201–212
Keay S (2010) Iberia and Italica: issues and challenges in the comparative study of Roman urbanism. In: Corsi C, Vermeulen F (eds) Changing landscapes. The impact of Roman towns in the Western Mediterranean. Proceedings of the international colloquium, Castelo de Vide – Marvão 15–17 May 2008. Antequem, Bologna, pp 27–45
Keay S, Millett M, Poppy S, Robinson J, Taylor J, Terrenato N (2000) Falerii Novi: a new survey of the walled area. PBSR 68:1–93
Keay S, Millett M, Paroli L, Strutt K (eds) (2005) Portus. An archaeological survey of the port of Imperial Rome, vol 15, Archaeological monograph. The British School at Rome and Soprintendenza Archeologica di Ostia, London
Keller DR, Rupp DW (1983) Archaeological survey in the Mediterranean area. B.A.R, Oxford
Macready S, Thompson FH (1985) Archaeological field survey in Britain and abroad. Society of Antiquaries of London, London
Martens F, Mušič B, Poblome J, Waelkens M (2012) The integrated urban survey at Sagalassos. In: Vermeulen F, Burgers G-J, Keay S, Corsi C (eds) Urban landscape survey in Italy and the Mediterranean. Oxbow, Oxford, pp 111–122
Palmer R (2005) If they used their own photographs they wouldn't take them like that. In: Brophy K, Cowley D (eds) From the air. Understanding aerial archaeology. Tempus, Stroud, pp 94–116
Patterson H (ed) (2004) Bridging the Tiber. Approaches to regional archaeology in the Middle Tiber Valley, vol 13, Archaeological monographs of the British School at Rome. British School at Rome, London
Schmiedt G, Castagnoli F (1957) L'antica città di Norba – documentazione aerofotogrammetrica. L'Universo 37:125–148
Scollar I (2002) Making things look vertical. In: Bewley R, Raczkowski W (eds) Aerial archaeology. Developing future practice. Ios Press, Ohmsha, pp 166–172
Scollar I, Tabbagh A, Hesse A, Herzog I (1990) Archaeological prospecting and remote sensing. Cambridge University Press, Cambridge
Slapšak B (2012) Towards integrated non-invasive research on complex urban sites: Ljubljana research in Tanagra and beyond. In: Vermeulen F, Burgers G-J, Keay S, Corsi C (eds) Urban landscape survey in Italy and the Mediterranean. Oxbow, Oxford, pp 53–61
Verhoeven G (2008) Imaging the invisible using modified digital still cameras for straightforward and low-cost archaeological near-infrared photography. J Archaeol Sci 35:3087–3100
Verhoeven G (2011a) Near-infrared aerial crop mark archaeology: from its historical use to current digital implementations. J Archaeol Method Theory. doi:10.1007/s10816-011-9104-5
Verhoeven G (2011b) Taking computer vision aloft: archaeological three-dimensional reconstructions

from aerial photographs with PhotoScan. Archaeol Prospect 18:67–73
Verhoeven G, Schmitt KD (2010) An attempt to push back frontiers – digital near-ultraviolet aerial archaeology. J Archaeol Sci 37:833–845
Verhoeven G, Smet P, Poelman D, Vermeulen F (2009) Spectral characterization of a digital still camera's NIR modification to enhance archaeological observation. IEEE Trans Geosci Remote Sens 47:3456–3468
Verhoeven G, Doneus M, Briese C, Vermeulen F (2012a) Mapping by matching – a computer vision-based approach to fast and accurate georeferencing of archaeological aerial photographs. J Archaeol Sci 39:2060–2070
Verhoeven G, Taelman D, Vermeulen F (2012b) Computer vision-based orthophoto mapping of complex archaeological sites: the ancient quarry of Pitaranha (Portugal-Spain). Archaeometry. doi:10.1111/j.1475-4754.2012.00667.x
Vermeulen F (2004) Fotografia aerea finalizzata nelle Marche Centrali: Un progetto integrato. Archeologia Aerea Studi di Aerotopografia Archeologica 1: 91–118
Vermeulen F (2011) Reviewing 10 years of aerial photography in the valley of the River Potenza (Le Marche). Archeologia Aerea Studi di Aerotopografia Archeologica 4:259–264
Vermeulen F (2012) Potentia: a lost new town. In: Christie N, Augenti A (eds) Urbes Extinctae. Archaeologies of abandoned classical towns. Ashgate, Farnham, pp 77–95
Vermeulen F, Verhoeven G (2004) The contribution of aerial photography and field survey to the study of urbanization in the Potenza valley (Picenum). J Rom Archaeol 17:57–82
Vermeulen F, Verhoeven G, Semey J (2005) The integration of aerial photography and GIS in the Potenza Valley survey. In: Bourgeois J, Meganck M (eds) Aerial photography and archaeology 2003. A century of information. Academia Press, Gent, pp 371–382
Vermeulen F, Hay S, Verhoeven G (2006) Potentia: an integrated survey of a Roman Colony on the Adriatic Coast. PBSR 74:203–236
Vermeulen F, Burgers G-J, Keay S, Corsi C (eds) (2012a) Urban landscape survey in Italy and the Mediterranean. Oxbow, Oxford
Vermeulen F, Slapšak B, Mlekuž D (2012b) Surveying the townscape of Roman Trea (Picenum). In: Johnson PS, Martin M (eds) Archaeological survey and the city, vol 2, University of Cambridge Museum of Classical Archaeology monographs. Oxbow, Oxford, pp 261–282
Vermeulen F, Verdonck L, Corsi C (2013) L'apport des méthodes non-destructives à l'étude de sites archéologiques complexes. Etude de cas MARIANA. In: Rebière J (ed) Conserver, Etudier, Protéger, Valoriser le patrimoine en milieu isolé (actes du colloque de Bastia-Lucciana 9–12 octobre 2008). Publication Bastia, Luciana. ISBN 978-2-906375-10-9 – EAN 9782906375109
Whimster R (1989) The emerging past: air photography and the buried landscape. RCHM, London
White RH, Gaffney C, Gaffney V (with Baker A) (2013) Wroxeter, the Cornovii and the urban process. Final Report of the Wroxeter Hinterland Project, 1994–1997. Volume 2: Characterizing the city. J Roman Archaeol, Supplementary series. Archaeopress, Oxford
Wilson D (2000) Air photo interpretation for archaeologists. Tempus, Stroud

5 Integrated Approach for Archaeological Prospection Exploiting Airborne Hyperspectral Remote Sensing

Rosa Maria Cavalli

Contents

5.1 Introduction

5.1.1 Overview of Remote Sensing for Archaeological Prospection

Since the end of the nineteenth century, remote sensing has been considered one of the techniques of archaeological prospection together with geophysical and geochemical methods. The first applications of passive satellite remote sensing were carried out one century later on Landsat Thematic Matter (TM) to identify former field divisions (Clark et al. 1998; Server 1998).

If aerial photography is considered as a remote-sensing technique – where remote sensing is the acquisition of information about a surface without any physical contact with it – then, to be precise, this remote-sensing technique (i.e. aerial photography, active and passive remote-sensing images) has, since the beginning of the twentieth century, been the tool most widely used for surveying, detecting and identifying buried archaeological remains (Bradford 1949, 1950; Miller 1957; Reeves 1936; Schmiedt 1966, 1968).

R.M. Cavalli
Institute of Atmospheric Pollution Research, CNR – IIA, Area della Ricerca di Roma1, Rome, Italy
e-mail: cavalli@iia.cnr.it

C. Corsi et al. (eds.), *Good Practice in Archaeological Diagnostics*, Natural Science in Archaeology,
DOI 10.1007/978-3-319-01784-6_5,

The reason why buried archaeological remains leave traces on the surface above them is that their presence can alter terrain characteristics (i.e. the surface and land cover) and can cause modification of the "natural" trend of the hidden terrain. For example, buried structures cause a difference of permeability in soil compared with soil where no remains exist.

Therefore, the reason why images taken over the surface of buried structures can detect these remains is that this modification is limited to the area of the image (i.e. pixels) and does not involve the surrounding area. This alteration is highlighted on the image with different degrees of brightness and/or patterns (called anomalies or marks or sites) with respect to the surrounding area.

In photo-interpretation, these alterations related to the presence of buried archaeological remains, which can be identified on the images, are classified in Piccarreta (1987): damp marks, grass-weed-crop marks, soil sites and shadow sites (see Chaps. 2 and 3, this volume).

Thus, the presence of buried archaeological remains can modify the morpho-geophysical characteristics of the soil and terrain (shadow sites, soil sites and damp marks) and/or modulate the trend of the vegetation growth (grass-weed-crop marks) over the buried structures, while leaving the surrounding area unaltered.

The anomaly related to the same buried archaeological remains on different images can be present or not and can differ in size and/or intensity and/or pattern for many reasons (Rowlands and Sarris 2007; Bassani et al. 2009). These reasons can be classified as follows:

- Environmental, atmospheric and lighting conditions at the time of the acquisition of the image (e.g. land cover and land use, ground water level related to level of archaeological structure, season, day and hour of the acquisition, the heaviness of the last rainfall)
- Characteristics of archaeological structures and positions of the archaeological sites (e.g. typology, building material, dimensions, depth of archaeological remains, depth of archaeological sites with respect to the water-bearing stratum)
- Imaging system characteristics used (e.g. active or passive sensors, spatial and spectral characteristics, instrumental noise)

Occasionally, the anomaly or mark related to buried archaeological remains has a characteristic spectral signature and/or particular pattern. In this case, the detection and extraction of the anomaly related to the buried structure can be easily seen on the remote-sensing images, because its spectral characteristic and pattern is evident from the surrounding area. Figure 5.1 shows an example of a mark (a soil site) which exhibits different chemical-physical characteristics from the surrounding area. The soil site is related to the Selinunte Archaeological Park (Italy, Sicily – see Fig. 5.4a), and the image and position of the soil site are shown on a Multispectral Infrared and Visible Imaging Spectrometer (MIVIS, Bianchi et al. 1996) image with a white arrow in Fig. 5.1a (the grey arrow indicates the surrounding area). These two spectra, one of the surfaces of the soil site and the other of the surrounding area, are shown in Fig. 5.1b. To be more precise, this soil site was partially excavated at the time of MIVIS acquisition.

However, the spectral signature of the anomaly is often comparable with the surrounding area, and the spectral signature of the pixels over the buried structure does not differ greatly from the spectral signature of the surrounding area. In this case, a problem arises regarding the extraction and classification of the archaeological spectral features related to the subsurface structure, because the anomaly related to that structure does not have a unique shape and/or spectral characteristics. Figure 5.2 shows an example of a mark (a damp mark) which shows different water content (soil moisture) from that of the surrounding area. The damp mark is related to the Arpi archaeological site (Italy, Puglia – see Fig. 5.4b), and the image and position of the damp mark are shown on the MIVIS image with a white arrow in Fig. 5.2a. These two spectra, one of the surfaces of the damp mark and another of the surface of the surrounding area, are shown in Fig. 5.2b, where the higher the water content is, the less bright the spectrum is. Figure 5.3, which is the last example, shows an example of a mark (grass-weed-crop marks) which exhibits a different trend of vegetation growth from that of the surrounding area. The grass-weed-crop mark is related to the Arpi archaeological site (Italy, Puglia – see Fig. 5.4b) too, and the image and the position of the grass-weed-crop mark are shown on the MIVIS image with white arrows in Fig. 5.3a, b shows the spectra of the different trends of growth of vegetation.

Fig. 5.1 An example of a soil site related to the Selinunte Archaeological Park (Italy, Sicily). (**a**) Shows the soil site on the MIVIS data and the positions of the soil site (*white arrow*) and the surrounding area (*grey arrow*) on the image. (**b**) Shows two spectra of the soil, one of the surfaces of the soil site and another of the surrounding area

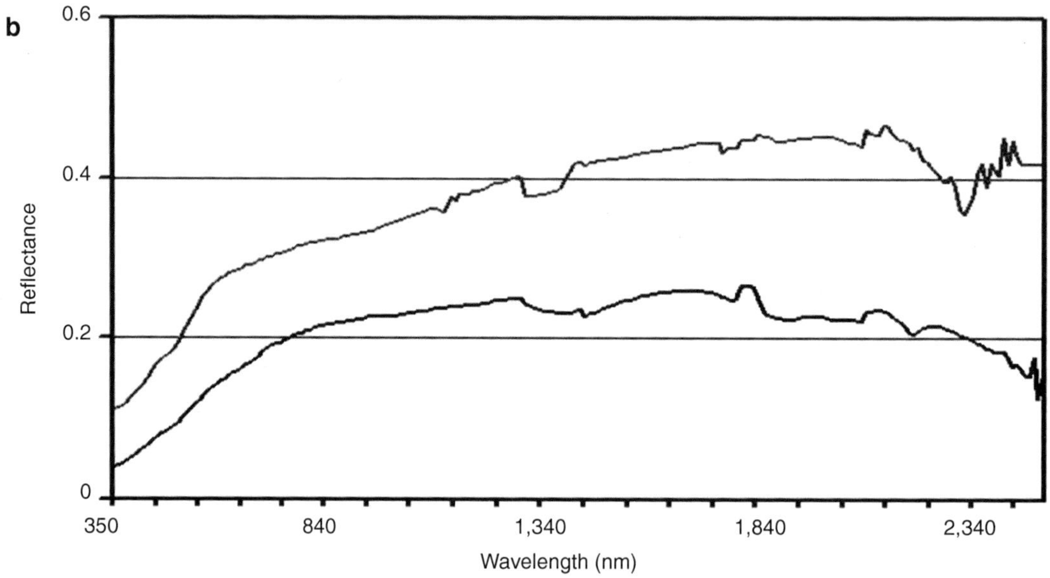

Fig. 5.1 (continued)

5.1.2 Background Literature on Hyperspectral Remote Sensing

In the literature many papers highlight the ability of the hyperspectral image to survey, detect and identify anomalies or marks related to buried archaeological remains. This capability has been confirmed by their contiguous and narrow bandwidth characteristics, which allow a better examination of the Earth's surface than discontinuous and broad bandwidths. In fact, there are many works on the subject of target and anomaly detection methods applied to hyperspectral data (Stein et al. 2002; Goudail et al. 2006; Zhang et al. 2010; Fowler and Du 2012) and which evaluate the ability of hyperspectral data to detect buried archaeological structures. These papers are introduced and discussed in the following pages.

The advent of hyperspectral technology has promoted the investigation of a variety of fields of remote sensing (Goetz 2009). In recent years it has been possible to acquire hyperspectral imaging data from airborne and satellite platforms. Hyperspectral data are currently available from several different types of airborne spectrometers (e.g. Airborne Imaging Spectrometer (AISA), Airborne Hyperspectral Scanner (AHS), Airborne Prism Experiment (APEX), Airborne Visible Infrared Imaging Spectrometer (AVIRIS), Compact Airborne Spectrographic Imager (CASI), Environmental Protection System (EPS-H), Digital Airborne Imaging Spectrometer (DAIS), Hyperspectral Digital Imagery Collection Experiment (HYDICE), Hyperspectral Mapper HyMap, Multispectral Infrared and Visible Imaging Spectrometer (MIVIS), Reflective Optics System Imaging Spectrometer (ROSIS)).

Meanwhile, in November 2000 NASA (National Aeronautics and Space Administration) launched Hyperion, the first civil hyperspectral sensor, aboard an Earth Observation satellite platform (EO-1) (Ungar et al. 2003). In October 2001 ESA (European Space Agency) launched its Compact High-Resolution Imaging Spectrometer (CHRIS), a hyperspectral satellite sensor, as part of the Project for On-Board Autonomy (PROBA) platform system (Barnsley et al. 2004).

The German, Japanese and Italian Space Agencies have recently started three new hyperspectral satellite missions (Environmental Mapping and Analysis Program EnMAP, Kaufmann et al. 2008; Hyperspectral Imager Suite (HISUI); and Hyperspectral Precursor and

Fig. 5.2 An example of a damp mark related to the Arpi archaeological site (Italy, Puglia). (**a**) Shows the damp mark on the MIVIS data and its position on the image (*white arrow*). (**b**) Shows these two spectra, one of the surfaces of the soil site and another of the surrounding area

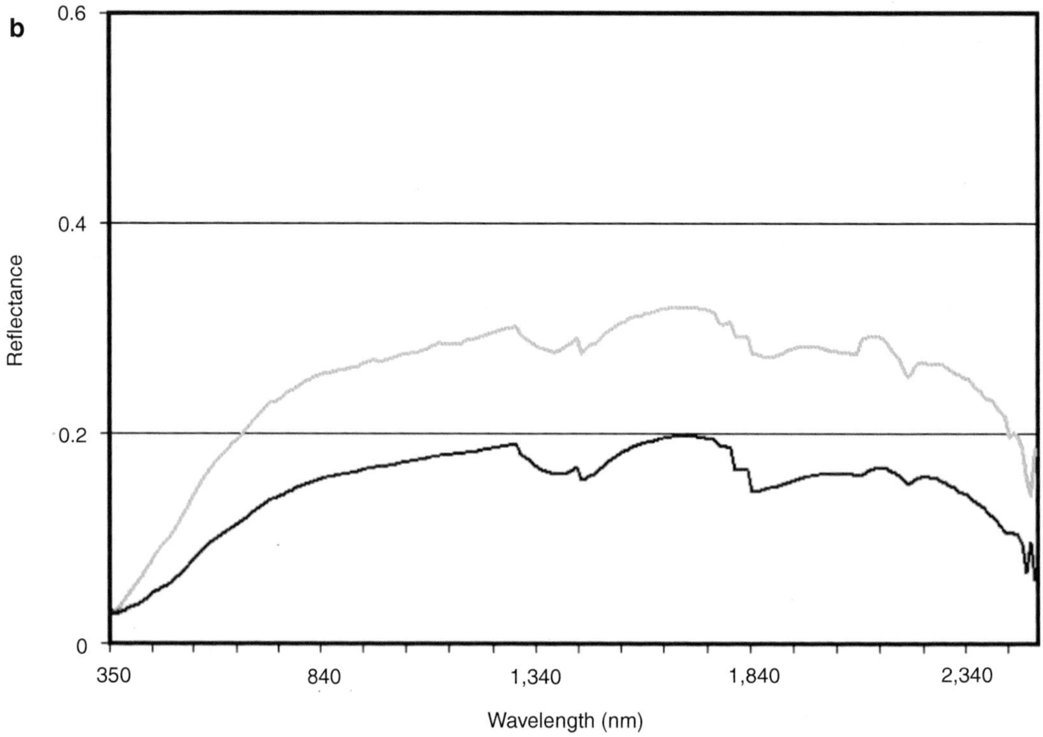

Fig. 5.2 (continued)

Application Mission (PRISMA), Galeazzi et al. 2009) and the launch of these satellites is programmed to take place in the next few years.

The concept of hyperspectral imaging, or imaging spectroscopy, was formulated by Goetz et al. (1985) in order to present the primary results of the technique of imaging spectrometry in mineral exploration (Goetz 2009). Hyperspectral imaging is recorded by sensors that can acquire a lot of very narrow, contiguous spectral bands by means of the visible, near-infrared, mid-infrared and thermal infrared portions of the electromagnetic spectrum, thereby enabling the construction of a continuous spectrum for every pixel in an image. The original definition given by the authors (Goetz et al. 1985) of hyperspectral imaging was "the acquisition of images in hundreds of contiguous, registered, spectral bands such that for each pixel a radiance spectrum can be derived". In the literature (Aspinall et al. 2002; Gianinetto and Lechi 2004; Goetz 2009), hyperspectral sensors are instruments that collect contiguous spectral bands with fixed narrow bandwidths, and the requirement for hundreds of spectral bands is of secondary importance with respect to the fixed narrow bandwidths. Contiguous spectral bands provide detailed information about individual elements in an image and increase the probability of finding a unique characteristic for any given element which better distinguishes it from other elements in the image (Jensen 1996).

5.2 Integrated Approach for Archaeological Prospection

All the activities of calibration (which is the process of quantitatively defining the system responses to known, controlled signal input, in accordance with the definition of the Working Group on Calibration and Validation (WGCV) of Committee on Earth Observation Satellites (CEOS) homepage http://www.ceos.org/index.php?option=com_content&view=category&layout=blog&id=75&Itemid=113) and validation (which is the process of assessing, by independent

a

Fig. 5.3 An example of grass-weed-crop marks related to the Arpi archaeological site (Italy, Puglia). (**a**) Shows the grass-weed-crop marks on the MIVIS data and their position on the image (*white arrows*). Different spectra of this vegetation are shown in (**b**)

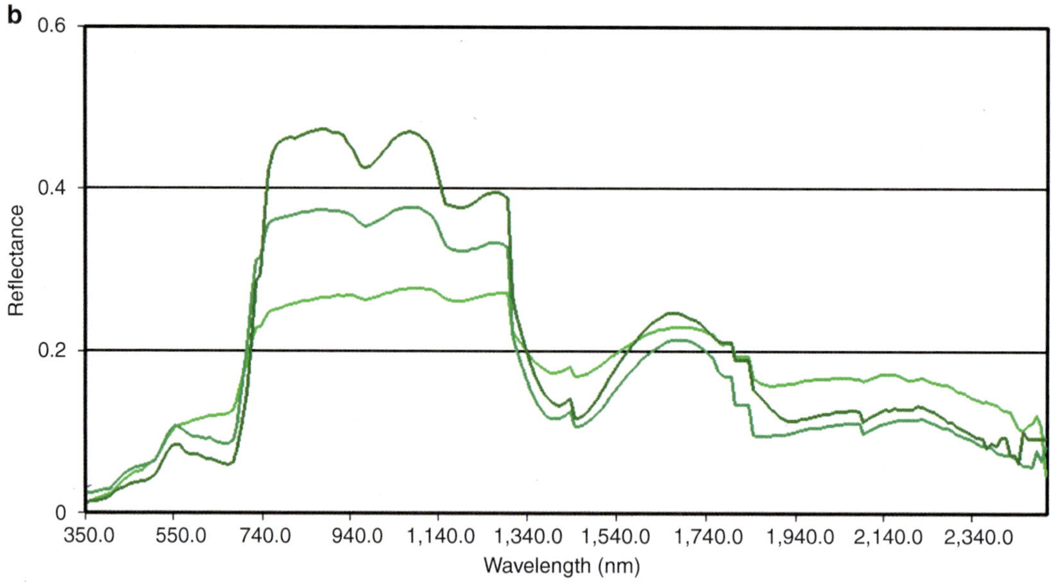

Fig. 5.3 (continued)

means, the quality of data products derived from the system output, in accordance with definition of the WGCV of CEOS homepage http://www.ceos.org/index.php?option=com_content&view=category&layout=blog&id=75&Itemid=113 – (Cal/Val) of Earth Observation (EO) data) are very important to improve the quality of data products of all EO applications (work plans of WGCV of CEOS (Greening 2012) and GEOSS (Global Earth Observation System of System; (Battrick 2005)). The integration of different kinds of data (satellite, airborne and in situ) allows Cal/Val activities to be carried out. The majority of EO activities, which analyse remote-sensing data in order to survey, detect and identify buried archaeological remains, are carried out by integrating multi-platform (satellite, airborne and in situ data), multi-sensor (active and passive, multi – and hyperspectral data) and frequency data. The integration of these data for Cal/Val activities is part of the research concerning the detection of buried archaeological remains by using remote-sensing data. In particular, the integration and comparison of the data improves and controls the atmospheric and geometric correction of remote images, which is essential in archaeological recognition and identification. Furthermore, different data are integrated in order to calibrate and validate the results. In particular, the integration of geophysical and geochemical in situ data with remote-sensing data improves archaeological interpretation and validates its results.

In addition, EO activities used to detect buried archaeological remains integrate multi-platform, multi-sensor and frequency data, not only to perform the Cal/Val activities but also to exploit every aspect of the integrated approach to archaeological prospection, as listed below:

- The integration of multi-platform, multi-sensor and frequency data gives merged images which allow a more detailed survey of the buried remains to be carried out with respect to a single datum.
- The comparison and appending of the results of the interpretation of these data allow a more complete survey of the buried structures to be obtained with respect to a single datum.
- The exportation of the characteristic methods of each datum, technique and application for use with other data, techniques and applications produces synthetic images which allow a more useful survey of the buried archaeological remains to be obtained.

- The collaboration of different researchers (experts in remote sensing, archaeologists, topographers, geophysicists, etc.) encourages the integration of further knowledge.

This integrated approach to detect buried archaeological remains and survey the archaeological landscape has been widely used to exploit the capability of hyperspectral remote-sensing data (Bianchi et al. 1998a, b; Cavalli et al. 1998, 2005, 2007, 2009, 2010; Malagoli and de Paolis 2001; Ardissone et al. 2003; Buck et al. 2003; Coren et al. 2005; Giardino and Haley 2006; Aqdus et al. 2007, 2012; Rowlands and Sarris 2007; Alexakis et al. 2009; Bassani et al. 2009; Cavalli and Pignatti 2009; Challis et al. 2009; Pascucci et al. 2010) and in situ hyperspectral data (Agapiou et al. 2010; Agapiou and Hadjimitsis 2011). In the literature there are also papers which have applied this integrated approach by utilising hyperspectral data in order to detect buried archaeological remains and which also manage and analyse archaeological data using Geographic Information Systems (GIS) (Boccardo et al. 2002; Emmolo et al. 2004; Traviglia 2011).

Some examples of this integrated approach using airborne hyperspectral data are introduced in the following paragraphs. The principal spectral characteristics of the airborne multispectral and hyperspectral data used in this research are

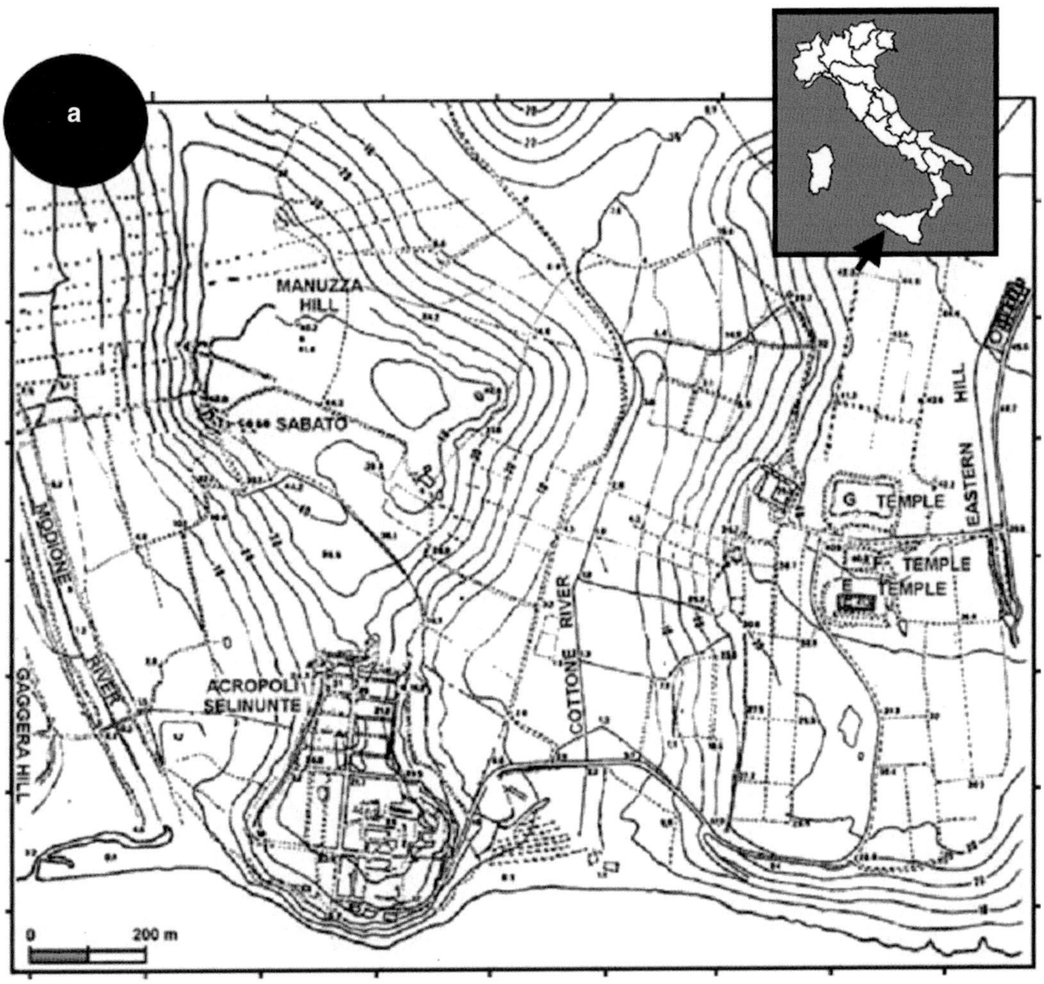

Fig. 5.4 The positions and the maps of the study areas: the Selinunte Archaeological Park in (**a**) the Arpi archaeological site in (**b**) the Marsala archaeological site in (**c**) the *Mothia* archaeological site in (**d**) and the Aquileia archaeological site in (**e**)

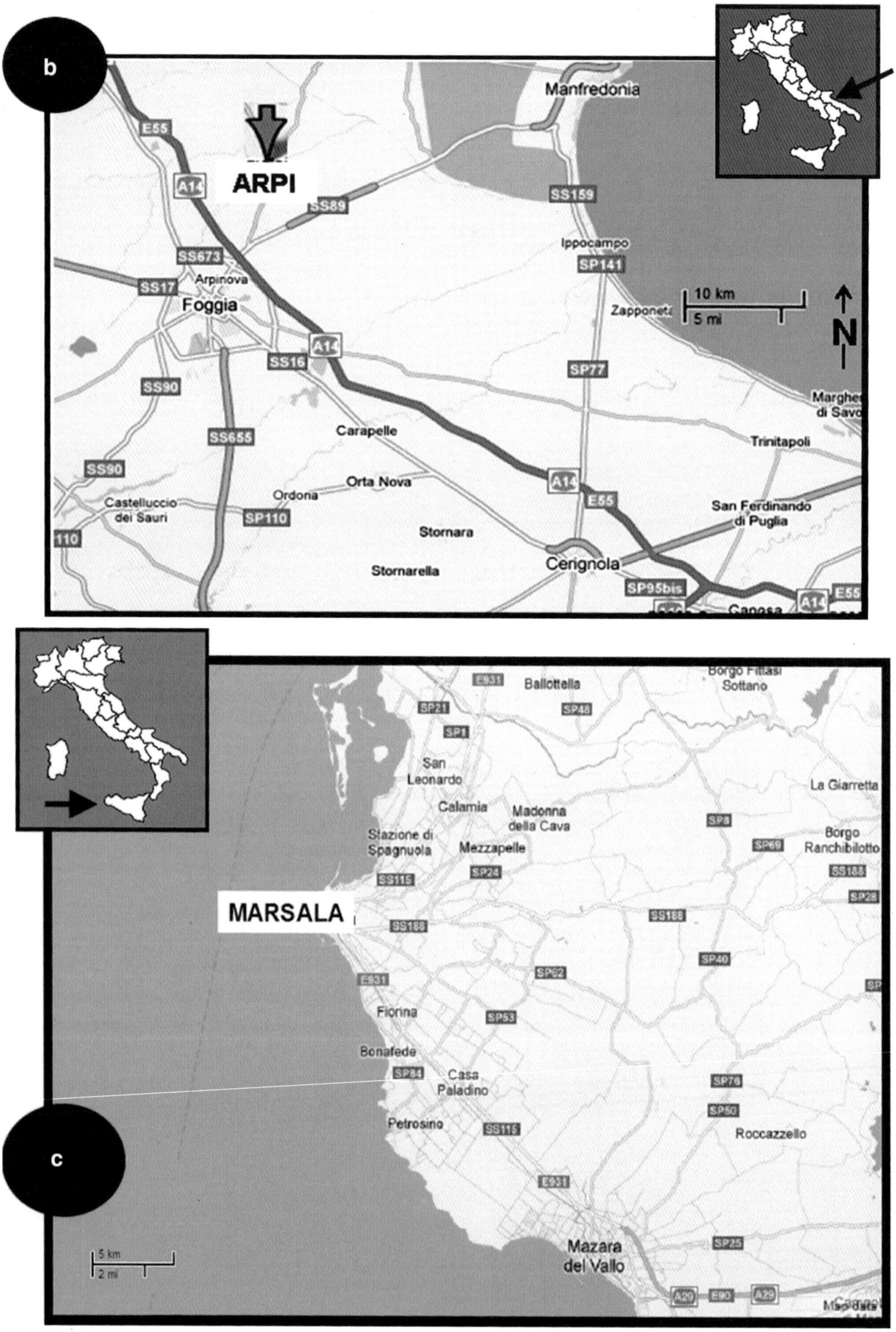

Fig. 5.4 (continued)

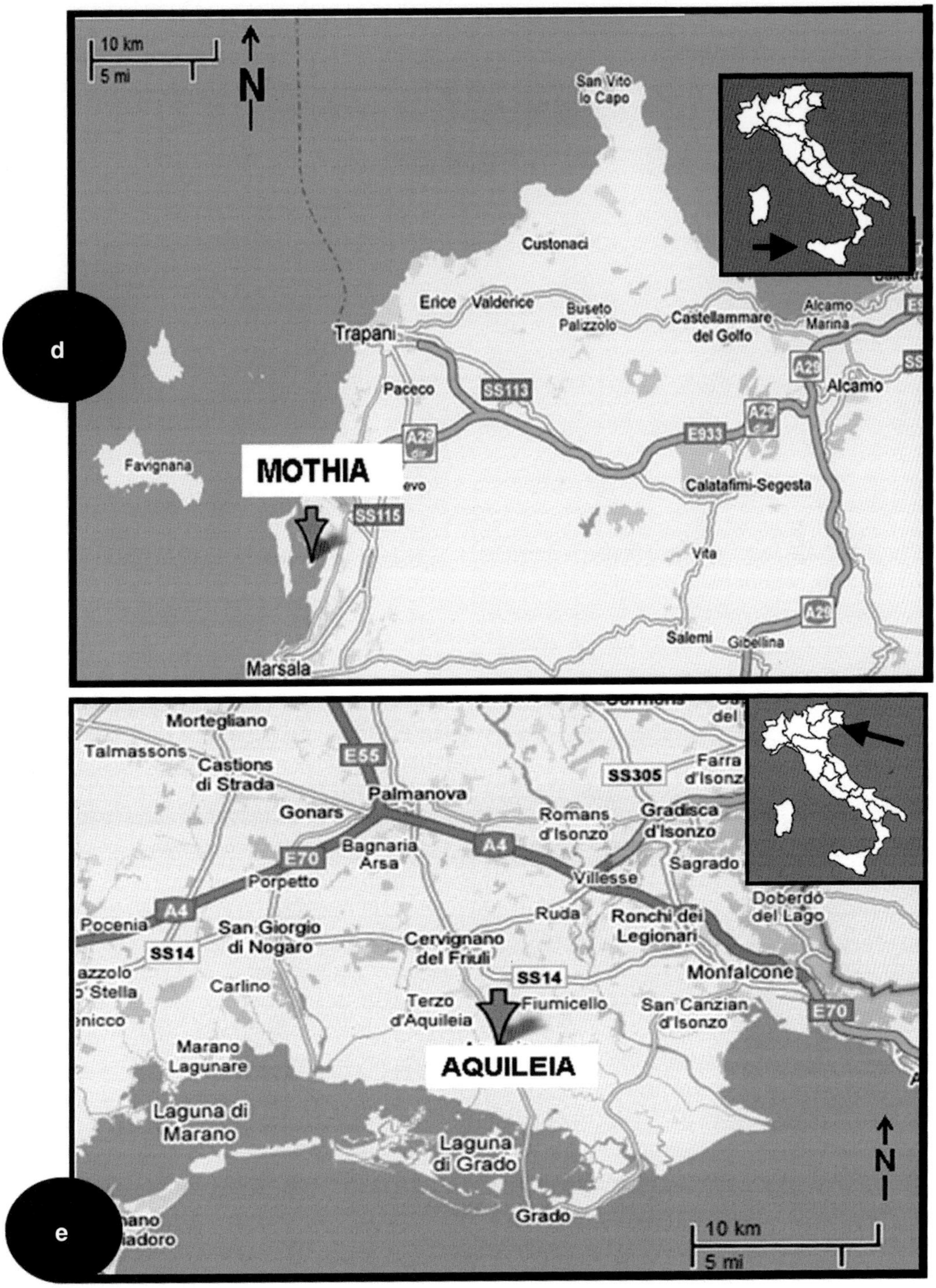

Fig. 5.4 (continued)

Table 5.1 Principal spectral characteristics of the airborne multispectral and hyperspectral data used in this research

	Spectral region	Spectral resolution (μm)	Spectral range (μm)
ATM	VIS-NIR SWIR-TIR (tot 12 ch.)	variable from 24 to 3,100	0.42–1,150
CASI	VNIR (48 ch.)	0.01	0.40–0.94
MIVIS	VNIR (28 ch.)	0.02 (VIS)	0.43–0.83(VIS)
		0.05 (NIR)	1.15–1.55 (NIR)
	SWIR (64 ch.)	0.09	1.983–2.478
	TIR (10 ch.)	0.34–0.54	8.180–12.700

summarised in Table 5.1. These research activities were initially performed, in particular, to evaluate the ability of hyperspectral data to detect buried archaeological remains and survey the archaeological landscape, and they were then improved, where necessary, by using this integrated approach.

5.2.1 Selinunte Archaeological Park

The Italian National Research Council (CNR) performed a research activity to detect and identify the buried archaeological remains of the Selinunte Archaeological Park (see Fig. 5.4a) by exploiting every feature of the integrated approach. This park is a cultural heritage site in Southern Sicily located along the SW coast. The city of Selinunte was founded in the seventh century BC by colonisers who came from *Megara Hyblaea*.

The integrated multi-platform, multi-sensor and frequency data and, consequently, the activities related to the applied integrated approach that were used for this research are briefly introduced.

The in situ survey was performed using a hyperspectral radiometer (FieldSpec spectroradiometer, Analytical Spectral Devices (ASD)) in order to carry out the Cal/Val activities used to:

- Improve and control the atmospheric correction of the airborne hyperspectral data (Bassani et al. 2006, 2010)
- Calibrate the land use and land cover classification and validate its results (Cavalli et al. 2007, 2009)

Two airborne hyperspectral surveys were performed using MIVIS on 23 May 1996, at 7.30 and 12.30 h local time with clear sky conditions, at an altitude of 1,500 m a.s.l. (3 m/pixel ground resolution) in order to:

- Evaluate the ability of the MIVIS data to detect and identify buried archaeological remains
- Compare and merge different acquisition times to detect buried archaeological remains (Cavalli et al. 1998, 2007)

Synthetic Aperture Radar (SAR) was used within the framework of the Southern Italy Demonstration of Remote Sensing Applications (SIDERA) project in order to:

- Compare and merge passive (MIVIS) and active (SAR) data in order to detect and identify buried archaeological remains (Bianchi et al. 1998a)

Historical aerial photos were collected in 1968 on a scale of 1:13,500, in 1971 on scales of 1:4,000 and 1:8,000, in 1973 on scales of 1:5,000 and 1:10,000, in 1975 and in 1987 on a scale of 1:10,000 as colour prints and in 1993 on a scale of 1:8,000 in order to:

- Evaluate the results of detecting archaeological remains (Cavalli et al. 2007)
- Compare and append the results with the MIVIS data (Cavalli et al. 2007)

Lastly, all the activities devoted to integrating previous archaeological knowledge were performed by an archaeologist to obtain an archaeological interpretation and validation of the anomalies detected and identified.

5.2.2 *Mothia* and Marsala Archaeological Areas

The CNR performed another research activity to detect and identify the buried archaeological remains of the *Mothia* (Fig. 5.4d) and Marsala (Fig. 5.4c) archaeological areas by exploiting any features of the integrated approach.

Mothia and Marsala are located on the western coast of Sicily (Fig. 5.4d, c). The *Mothia* archaeological area covers an ancient Phoenician colony that was founded at the end of the seventh century BC on the island of San Pantaleo. *Mothia* became one of the most prosperous Western Phoenician colonies as a result of its favourable location for maritime trade. Marsala was founded by the Phoenicians under the name of Lilibeo and was inhabited intensely during the Punic, Roman, Arab and Norman dominations.

For this research, the acquired data and, therefore, activities of the integrated approach performed are briefly described in the following:

The in situ survey was performed using a hyperspectral radiometer (FieldSpec spectroradiometer, Analytical Spectral Devices (ASD)) in order to carry out the Cal/Val activities used to:

- Improve and control the atmospheric correction of the airborne hyperspectral data (Bianchi et al. 1998b; Cavalli et al. 2009)
- Calibrate the land use and land cover classification and validate its results (Cavalli et al. 2009)

An airborne hyperspectral survey was performed on *Mothia* and Marsala by using MIVIS on 20 July 1994 at 10.15 and 10.30 h local time, respectively, with clear sky conditions, from an altitude of 2,000 m a.s.l. (4 m/pixel ground resolution) in order to:

- Evaluate the ability of MIVIS data to detect and identify the buried archaeological remains (Bianchi et al. 1998b; Merola et al. 2008; Cavalli et al. 2009)

All the activities devoted to integrating the previous archaeological knowledge were performed by archaeologists in order to obtain an archaeological interpretation and validation of any anomaly detected and identified (Bianchi et al. 1998b; Merola et al. 2008; Cavalli et al. 2009).

5.2.3 Arpi Archaeological Site

After the encouraging results obtained by research activity over the ancient sites of Selinunte, *Mothia* and Marsala, the CNR performed a further research activity to detect and identify the buried archaeological remains of the Arpi archaeological site (see Fig. 5.4b). This activity was carried out in collaboration with the Natural Environment Research Council (NERC, UK) as part of a Mediterranean Flight Campaign to exploit any features of the integrated approach by using two airborne hyperspectral sensors and one airborne multispectral sensor.

The site of Arpi, situated northeast of Foggia (Puglia, Italy – see Fig. 5.4b), was one of the largest and most important pre-Roman sites between the seventh and the third centuries BC, from the Late Iron Age to the Roman conquest.

The multi-platform, multi-sensor and multi-frequency time data and, therefore, the activities of the integrated approach performed for this research are described in brief.

The in situ survey was performed using a hyperspectral radiometer (FieldSpec spectroradiometer, Analytical Spectral Devices (ASD)) in order to carry out the Cal/Val activities used to:

- Improve and control the atmospheric correction of the airborne hyperspectral data (Bassani et al. 2006, 2009, 2010; Cavalli et al. 2005, 2009; Pascucci et al. 2010)
- Calibrate the land use and land cover classification and validate its results (Cavalli et al. 2005, 2009; Bassani et al. 2009; Pascucci et al. 2010)

Two airborne hyperspectral surveys were performed using:

- MIVIS on 27 June 2002 at 10:55 GMT with clear sky conditions from an altitude of 1,500 m a.s.l. (3 m/pixel ground resolution) in order to:
  - Evaluate the ability of MIVIS data to detect and identify the buried archaeological remains (Cavalli et al. 2005, 2009; Bassani et al. 2009)
  - Compare the results of CASI and ATM to detect the buried archaeological remains (Pascucci et al. 2010)
- CASI on 25 April 2005, at 13:50 local time with clear sky conditions with 2 m/pixel ground resolution in order to:
  - Evaluate the ability of CASI data to survey, detect and identify the buried archaeological remains (Pascucci et al. 2010)
  - Compare the results of MIVIS and ATM to detect the buried archaeological remains (Pascucci et al. 2010)

Two airborne multispectral surveys were performed using ATM on 25 April 2005 at 13:50

local time and on 26 April 2005 at 03:06 local time, with clear sky conditions with 2 m/pixel ground resolution in order to:

- Evaluate the ability of ATM data to detect and identify the buried archaeological remains (Pascucci et al. 2010)
- Compare the results of MIVIS and CASI to detect the buried archaeological remains and integrate the different acquisition times (Pascucci et al. 2010)

The vertical photos of the site taken by J. Bradford (1957) in 1944–1945 and the IGMs of 1954 were examined in order:

- To evaluate the results of the detection of related archaeological remains (Bassani et al. 2009)
- To compare the results with the MIVIS results (Bassani et al. 2009)

Lastly, all the activities to integrate the previous archaeological knowledge were performed by archaeologists in order to make an archaeological interpretation and validation of any anomaly detected and identified.

5.2.4 Albanian Archaeological Site

These research activities highlighted the benefits of the integrated approach in detecting and identifying buried archaeological remains of well-known archaeological sites by using airborne hyperspectral data. Therefore, the CNR performed a further research activity to evaluate the ability of this integrated approach to survey archaeological sites and detect and identify buried archaeological remains of relatively unknown archaeological sites. In this framework, the CNR in cooperation with the Montenegrin and Albanian Institutions were involved in the HYPerspectral for Adriatic Coastal Monitoring (HYPAD.COM) international project, funded by the Italian Ministry for the Environment and Territory. The Albanian-Montenegrin coastal and transition areas were selected as study sites in accordance with the requests of the local institutions. Four archaeological sites (Butrint, *Phoinike*, *Hadrianopolis* and *Antigonea*) were identified in collaboration with the Archaeological Institute of Tirana and University of Bologna, Department of Archaeology, within the framework of the SITARC (Sistema Informativo Territoriale Archeologico della Regione Caona) project (Cavalli and Pignatti 2009).

The data acquired and, therefore, the activities of the integrated approach are described in brief.

The in situ data was obtained by using a hyperspectral radiometer (FieldSpec spectroradiometer, Analytical Spectral Devices (ASD)) in order to carry out the Cal/Val activities in order to:

- Improve and control the atmospheric correction of the airborne hyperspectral data
- Calibrate the land use and land cover classification and validate its results
- Survey archaeological sites

Airborne data was obtained by using the MIVIS sensor on 27 July 2008 with clear sky conditions from an altitude of 1,500 m a.s.l. (3 m/pixel ground resolution) in order to:

- Evaluate the ability of MIVIS data to survey archaeological sites
- Evaluate the ability of MIVIS data to detect and identify the buried archaeological remains

Furthermore, an aerial photo survey was carried out over the archaeological site of *Phoinike*.

5.3 Integration Data, Results and Methodologies for Archaeological Prospection

To start with, this integrated approach was carried out to improve the performance the Cal/Val activities, and then, to enhance the capability of airborne hyperspectral data to detect buried archaeological remains and survey the archaeological landscape. These research activities were undertaken first to evaluate this capability, but now, the objective was to exploit this integrated approach in order to optimise this capability.

5.3.1 Integration Data for Archaeological Prospection

Therefore, the integration of multi-platform (satellite, airborne and in situ data), multi-sensor (active and passive, multi – and hyperspectral data) and frequency data was initially applied in order to perform Cal/Val activities. In particular, the Cal/Val activities in this research were used to improve and control the atmospheric correction of the airborne hyperspectral data, calibrate the land use and land cover classification and validate its results (Bassani et al. 2006, 2010).

Afterwards, the multi-platform, multi-sensor and frequency data was integrated to create merged images which integrate information from each data source and allow a more detailed survey of the buried remains to be carried out with respect to a single datum.

It is important to note that the merging of more images, collected by the same sensor but acquired at a different time or by different sensors, needs to be perfectly matched (i.e. co-registered). An example of a merged image is the result of Apparent Thermal Inertia (ATI) performed on images collected by the same sensor (MIVIS). ATI of the Selinunte Archaeological Park was obtained by MIVIS data recorded on 23 May 1996 at 7.30 and 12.30 h local time and is shown in Fig. 5.5 (Cavalli et al. 1998, 2007). ATI is a physical parameter related to thermal conductivity, density and thermal capacity, and it is therefore sensitive to change in porosity and, subsequently, to the content of soil moisture (Kahle 1987; Gupta 1991; Bendor et al. 1999). ATI is defined as the ratio, within a given time range, between the energy absorbed by surface materials and the corresponding temperature changes (i.e. daytime temperature less night-time temperature).

Other examples are merged images between SAR images and MIVIS data over the Selinunte Archaeological Park (Bianchi et al. 1998a, b). These images were performed on data collected by different sensors in order to merge and enhance the information of the SAR images and MIVIS data over Selinunte Archaeological Park. Figure 5.6 shows the MIVIS image in RGB (red represents channel 13, green represents channel 7 and blue represents channel 1) merged with the SAR image of the park (Bianchi et al. 1998a, b).

5.3.2 Comparison and Appending of the Results for Archaeological Prospection

Comparison and appending of the results obtained from the multi-platform, multi-sensor and multifrequency time analysed data were performed to allow a more complete survey of the buried structures to be obtained with respect to a single datum. An example of this is shown in Fig. 5.7, which describes the anomalies detected and archaeologically interpreted over the Selinunte Archaeological Park by a MIVIS image recorded on 23 May 1996 at 7:15 (grey lines) and by a MIVIS image recorded the same day at 12:30 (white lines).

5.3.3 Exportation of the Characteristic Methods of Each Datum, Technique and Application for Employing with Other Datum, Techniques and Applications for Archaeological Prospection

Exportation of the characteristic methods of each datum, technique and application for employing with other data, techniques and applications allows a more useful survey of the buried archaeological remains to be obtained.

For example, photo-interpretation, used to detect and identify buried archaeological remains by aerial photography, is applied to all kinds of image to detect and identify buried archaeological structures.

These creative integrations have been involved in exporting methods used to other data, techniques and applications in order to detect and identify buried archaeological structures. For example, principal components analysis (PCA) was used to produce uncorrelated output bands, segregate noise components and reduce the dimensionality of data sets. Another example is vegetation indexes designed to accentuate a particular vegetation property. These methods were applied to enhance the identification of buried archaeological remains from multispectral data (Argote-Espino and

Fig. 5.5 An example of a merged image, Apparent Thermal Inertia of Selinunte Archaeological Park obtained by the MIVIS data recorded on 23 May 1996 at 7.30 and 12.30 h local time (Cavalli et al. 1998, 2007)

Chavez 2005; Winterbottom and Dawson 2005) and also hyperspectral data (Bianchi et al. 1998b; Malagoli and de Paolis 2001; Cavalli et al. 2007; Merola et al. 2008; Traviglia 2011).

Moreover, principal classification of the surface-cover (i.e. Spectral Angle Mapper and Minimum Distance) was performed to highlight and detect buried archaeological structures. In particular, the rule images of the Spectral Angle Mapper and Minimum Distance classifier were performed on hyperspectral data to detect the structures (Ardissone et al. 2003; Coren et al.

Fig. 5.6 An example of a merged image, the combination of MIVIS image in RGB (*red* represents channel 13, *green* represents channel 7 and *blue* represents channel 1) and SAR image of the Selinunte Archaeological Park (Bianchi et al. 1998a, b)

2005; Cavalli et al. 1998, 2007; Merola et al. 2008).

Therefore, a dimensionality reduction technique, nonlinear principal component analysis, referred to as an Auto-associative Neural Network, was performed on MIVIS data collected over the Selinunte Archaeological Park to detect the buried archaeological structures (Cavalli et al. 2013).

It is important to note that the application of these methods was led by a research team (i.e. an expert in remote sensing, archaeologist) in order to select, follow the calibration and validation procedure and verify the results.

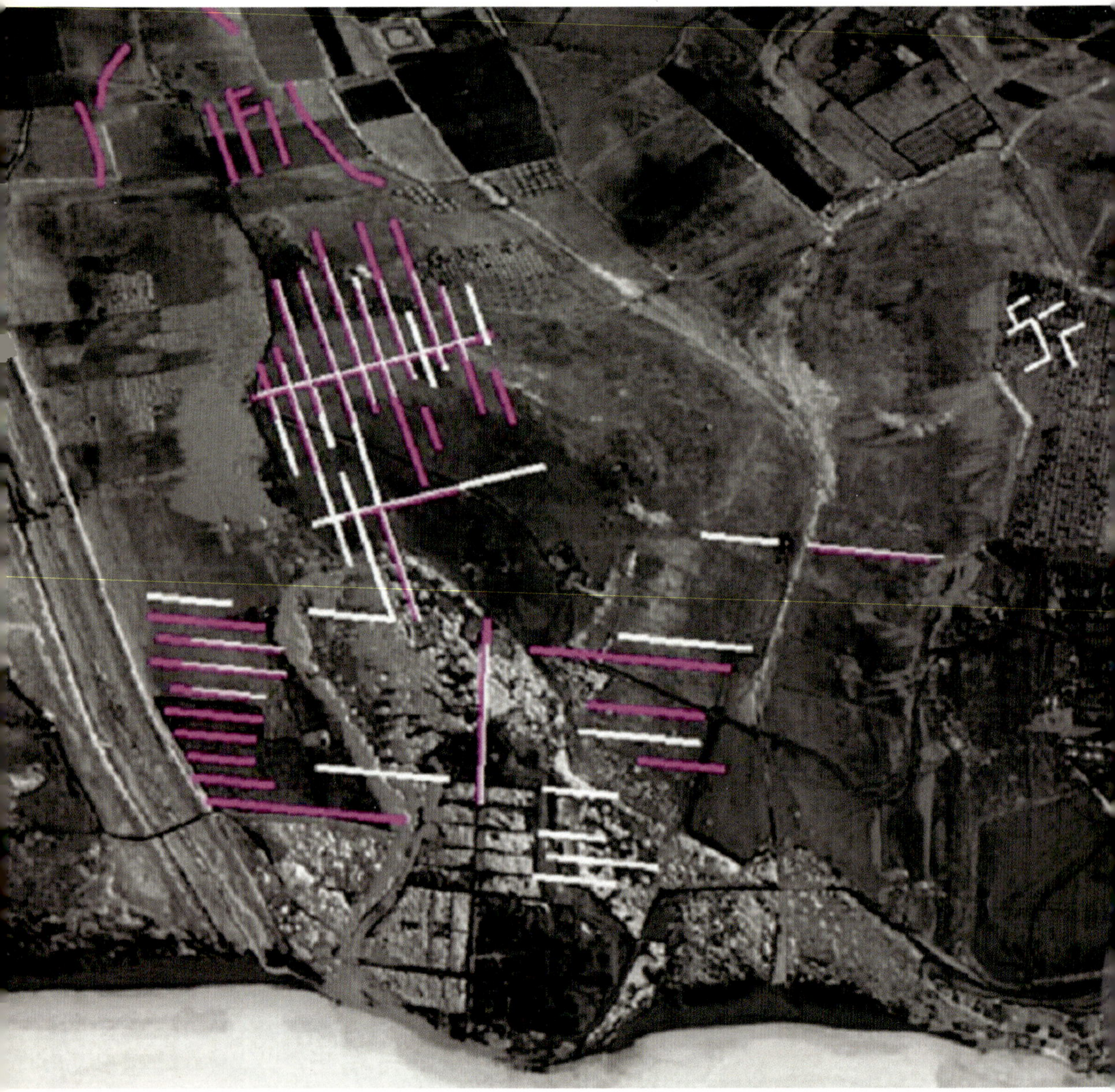

Fig. 5.7 An example of the comparison and appending of the results. Anomalies detected and archaeologically interpreted of two MIVIS images carried out over the Selinunte Archaeological Park. One image recorded on 23 May 1996 at 7:15 (*grey lines*) and another at 12:30 (*white lines*)

5.4 Evaluation and Ranking the Results of Integrated Approach

5.4.1 Image Obtained by Integrated Approach

These researches for archaeological prospection thus obtained different kinds of image, which can be separated into:

- images collected by each sensor;
- merged images obtained by the integration of data for archaeological prospection;
- synthetic images achieved by exporting the characteristic methods of each datum, technique and application for use with other data, techniques and applications for archaeological prospection.

Each image obtained using the integrated approach was examined to detect and identify the anomalies, and then an archaeological interpretation was performed on each image in order to associate each anomaly with the related buried archaeological structure.

In particular, each image collected with hyperspectral and multispectral data was examined by the researchers. The results of the archaeological prospection and their discussions are presented in Bianchi et al. (1998b), Cavalli et al. (1998, 2005, 2007) and Pascucci et al. (2010). The results of the detection and archaeological interpretation of the merged images achieved by SAR and MIVIS data are presented in Bianchi et al. (1998a), while the results of ATI obtained by two acquisitions of MIVIS data, one in daytime and another at night, are discussed in Cavalli et al. (1998, 2007). Finally, the results of the detection and archaeological interpretation of the synthetic images obtained by principal component analysis, vegetation indices and rule images of Spectral Angle Mapper and Minimum Distance are shown in Cavalli et al. (2007); the results of nonlinear principal component analysis are presented in Cavalli et al. (2013).

5.4.2 Methods to Evaluate and Rank the Image Obtained by the Integrated Approach

These images represented the anomaly with a typical size and degree of brightness due to characteristics of the preprocessing, elaboration, image, sensor, time of acquisition, etc. Different images highlight the same anomaly with diverse sizes and diverse degrees of brightness. Thus, each image has a typical capability in detecting each buried archaeological structure.

Comparison and evaluation of the capability of each image to detect and identify buried archaeological structures were performed by the photo-interpreter. However, this evaluation can be considered descriptive and subjective, because the visual interpretation depends on the person and his knowledge of the area. In order to remedy this and to evaluate the ability of each image to enhance the anomalies related to the buried structures, two indices were proposed: the Detection Index (DI) and Separability Index (SI) (Cavalli et al. 2007)

The first parameter, the Detection Index (DI), quantifies the results of the photo-interpretation work (i.e. the number of detected pixels related to the mark) for each image with respect to the total number of pixels related to the same mark detected by interpreting the whole analysed data set. The DI is expressed by

$$\mathrm{DI} = \frac{\mathrm{Npixel}_{\mathrm{mark(image)}}}{\mathrm{Npixel}_{\mathrm{tot\_mark(data\_set)}}} \times 100 \qquad (5.1)$$

where, for a given area, $N\mathrm{pixel}_{\mathrm{mark}}$ is the number of pixels pertaining to the mark identified by the photo-interpreter on the image, while $N\mathrm{pixel}_{\mathrm{tot\_mark(data\_set)}}$ corresponds to the total number of pixels related to the same mark detected by interpreting the whole analysed data set.

The second parameter, the Separability Index (SI), describes, for a given image, the tonal differences between the pixels of a mark ($N\mathrm{pixel}_{\mathrm{tot\_mark(data\_set)}}$) and the pixels of its surrounding area. The SI corresponds to a normalised scalar product expressed as follows:

$$SI = \frac{1 - \int D_{mark} D_{surrounding} dx}{\sqrt{\int D^2_{mark} dx \int D^2_{surrounding} dx}} \times 100 \quad (5.2)$$

where D_{mark} represents the frequency distribution of the digital values of those pixels belonging to the archaeological anomaly in every image, while $D_{surrounding}$ corresponds to the frequency distribution of those pixels selected as the surrounding area. According to the index definition, the SI is a parameter related to the similarity of the brightness of the marks to that of the surrounding area.

The DI and SI combine visual interpretation with statistical variables related to tonal marks. It is important to note that the evaluation is performed not on the anomaly but on the marks. That is, a mark is an archaeological anomaly interpreted as being related to a buried archaeological structure.

In fact, when calculating DI and SI, visual interpretation is necessary to identify the mark related to the buried structure and the surrounding area and to select and draw these marks on the image.

5.5 Results

For these researchers, the DI and SI evaluated and ranked the ability of the image obtained by using the integrated approach to detect buried archaeological remains. In particular, they assessed the quality of:

- Images collected by each sensor
- Merged images obtained from the integration of data for archaeological prospection
- Synthetic images achieved by exporting the characteristic methods of each datum, technique and application for use with other data, techniques and applications for archaeological prospection

5.5.1 The Quality of Images Collected by Each Sensor

The results of the evaluation of the images collected by each sensor are shown in Figs. 5.8, 5.9, and 5.10. Figure 5.8 shows two scatter plots of DI versus SI computed from each MIVIS band related to two different areas of the archaeological park (Manuzza Hill in Fig. 5.8a and the western slope of Manuzza Hill and the Acropolis in Fig. 5.8b). The bands are shown with different symbols and colours (see legend) according to the MIVIS optical port (see Table 5.1). The spectral ranges which obtained the highest values of DI and SI, and therefore the greatest ability to detect and identify the buried archaeological structures, are highlighted with coloured arrows and their wavelengths (Cavalli et al. 2007). These results show that the ability to detect and identify the buried archaeological structures is a function of the land cover. Therefore, 97 archaeological areas were selected from the MIVIS images taken over Aquileia (see Fig. 5.4e), Arpi, Marsala, *Mothia* and Selinunte archaeological sites (Cavalli et al. 2009) according to the following considerations:

- Archaeological structures were not excavated at the time of the remote acquisition.
- Archaeological structures show a sharp geometry.
- Archaeological structures are not too deeply buried.
- Geophysical surveys were performed.

In order to confirm that the ability of each band to detect the buried archaeological structures is a function of the land cover, the following methodology was used with the MIVIS images (Cavalli et al. 2009):

- Spectral Angle Mapper (SAM) classification to perform land cover classification
- Spectral unmixing (LSU) to verify the homogeneity of the land cover and to select only the anomaly and surrounding area which shows more than 75 % abundance of end-members
- Separability Index (SI) for each MIVIS band to rank their capability in detecting the buried archaeological structures
- Scatter plot analysis of the SI versus the MIVIS bands in function of the land cover fractional abundances (i.e. LSU results)

Figure 5.9 shows the trend (i.e. mean and +/– deviation standard) of SI for each MIVIS band (represented with their wavelengths in microns) of all the test sites showing more than 75 % of (a)

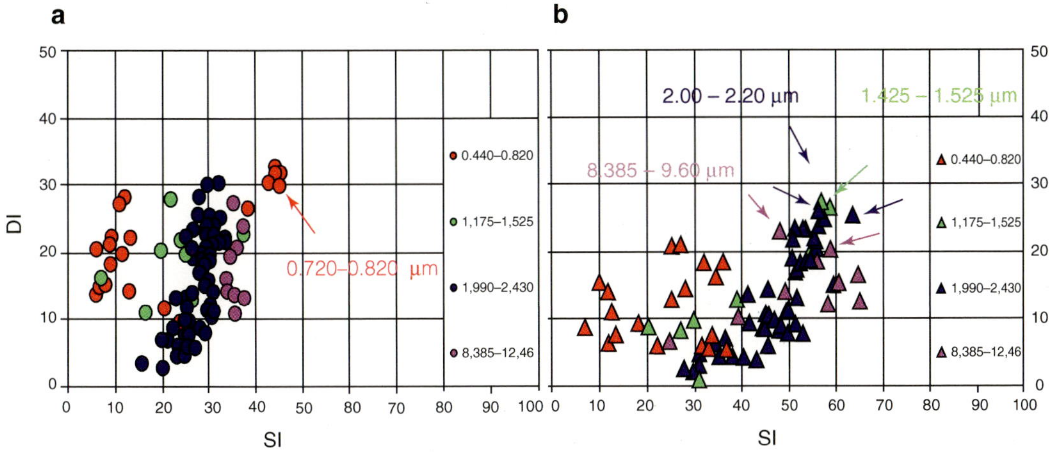

Fig. 5.8 Two scatter plots of DI versus SI computed from each MIVIS band related to different areas of the archaeological park. (**a**) Shows scatter plots of the Manuzza Hill and (**b**) Shows scatter plots of the western slope of Manuzza Hill and Acropolis Hill. Each band is shown with different *symbols and colours* (see legend) according to their optical port (see Table 5.1)

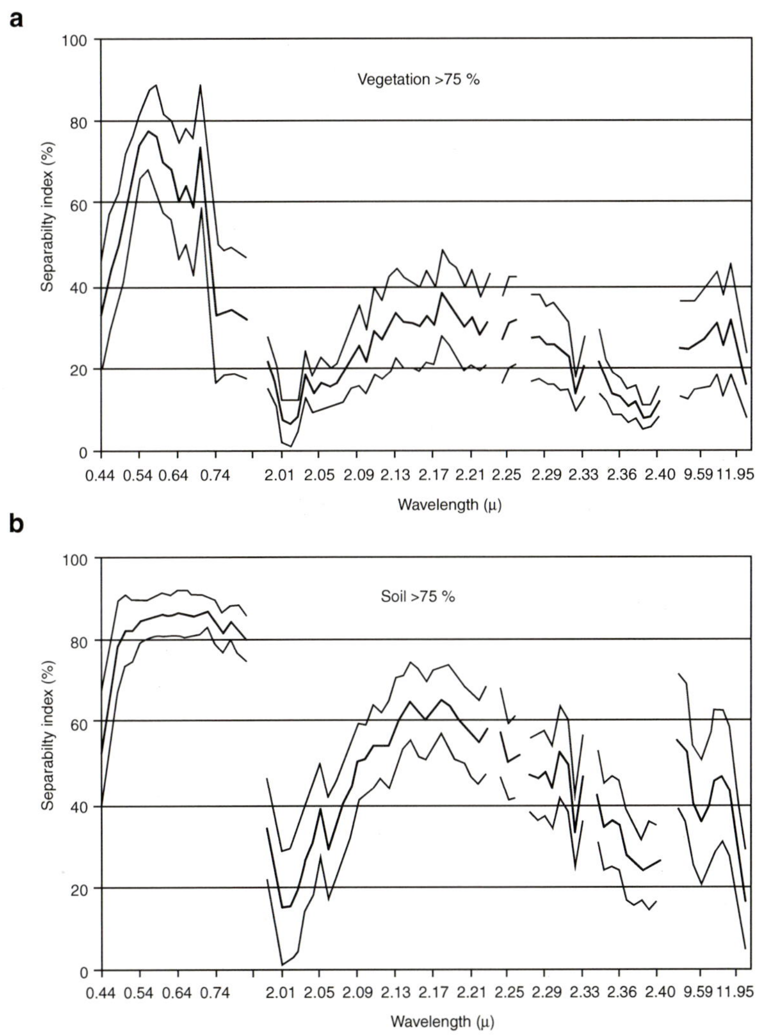

Fig. 5.9 The trend of SI for each MIVIS band (represented with the wavelength in microns) of all the test sites showing more than 75 % of (**a**) green crop (vegetation) and (**b**) bare soil (soil) end-members. *Black lines* show the values of the mean and *grey lines* the values of mean +/− deviation standard (Cavalli et al. 2009)

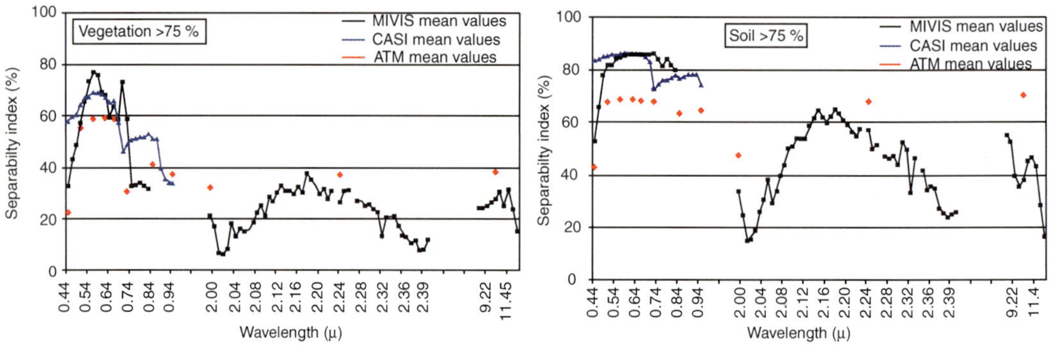

Fig. 5.10 The mean values of SI for each ATM, CASI and MIVIS band (represented with the wavelength in microns) of all the test sites showing more than 75 % of (**a**) green crop and (**b**) bare soil end-members (Pascucci et al. 2010)

green crop and (b) bare soil end-members (Cavalli et al. 2009). These results confirmed that the ability of MIVIS bands to detect and identify buried archaeological structures is a function of the land cover.

The same procedure was performed on 30 archaeological areas selected on the ATM, CASI and MIVIS images collected over the Arpi archaeological site (Pascucci et al. 2010) in order to prove that the ability to detect buried archaeological structures is a function of the land cover, not only for MIVIS data but also for ATM and CASI data.

Figure 5.10 shows the mean values of SI for each ATM, CASI and MIVIS band (represented with their wavelengths in microns) of all the test sites showing more than 75 % of (a) green crop and (b) bare soil end-members (Pascucci et al. 2010). This again showed that the ability of the hyperspectral and multispectral data to detect buried archaeological structures is a function of the land cover.

5.5.2 The Capability of the Merged Images

The results of evaluation, performed on the merged images obtained from the integrated data for archaeological prospection, are inferior to or comparable with the results performed on the images collected by each sensor. These low capabilities are due to a reduction of geometric precision. In particular, the 3 m spatial resolution of the MIVIS data was increased to match different data.

5.5.3 The Capability of the Synthetic Images

Finally, an evaluation of the capability of the synthetic images achieved by exporting the characteristic methods of each datum, technique and application for use with other datum, techniques and applications for archaeological prospection was performed on the MIVIS data acquired on the Selinunte Archaeological Park. Figure 5.11 shows two scatter plots of DI versus SI computed from each synthetic image related to two different areas of the park (Manuzza Hill in Fig. 5.11a and the western slope of Manuzza Hill and the Acropolis in Fig. 5.11b). The synthetic images are shown with different symbols and colours (see legend) (Cavalli et al. 2009). The grey squares represent the area of two scatter plots of DI versus SI computed from each MIVIS band related to the same area and the same anomaly (see Fig. 5.8). The synthetic images produced for these researchers were obtained by the whole hyperspectral data set (i.e. rule images from the Spectral Angle Mapper and Minimum Distance) and by a fraction of the hyperspectral data set (i.e. principal component analysis and vegetation indices). The results show that the synthetic images obtained from the whole hyperspectral

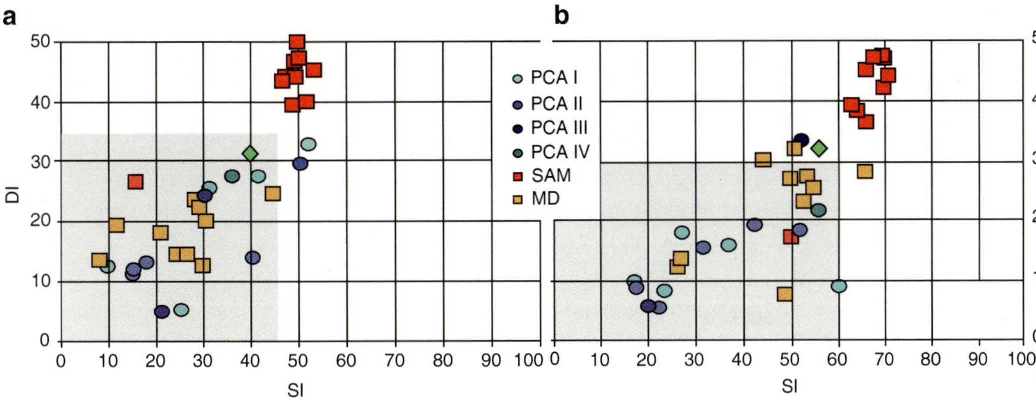

Fig. 5.11 The two scatter plots of DI versus SI computed from each synthetic image related to different areas of the archaeological park. (**a**) Shows scatter plots of the Manuzza Hill and (**b**) Shows scatter plots of the western slope of Manuzza Hill and Acropolis Hill. The synthetic images are shown with different *symbols and colours* (see legend) (Cavalli et al. 2009). The *grey squares* represent the limits of two scatter plots of DI versus SI computed from each MIVIS band related to same area (see Fig. 5.8)

data set have the greatest values of DI and SI and are therefore the best at detecting buried archaeological structures.

Furthermore, the synthetic images achieved by nonlinear principal component analysis show values of SI greater than all the other synthetic images obtained over the Selinunte Archaeological Park (Cavalli et al. 2013). In this paper a new data processing flow chart for the retrieval of buried archaeological structures, based on the use of nonlinear principal component analysis applied to airborne hyperspectral images, was presented. This flow chart, including nonlinear principal component analysis and SI techniques, data resampling criteria and anomaly evaluations criteria, applied to MIVIS airborne hyperspectral data collected over Selinunte Archaeological Park, emphasises the anomalies that are related to the presence of buried structures. The results show that the capability of the nonlinear principal component analysis approach as a preprocessing technique is better than the results of other synthetic images. Therefore, two criteria adopted for resampling the input bands obtained good results: (1) selection of the bands containing the most archaeological content (the highest value of SI), identified a greater number resulting from nonlinear principal component analysis, with SI values higher than the images used as the data set and the images of the whole data set; (2) selection of the bands with a low signal-to-noise ratio, identified a few nonlinear principal component analysis, with SI values higher than the images used as the data set and the images of the whole data set (Cavalli et al. 2013).

In conclusion, the results show that the ability to detect buried archaeological structures and survey the archaeological landscape of the synthetic images obtained by exporting the characteristic methods of each datum, technique and application to other data, techniques and applications is greater than both the capability of each single band of each sensor employed and the capability of each merged image. The best capability was evaluated by the application of nonlinear principal component analysis.

Conclusion

This paper described the advantages of applying an integrated approach to detect buried archaeological remains and survey the archaeological landscape. This approach exploits the integration of multi-platform (satellite, airborne and in situ), multi-sensor (active and passive, multi – and hyperspectral) and frequency data in order to carry out calibration and validation activities, compare and append the results of each datum, integrate different data and export the characteristic methods of each datum, technique and application for use with other data, techniques and applications. This approach was adopted not only to improve the performance of the calibration and validation activities, but also to enhance the ability of airborne hyperspectral data to detect buried archaeological remains and survey the archaeological landscape. This integrated approach promoted the collaboration of different researchers (experts in remote sensing, archaeologists, topographers, geophysicists etc.).

To start with, CNR performed these research activities to evaluate the ability of airborne hyperspectral data to detect buried archaeological remains and survey the archaeological landscape and then to improve this capability by using the integrated approach. This improvement was evaluated by comparing the capability of each band of each collected sensor, the capability of each merged image and the capability of each synthetic image obtained to export the characteristic methods of each datum, technique and application to other data, techniques and applications.

This evaluation was performed using two indexes, DI and SI, which allow us to rank the ability of each image of each mark to detect buried archaeological remains. For each image for each mark, these indexes quantify the number of pixels which highlighted the anomaly and the tonal differences between the pixels of the mark and the pixels of the surrounding area.

The results of this archaeological research highlighted that this integrated approach improves the capability of hyperspectral data in detecting buried archaeological structures and surveying the archaeological landscape. In fact, the results of this evaluation were that the capability (in detecting buried archaeological structures and surveying the archaeological landscape) of the synthetic image (obtained to export the characteristic methods of each datum, technique and application to other data, techniques and applications) is greater than the capability of each single band of each collected sensor and the capability of each merged image. In particular, the best capability was obtained by the application of nonlinear principal component analysis. Moreover, the values of DI and SI obtained by each band of each collected sensor were greater or comparable with their values obtained by each merged image (i.e. ATI integration SAR and MIVIS data).

References

Agapiou A, Hadjimitsis DG (2011) Vegetation indices and field spectroradiometric measurements for validation of buried architectural remains: verification under area surveyed with geophysical campaigns. J Appl Remote Sens 5(1):053554. doi:10.1117/1.3645590

Agapiou A, Hadjimitsis DG, Sarris A, Themistocleous K, Papadavid G (2010) Hyperspectral ground truth data for the detection on buried architectural remains. In: Ioannides M et al (eds) Digital heritage. Third international conference, EuroMed 2010. Springer, Berlin/Heidelberg, pp 318–331, LNCS 6436

Alexakis D, Sarris A, Astaras T, Albanakis K (2009) Detection of Neolithic settlements in Thessaly (Greece) through multispectral and hyperspectral satellite imagery. Sensors 9(2):1167–1187

Aqdus SA, Hanson WS, Drummond J (2007) A comparative study for finding archaeological crop marks using airborne hyperspectral, multispectral and digital photographic data. In: Challenges for earth observation: scientific, technical and commercial. Proceedings of the 2007 annual conference of the remote sensing & photogrammetry society, RSPSoc2007, Newcastle upon Tyne, 11–14 Sept 2007. Remote Sensing and Photogrammetry Society, Nottingham

Aqdus SA, Hanson WS, Drummond J (2012) The potential of hyperspectral and multi-spectral imagery to enhance archaeological cropmark detection: a comparative study. J Archaeol Sci 39(7):1915–1924

Ardissone P, Boccardo P, Borgogno Mondino E (2003) Digital images processing of hyperspectral airborne data: a cultural example. In: Proceedings of CIPA XIX international symposium, Commission V, WG V/5, Antalya, 30 September–4 October 2003, pp 202–205

Argote-Espino D, Chávez RE (2005) Detection of the possible archaeological pathways in Central Mexico through digital processing of remote sensing images. Archaeol Prospect 12(2):105–114

Aspinall RJ, Andrew Marcus W, Boardman JW (2002) Considerations in collecting, processing, and analyzing high spatial resolution hyperspectral data for environmental investigations. J Geogr Syst 4(1):15–29

Barnsley MJ, Settle JJ et al (2004) The PROBA/CHRIS mission: a low-cost smallsat for hyperspectral, multi-angle, observations of the earth surface and atmosphere. IEEE Trans Geosci Remote Sens 42(7):1512–1520

Bassani C, Cavalli RM, Palombo A, Pignatti S, Madonna F (2006) Laboratory activity for a new procedure of MIVIS calibration and relative validation with test data. Ann Geophys 49(1):45–56

Bassani C, Cavalli RM, Goffredo R, Palombo A, Pascucci S, Pignatti S (2009) Specific spectral bands for different land cover contexts to improve the efficiency of remote sensing archaeological prospection: the Arpi case study. J Cult Herit 10(S1):41–48

Bassani C, Cavalli RM, Pignatti S (2010) Aerosol optical retrieval and surface reflectance from airborne remote sensing data over land. Sensors 10(7):6421–6438

Battrick B (2005) Global Earth Observation System of Systems: (GEOSS): 10-Year Plan Reference Document, pp.197–210; ESA Publications Div

Ben-Dor E, Portugali J, Kochavi M, Shimoni M, Vinitzky L (1999) Airborne thermal video radiometry and excavation planning at Tel Leviah, Golan Heights, Israel. J Field Archaeol 26(2):117–127

Bianchi R, Cavalli RM, Fiumi L, Marino CM, Pignatti S, Pizzaferri G (1996) 1994–1995 CNR LARA Project airborne hyperspectral campaigns. In: Proceedings Of The Thematic Conference On Geologic Remote Sensing 1996, 27–29 February, Las Vegas USA. Environmental Research Institute Of Michigan, pp. II 301–II 311

Bianchi R, Cavalli RM, Marino CM, Pignatti S (1998a) Il telerilevamento aereo per lo studio dei beni archeologici. Le campagne di acquisizione dati sul Parco Archeologico di Selinunte. In: Bianchi R (ed) Selinunte, 4: Intesa di programma CNR-MISM. Progetto strategico tecnologie moderne per la conservazione dei beni culturali. Bulzoni, Roma, pp 321–335

Bianchi R, Cavalli RM, Corsi C, Marino CM, Pignatti S (1998b) Ricerche topografiche in Sicilia: integrazione tra metodi tradizionali e dati iperspettrali da piattaforma aerea. In: Proceedings of the XVth international congress of classical archaeology, Amsterdam, 12–17 July 1998. Allard Pierson, Amsterdam, pp 15–16

Boccardo P, Tonolo FG, Spanò A (2002) GIS design using high geometric resolution satellite images and hyperspectral airborne data. Int Arch Photogramm Remote Sens Spat Inf Sci 34(5):309–314

Bradford JSP (1949) Buried landscapes in Southern Italy. Antiquity 23:58–72

Bradford JSP (1950) The Apulia expedition. Antiquity 24:84–95

Bradford JSP (1957) The ancient city of Arpi in Apulia. Antiquity 31:167–169

Buck PE, Sabol DE, Gillespie AR (2003) Sub-pixel artifact detection using remote sensing. J Archaeol Sci 30:973–989

Cavalli RM, Pignatti S (2009) Telerilevamento iperspettrale per i rilievi archeologici. In: Giorgi E (ed) Groma 2. In profondità senza scavare. Metodologie di indagine non invasiva e diagnostica per l'archeologia. BraDypUS, Bologna, pp 159–169

Cavalli RM, Colosi F, Pignatti S, Poscolieri M (1998) Il telerilevamento aereo per lo studio dei beni archeologici. Applicazione dei dati iperspettrali sul Parco Archeologico di Selinunte. In: Bianchi R (ed) Selinunte, 4: Intesa di programma CNR-MISM. Progetto strategico tecnologie moderne per la conservazione dei beni culturali. Bulzoni, Roma, pp 339–358

Cavalli RM, Merola P, Pignatti S, Poscolieri M (2005) Telerilevamento iperspettrale MIVIS per lo studio delle testimonianze antropiche nell'area archeologica di Arpi (FG) Italia. Rivista Italiana di Telerilevamento 33(34):109–117

Cavalli RM, Colosi F, Palombo A, Pignatti S, Poscolieri M (2007) Remote hyperspectral imagery as a support to archaeological prospection. J Cult Herit 8(3):272–283

Cavalli RM, Pascucci S, Pignatti S (2009) Optimal spectral domain selection for maximizing archaeological signatures: Italy case studies. Sensors 9(3):1754–1767

Cavalli RM, Masini N, Pascucci S, Palombo A, Pignatti S (2010) Integration of thermal and hyperspectral VNIR imagery for architectural and artistic heritage analysis and monitoring. In: EGU General Assembly 2010, 2–7 May 2010, Vienna, Austria, vol 12, Geophysical research abstracts. EGU General Assembly, Vienna, p 4919

Cavalli RM, Licciardi GA, Chanussot J (2013) Detection of anomalies produced by buried archaeological structures using nonlinear principal component analysis applied to airborne hyperspectral image. IEEE J Sel Top Appl Earth Obs Remote Sens 6(2):659–669

Challis K, Kincey M, Howard A (2009) Airborne remote sensing of valley floor geoarchaeology using daedalus ATM and CASI archaeological prospection. Archaeol Prospect 16:17–33

Clark CD, Garrod SM, Parker Pearson M (1998) Landscape archaeology and remote sensing in southern Madagascar. Int J Remote Sens 19(8):1461–1477

Coren F, Visintini D, Fales FM, Sterzai P, Prearo G, Rubinich M (2005) Integrazione di dati laserscanning ed iperspettrali per applicazioni archeologiche. In: Atti della 9a Conferenza Nazionale ASITA, 15–18 novembre 2005, vol 1. ASITA, Catania, pp 793–798

Emmolo D, Franco V, Lo Brutto M, Orlando P, Villa B (2004) Hyperspectral techniques and GIS for archaeological investigation. In: International archives of the photogrammetry, remote sensing and spatial information sciences, vol XXXV, part B7, XXth ISPRS Congress, Istanbul, 12–23 July 2004, pp 492–497

Fowler JE, Du Q (2012) Anomaly detection and reconstruction from random projections. IEEE Trans Image Process 21(1):184–195

Galeazzi C, Ananasso C, Loizzo R (2009) The PRISMA mission. In: 6th EARSeL workshop, Tel Aviv, Israel, 16–18 Mar 2009

Gianinetto M, Lechi G (2004) The development of superspectral approaches for the improvement of land cover classification. IEEE Trans Geosci Remote Sens 42:2670–2679

Giardino M, Haley BS (2006) Airborne remote sensing and geospatial analysis. In: Johnson JK (ed) Remote sensing in archaeology: an explicitly North American perspective. University of Alabama Press, Tuscaloosa, pp 47–77
Goetz AFH (2009) Three decades of hyperspectral remote sensing of the earth: a personal view. Remote Sens Environ 113:5–16
Goetz AFH, Vane G, Solomon JE, Rock BN (1985) Imaging spectrometry for earth remote sensing. Science 228:1147–1153
Goudail F, Roux N, Baarstad I, Løke T, Kaspersen P, Alouini M, Normandin X (2006) Some practical issues in anomaly detection and exploitation of regions of interest in hyperspectral images. Appl Optics 45(21):5223–5236
Greening M-C (2012) Five-year plan 2011–2016. Working group on calibration and validation. Committee on Earth Observation Satellites, Version 5.4, July 2012. Available at http://www.ceos.org/images/WGCV/WGCV35/WGCV_work_plan_v5.4.pdf
Gupta RP (1991) Interpretation of data in the thermal infrared region. In: Gupta RP (ed) Remote sensing geology. Springer, Berlin, pp 183–216
Jensen JR (1996) Introductory digital image processing: a remote sensing perspective. Prentice-Hall, Upper Saddle River
Kahle AB (1987) Surface emittance, temperature, and thermal inertia derived from Thermal Infrared Multispectral Scanner (TIMS) data for Death Valley, California. Geophysics 52(7):858–874
Kaufmann H, Segl K, Guanter L, Hofer S, Förster K-P, Stuffler T, Mueller A, Richter R, Bach H, Hostert P, Chlebek C (2008) Environmental Mapping and Analysis Program (EnMAP) – recent advances and status. In: Geoscience and remote sensing. IEEE international symposium – IGARSS, 7–11 July 2008, vol 4. NJ IEEE Service Center, Piscataway, pp 109–112
Malagoli P, De Paolis R (2001) MIVIS dalla sperimentazione alle applicazioni. Rivista Italiana di Telerilevamento 20(21):145–149
Merola P, Guglietta D, Sampieri S, Allegrini A (2008) Lylibaeum reconstruction by remotely data. In: Remote sensing for archaeology and cultural heritage management, Proceedings of the 1st international EARSeL workshop. CNR, Rome, pp 71–74
Miller WC (1957) Uses of aerial photographs in archaeological field work. Am Antiquity 23(1):46–62
Pascucci S, Cavalli RM, Palombo A, Pignatti S (2010) Suitability of CASI and ATM airborne remote sensing data for archaeological subsurface structure detection under different land cover: the Arpi case study (Italy). J Geophys Eng 7:183–189
Piccarreta F (1987) Manuale di fotografia aerea: uso archeologico, vol 42, Studia Archaeologica. L'Erma Di Bretschneider, Roma
Reeves DM (1936) Aerial photography and archaeology. Am Antiquity 2(2):102–107
Rowlands A, Sarris A (2007) Detection of exposed and subsurface archaeological remains using multi-sensor remote sensing. J Archaeol Sci 34:795–803
Schmiedt G (1966) Contributo della foto-interpretazione alla ricostruzione del paesaggio agrario altomedievale. In: Settimane di Studio del Centro Italiano di Studi sull'AltoMedioevo XIII, 22–28 April 1965. Centro italiano di studi sull'alto medioevo, Spoleto, pp 771–837, tables I-XLVIII
Schmiedt G (1968) Le fortificazioni altomedievali viste dall'aereo. In: Ordinamenti militari in Occidente nell'Alto Medioevo: Settimane di Studio del Centro Italiano di Studi sull'Alto Medioevo XV, 30 March – 5 April 1967, vol II. Centro italiano di studi sull'alto medioevo, Spoleto, pp 860–927, tables I-XL
Server TL (1998) Validating prehistoric and current social phenomena upon the landscape of Peten, Guatemala. In: Liverman D, Moran EF, Rindfuss RR, Stern PC (eds) People and pixels: linking remote sensing and social science. National Academy Press, Washington, DC, pp 145–163
Stein DJW, Beaven SG, Hoff LE, Winter EM, Schaum AP, Stocker AD (2002) Anomaly detection from hyperspectral imagery. IEEE Signal Process Mag 19(1):58–69
Traviglia A (2011) Integrated archaeological investigations for the study of the Greater Aquileia Area. In: The new technologies for Aquileia. Proceedings of the 1st workshop, Aquileia, Italy, 2 May 2011, pp 1–14
Ungar SG, Pearlman JS, Mendenhall JA, Reuter D (2003) Overview of the earth observing one (EO-1) mission. IEEE Trans Geosci Remote Sens 41:1149–1153
Winterbottom SJ, Dawson T (2005) Airborne multispectral prospection for buried archaeology in mobile sand dominated systems. Archaeological prospection. Archaeol Prospect 12:205–219
Zhang L, Du B, Zhong Y (2010) Hybrid detectors based on selective endmembers. IEEE Trans Geosci Remote Sens 48(6):2633–2646

Skin Deep: LiDAR and Good Practice of Landscape Archaeology

6

Dimitrij Mlekuž

Contents

D. Mlekuž
Department of Archaeology,
University of Ljubljana, Ljubljana, Slovenia

Institute for the Protection of the Cultural Heritage of Slovenia, Centre for Preventive Archaeology, Ljubljana, Slovenia
e-mail: dmlekuz@gmail.com

6.1 Introduction

What makes complex sites complex? It is the same properties that make landscapes complex. Complex sites are packed with traces of past practices that combine in complex ways, the same way traces amalgamate into landscapes. Complex sites are palimpsests, the same way landscapes are palimpsests. Complex sites are not isolated entities, they are integral parts of landscapes, nexuses in a continuum of features and traces that constitute landscapes. Study of complex sites is therefore same as study of landscapes, as complex sites are integral part of landscapes and landscapes (site-scapes) in themselves.

The term "laser scanning" describes any technology which accurately and repeatedly measures distance using laser pulse, by precise measurement of time needed for the laser pulse to travel from the object and back and transforms these measurements into a series of points, or a point cloud, from which information on the morphology of the object being scanned may be derived. Airborne LiDAR (Light Detection and Ranging), ALS (airborne laser scanning), or ALSM (Airborne Laser Swath Mapping) is an active remote-sensing technique, which records the surface of the earth using laser scanning (Opitz 2012, p. 13)

LiDAR allows very precise three-dimensional mapping of the surface of the earth, producing high-resolution topographic data, even where surface is obscured by forest and

C. Corsi et al. (eds.), *Good Practice in Archaeological Diagnostics*, Natural Science in Archaeology,
DOI 10.1007/978-3-319-01784-6_6, © Springer International Publishing Switzerland 2013

vegetation. The level of detail on digital surface and terrain models produced from high-resolution LiDAR topographic data helps us enormously in the identification of past events which reworked and modified the surface of the earth. However, interpretation of LiDAR data poses much more than technical challenges. LiDAR does not provide only a layer of data but offers a different view of landscape. What do we see with LiDAR? What is good practice of working with LiDAR-derived high-resolution topographic data?

This chapter is not a series of prescriptions on what to do with LiDAR but an attempt at setting up a theoretical underpinning for the interpretation of LiDAR-derived high-resolution topographic data in landscape archaeology and study of complex sites.

6.2 What Do We Do?

LiDAR—like photography and other visual technologies—not only produces images but extends our power to detect, record, and imagine landscapes.

In archaeology, as in any other science, knowledge is not discovered, but constructed, produced, and crafted through scientific practice. If we want to understand what good practice is, we have first to understand what a practice consists of and how the knowledge is produced.

Science, including landscape archaeology, is a practice which, rather than touching directly the messy "real world," deals with signs and symbols. Since it is impossible to deal with the real world every time you want to make a statement about it, the work of the landscape archaeologists involves the creation of maps, sketches, illustrations, photographs, point clouds, papers, and books. Bruno Latour calls these accounts "inscriptions" or "immutable mobiles," which "refers to all the types of transformations through which an entity becomes materialized into a sign, an archive, a document, a piece of paper, a trace" (Latour 1986, 1987, p. 306, 1999). Inscriptions thus keep some types of relations of reality intact—therefore immutable, but can circulate around—thus mobiles, allowing new associations, translations, and articulations to take place (Latour 1999).

No serious scientists face nature with their bare eyes and hands, whether in the laboratory or in the field. Without instruments and tools, they are no different from nonscientists who have basic training about science. Bruno Latour defines scientific instruments as "inscription devices," as "any setup, no matter what its size, nature and cost, that provides a visual display of any sort in a scientific text" (Latour 1987, p. 68).

The instruments are the interface between the real world and the landscape archaeologist, where inscriptions are produced. If you want to find out what landscape archaeologist does, find out first what kind of instrument he or she is using, and then observe what he or she does to the instrument and with the instrument.

In order to be examined, the real world needs to be scaled down and inscribed. But how to establish a nonarbitrary, meaningful connection between the real world out there and the inscribed world on paper or computer file? One way is by "mathematicizing" the real world, by quantifying the observation and coding every sighting in coordinates. Once all sets of coordinates are collected, the shape of the observed phenomena may be redrawn by those who have not sighted them (Latour 1986, 1987, p. 224). Map can be then redrawn on paper or within geographic information system. LiDAR as inscription device does exactly this.

The practice of landscape archaeology is therefore transformation of material, real world into inscriptions and translations of these inscriptions, which are processes of transformation, conversion, juxtaposition, simplification, and combination of inscriptions (Latour 1986, 1987, 1999). With each step, we lose "locality, particularity, materiality, multiplicity, and continuity" yet we gain "compatibility, standardisation, text, calculation, circulation, and relative universality" (Latour 1999, p. 70). The series of references established along this "cascade" ends up as final publication, report, paper, book, and map, with the references maintained in the form of a project archive.

The purpose of the chain of references is to facilitate the way we retrace the process of transformation and translation. The process is reversible, it circulates: faced with the final publication, we can move backwards along this chain of mediations and transformations documented in the project archive, follow it ultimately back to the landscape studied.

6.3 LiDAR as an Inscription Device

LiDAR produces an inscription, trace of the landscape in a similar way as photography. Photograph is the result of light rays falling on an optically sensitive film, resulting in an irreversible chemical reaction on the surface of the film. A negative image is produced on the film, which can later be reproduced as a positive image on photographic paper. In this way the world inscribes itself upon the surface of the photographic image (Hauser 2007, p. 78). Photography is thus carefully manipulated at every stage. The technologies of photography, camera, and darkroom harness to produce a recognizable image. The technologies of photography, camera, and darkroom harness the indexical qualities of chemical photosensitivity in order to produce a lasting image (Hauser 2007, p. 88).

LiDAR is a much more complex way to produce trace of the landscape. Airborne laser scanning systems are assemblages of technologies, including a laser scanner, positioning and georeferencing equipment (GPS and inertial measurement unit, IMU), and data recording system, located on an airborne platform, aircraft, or helicopter that produce the trace of the surface of the earth.

Platform choice plays a role as it affects the flying height and speed, resulting in vegetation penetration and point density and area covered.

Laser transmitter sends a pulse of laser light. Most airborne LiDAR systems work at near-infrared range (NIR), although bathymetric LiDAR systems operate in the green range. Solid-state lasers emit very short, high-powered pulses at high repetition rate. Typical pulse duration ranges from 4 to 10 ns resulting in the scan rage of 100–150 kHz, or 100,000–150,000 pulses per second. Laser is coupled with the beam director which scans the laser pulses over a swath of terrain, usually centered on, and collinear with the flight path of the platform, the scan direction being orthogonal to the flight path. It uses a variety of systems, including oscillating mirrors, rotating mirrors, rotating prisms, and rocking or swaying mirrors; the system produces distinct pattern of measures on the ground, known as scan pattern.

When laser pulse reaches the ground, it has a finite diameter (around 10 cm and larger). It is possible that a part of the diameter interacts with an object above the ground, for example, a tree. Part of the pulse will interact with the tree canopy and then reflect back, while the rest of the pulse keeps travelling through the gaps of the canopy till it encounters other objects (branches, leaves, ground) which result in reflection of other parts of the pulse.

The sampling of the received laser pulse can be carried out in different ways. Airborne LiDAR systems may be divided into discrete return and full waveform systems. Discrete return systems operate so that the detector triggers when the incoming pulse amplitude (intensity or energy) reaches a set threshold, thus measuring the time of flight. Discrete return LiDAR usually records two to four returns per pulse.

Full waveform systems sample the intensity of the return signal at regular intervals, thus recording the form of the entire returned pulse. In this way, pulse interaction with the object can be precisely recorded. This can be useful for further processing, for example, in analyses of vegetation canopy and in differentiating between low vegetation and terrain (Doneus et al. 2008).

The round trip travel times of the laser pulses from the aircraft to the ground are measured with a precise interval timer and the time intervals are converted into range measurements based on the velocity of light. The position of the aircraft at the time of each measurement is determined by a phase difference kinematic GPS. Rotational positions of the beam director are combined with aircraft roll, pitch, and heading values determined with an inertial measurement system, and with the range measurements, to obtain vectors from the aircraft to the ground points. When these vectors are added to the aircraft locations, they yield accurate coordinates of points on the surface of the terrain.

LiDAR raw data is thus a series of measurements of times and intensities of returned laser pulses. Adjacent strips of points are aligned to improve accuracy within the dataset, and the final point cloud is adjusted to fit the coordinates of ground surveyed control points. These positions are typically stored as a point cloud, where each point contains attribute information such as GPS time, intensity, scan angle, and flightstrip number along with its X, Y, and Z coordinates. Due to their large volume, point clouds are usually manipulated and stored in the binary LAS format (Samberg 2007).

Airborne LiDAR technology is still developing rapidly in both sensor capability and data processing. The competition between LiDAR sensor manufactures is mostly focused on increasing laser pulse repetition rates to collect more data points. High-density data make it possible to represent terrain in much detail. However, high-density data lead to a significant increase in the data volume, imposing challenges with respect to data storage, processing, and manipulation. Although the cost of LiDAR data collection has become more affordable in the last decade, the processing and interpretation of raw data still remains a considerable challenge.

The essential and critical phase in the post-processing chain is classification. Points in the point cloud must be classified to differentiate between returns from the ground and those from vegetation (Fig. 6.1a). The processing implies assumptions to be made about the properties of the ground, either explicitly by the analyst of the data, or within the software systems applied. The assumptions have a major impact on the quality of derived data. Thus, the ground points are the measurements from bare-earth terrain that are usually the lowest surface features in a local area. Non-ground points are the measurements from the objects above the bare-earth terrain, such as trees, shrubs, and buildings. In order to appropriately identify ground points, it is important to understand the physical characteristics of ground points that differentiate them from non-ground points. Ground surfaces can be divided into four categories based on their physical characteristics (Meng et al. 2010).

Ground surfaces are usually the lowest features in a local neighborhood. Many ground-filtering methods are based on this important characteristic and search for the lowest elevations in a neighborhood to initialize the ground-filtering process.

Surface slope is generally lower between two neighboring bare ground points than between one bare ground and one non-ground point. Therefore many ground filters define a point with slope larger than the maximum ground slope as a non-ground point. Ground points can be distinguished from non-ground points on the basis of slope threshold value. Threshold differs for each surface type: relatively flat urban surfaces may have a low-threshold value, while complex surfaces such as mountain terrain or high-relief forest canopy surfaces will have steeper slopes and may require a higher threshold to accurately distinguish ground from non-ground.

Because most bare-earth surfaces have few sharp changes in elevation, the elevation difference from a ground point to neighboring ground points is usually lower than the difference to neighboring non-ground points. Therefore, points having an elevation difference higher than a location-specific threshold are probably non-ground points, such as shrubs, trees, or buildings.

Ground surfaces are relatively continuous and smooth. Objects such as trees and buildings are usually less smooth in texture than bare ground and may be removed on the base of morphological characteristics. Poor choice of filtering parameters will lead to non-ground points being classified as ground and ground points classified as non-ground. This can have serious consequences for the detection of archaeological features in the final terrain model (Opitz 2012, p. 20).

Classified point cloud allows interpolation of different digital elevation models. Digital terrain model (DTM) is a representation of bare-terrain

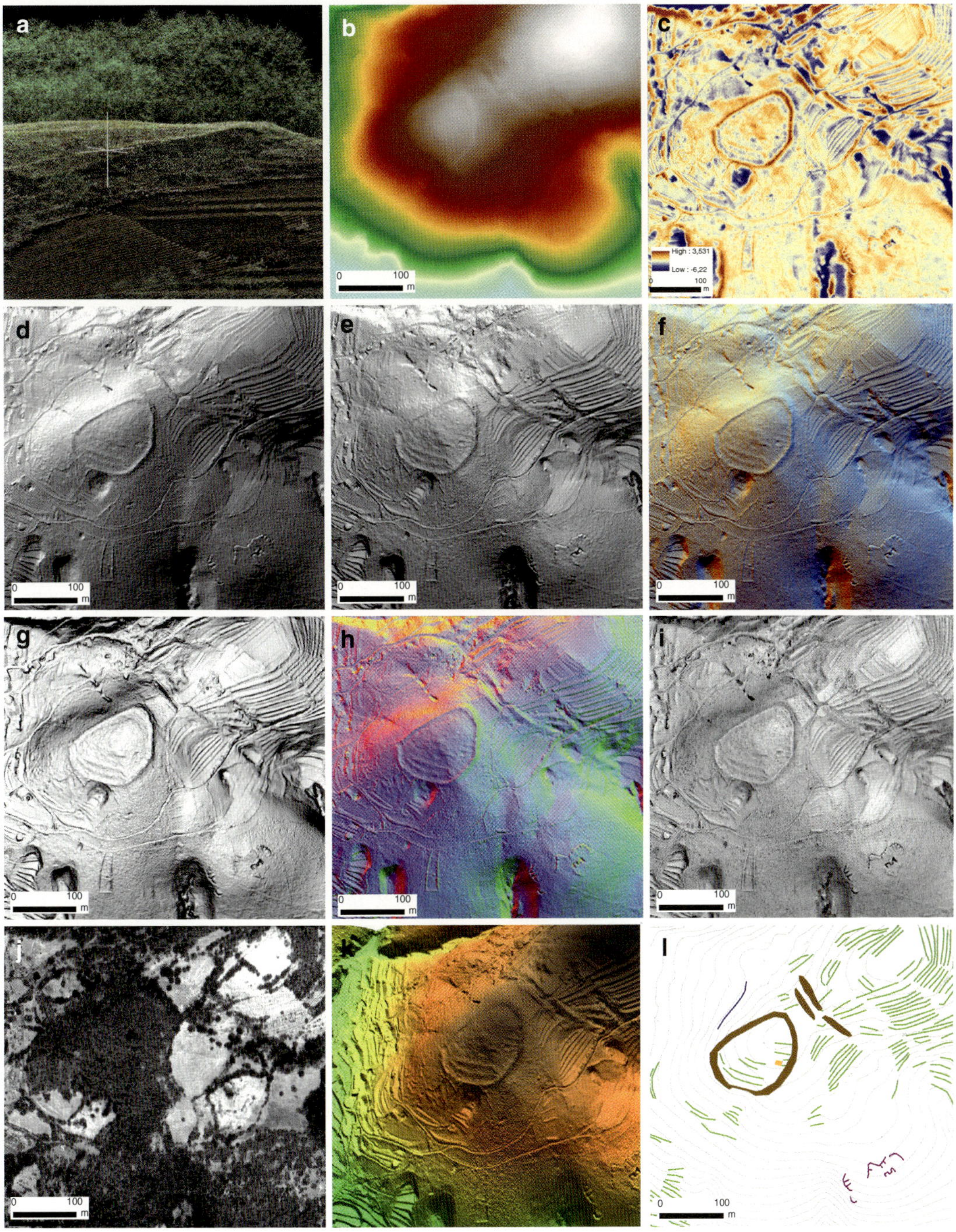

Fig. 6.1 Some typical inscriptions produced in the process of interpretation of high-resolution topographic data. Classified point cloud (**a**); color-coded digital terrain model (**b**); local relief (**c**); hill-shaded digital terrain model with illumination from 315° azimuth and 45° elevation (**d**); minimum of hill-shaded images illuminated from three consecutive directions (315, 0, 45°, light source height 45°) (**e**); RGB composite of three hill-shaded images from consecutive directions (315, 0, 45°, light source height 45°) (**f**); third principal component of hill-shaded images illuminated from 16 directions with 45° elevation of light source (**g**); RGB composite of first three principal components of hill-shaded images illuminated from 16 directions with 45° elevation of light source (**h**); sky-view factor visualization (5 m search radius in 24 directions) (**i**); LiDAR intensity image (**j**); 3D view of digital terrain model, draped with color-coded elevations (**k**); digitized features draped over contour map (**l**)

surface, free of any object, such as trees, buildings, etc. Digital surface models (DSM) include tops of the buildings, trees, power lines, and other "landscape clutter." For archaeology a combination between DTM with some landscape clutter is usually preferable.

While classification of point cloud is usually performed using specialist software, the interpolation and visualizations of LiDAR data are usually carried out in a geographic information system (GIS), which allows not only interpolation and visualization of DTMs and DSMs but also integration of other georeferenced datasets.

Digital elevation models are then visualized. There are number of different visualization methods at hand (see Devereux et al. 2008; Kokalj et al. 2012).

Digital elevation model can be represented by assigning a color to a pixel based on the elevation of that pixel (Fig. 6.1b). Since we are usually interested in a small range of height values, the image histogram must be manipulated to enhance contrast and emphasize details. This includes nonlinear enhancements such as logarithmic stretch, square root enhancement, exponential stretch, and histogram equalization. If the relative differences between the values (elevations) are important, a histogram is manipulated by cutting of the extreme values.

Archaeological features are generally of a much smaller scale than the landforms on which they lie. Trend removal separates local small-scale features from large-scale landforms by removing the height variation of large-scale forms (trend) and produces a local relief model (LRM) (Fig. 6.1c). The trend can be computed by generalizing a detailed DEM, usually using low-pass filters or by resampling to lower resolution.

However, since small-scale features are smoothed rather than eliminated by computing trend surface, the local relief model is biased towards small features. Hesse (2010) improved the procedure by replacing trend surface with a "purged DEM," resulting in a less biased representation of small-scale topographic features.

Most common visualization is a representation of terrain by hill-shaded images (Fig. 6.1d). Analytical hill-shading simulates illumination by direct light from a light source at an infinite distance, as rays have constant azimuth and elevation angle for the entire area. This represents relief with all the features in a natural in intuitive way, straightforward to interpret. However, single analytical hill-shading fails to reveal linear features that lie parallel to the direction of light (Devereux et al. 2008).

This can be addressed by creation of different hill-shaded images by changing the direction and height of light source. They can be combined using mathematical functions (either mean, minimum, or range of values, Fig. 6.1e), or RGB composites can be produced from images hill-shaded from three different directions (Hobbs 1999; Devereux et al. 2008; Kokalj et al. 2012; Fig. 6.1f).

Images hill-shaded from different directions are highly correlated and can be summarized using multivariate statistical analyses, such as principal component analysis (PCA) (Devereux et al. 2008). The first three components (Fig. 6.1g) computed from multiple (e.g., 16) directions usually contain a high percentage (typically over 99 %) of the information in the original dataset. The RGB composites of the first three components simplify the interpretation of multiple hill-shaded images (Fig. 6.1h).

A relatively new way of visualizing digital elevation model is the sky-view factor which overcomes the drawbacks of the directional hill-shading (Kokalj et al. 2012; Zakšek et al. 2011). Sky-view factor measures the portion of the sky visible from a certain point and thus represents diffuse illumination (Fig. 6.1i).

Another product of LiDAR survey is the intensity image, a monochromatic image of the illumination (energy) returns from the LiDAR system (Fig. 6.1j). Intensity images thus provide active illuminated scene in the near-infrared spectrum, which can provide information on the character of soils, sediments, and vegetation (Challis et al. 2011; Challis and Howard 2012).

Unlike photography, where the result is usually a single image, the result of LiDAR processing chain is a whole cascade of inscriptions (Fig. 6.1). These accounts are produced, constructed, and crafted using different and complicated methods, but that does not make them less real. As Bruno Latour observes (2001, p. 19) the "more instruments, the more mediation, the better the grasp of reality is ... the more human-made images are generated, the more objectivity will be collected."

6.4 LiDAR as Topography

Ancient topography can be defined as the study of places, their shape and features that can hint at the nature of an archaeological site and the potential existence of structures buried beneath the soil. Many sites are visible on the ground as a series of "humps and bumps." An accurate plan produced by topographical survey can reveal the outlines of features that had not previously been recognized. The English topographic school, for example, developed topography to the level of art, producing beautiful and rich inscriptions of places using hachure depictions (Johnson 2007).

Development of new tools such as differential GPS or laser theodolite allows places to be recorded in a more metrically accurate way, using three-dimensional representations. This generally entails the recording of elevations across a grid of a certain resolution, for instance at 5 or 10 m intervals, but also the recording of points on the visible breaks of slope, to emphasize archaeological features in the landscape. This allows us to produce even more inscriptions, using different visualization techniques.

LiDAR produces metrically accurate high-resolution topographic data as well and can be understood as an extension of these technologies.

What is the difference between LiDAR topographic data and data produced by topographic survey? There are obvious differences in spatial cover, acquisition speed, and sheer volume of data, but there is more. What makes LiDAR different from other topographic techniques is the absence of selectiveness. LiDAR data is typically gathered across complete landscape blocks, not limited to selected places, and does not record only important "humps and bumps," recognized as such, but the whole landscapes, all the mess of traces, humps, and bumps. LiDAR records landscape in an indiscriminate way, every place, every feature, every trace, and every square meter is in principle treated with the same attention and resolution.

6.5 How to Read Traces?

Photographs, laser scans, and other measurements are traces of the real world (Hauser 2007, p. 72). LiDAR-derived high-resolution topographical data are full of marks and traces, however legible or illegible those marks may be. LiDAR produces a trace of traces.

A trace is a mark of something, a material residue of an occurrence or an existence. Footprints, cuts, scrapes, scuffs, scratches, and scars are all traces. So, too, are piles, heaps, accumulations, mounds, and banks. One class of traces is thus the imprint of something on a surface, in which nothing of the object that made the imprint remains, merely a negative of its contours. It is created by removing, scratching, and cutting into the surface, such as footprints, holloways, ditches, and cuts. On the other hand, traces can be created by accumulation, by bringing things and substances together. Blood stains are an example of this kind of traces, same as mounds, walls, and cairns.

When it comes to leaving traces, people are in no way privileged. Natural processes leave traces too, same as animals and humans. Water can accumulate levees, banks, and bars and can erode gullies, channels, and valleys. Wind can throw trees, creating scars on the forest surface. Animals leave traces on their daily routines as well: badger sets, boar rootling, deer digs, and animal trails are all traces of animal practices.

Traces are also thus signs of past action, event that made them happen. Traces deposited by humans, animals, and natural processes, such as the weather, need not be purposefully fabricated signs, and in fact most are not. Those features are indices of daily routines, non-discursive practices that left marks on the surface of the earth, and material ripples of the practices that occurred on it.

People, things, animals, and places are mutually constituted. We act with and through material culture and material culture acts through us (Latour 1994, 2004; Knappett 2005; Olsen 2010). Tools, equipment, and machines change the ways we interact with the world around us and leave traces. Tim Ingold (2004), for example, writes how the development of the technology of the shoe changed the ways people moved through landscapes. Development of tools and machines and new historically specific ways of interacting with animals resulted in assemblages such as plows and ox carts that produce different traces than before.

And this is what we see on the LiDAR-derived high-resolution topographic data: multiplicity and richness of past things, traces of past activities and tasks, and human and non-human materialized in a landscape. Landscapes are full of these traces—lime kilns, charcoal-burning platforms, fields, holloways, tracks, lynchets, and quarry pits, but also animal trails, paleochannels, tree throws, landslides, etc. (Fig. 6.2). These features overlap, crisscross, and are destroyed, reworked, or incorporated into other features.

But some traces are intentionally constructed as signs: buildings, monuments, barrows, roads, parks, gardens, etc., signifying the idea of durability, control, aesthetic beauty, monumentality, or symbolic power. They were deliberately built to change the way people move, interact, access, see, and understand the landscape.

Landscape is therefore full of traces of past practices. These traces are not isolated discrete "features," but a material residue of a web of interrelated practices. Tim Ingold calls this web a "taskscape" and points to the relations between the landscape and the activities performed within it: "Just as the landscape is an array of related features, so — by analogy — the taskscape is an array of related activities" (Ingold 2000, p. 195).

Practices occur at places that are shaped by traces of previous activities. People, animals, stuff, and substances move around landscape and stop at places. Movement is an essential and ubiquitous practice that connects practices and places in a web of interrelated activities. Movement leaves trace related to movement such as tracks, holloways, and paths.

Landscape can then be viewed as a product of practices, movements, trajectories, interrelations, and flows. And this sense of landscape as a continuous weaving, relating, and associating, forever in the making, is in my opinion much more productive than static notions of landscape in terms of territory, boundedness, area, scale, and so on (Allen 2011).

This prompts us to shift to thinking about past landscape in practical and processual terms, as something that was in a perpetual state of becoming, made, and remade by people, animals, natural forces, and things. We must focus on the ways in which people routinely and creatively interact—with landscapes in their daily lives, along with associated embodied and technologized practices (Wylie 2007).

In order to operate as a sign, the trace must be visible and recognizable (Hauser 2007, p. 73). As marks of something, as the signs of past events and processes, intelligible only as such, these traces need to be interpreted. Interpreted means correlating trace with the event that produced it, supplementing the trace with a mental image of what is missing from it. The archaeological record is full of absences (Lucas 2010). Thus looking at the trace of holloways, we see something that is not there, people moving along the path (Hauser 2007, p. 93). There is a gap between trace and the past action or event that produced it, and it has to be crossed by archaeological imagination. This is not always possible, interpretation therefore includes many uncertain categories. Interpretation is a highly subjective process and there can be several "right" interpretations of the same traces, depending on experience of the interpreter and the questions asked.

Alison Wylie refers to this process as one of "tacking" between different hypotheses and evidence, scales of analysis, traces and landscape, times, space, and entities, gradually refining hypotheses over time (Wylie 2002). Interpretation thus implies weaving together different arguments, strands of evidence, and different frames of reference, and good interpretations depend on their strength on "concatenation of (many) cables of arguments" (Wylie 2002, p. 165) rather than just one.

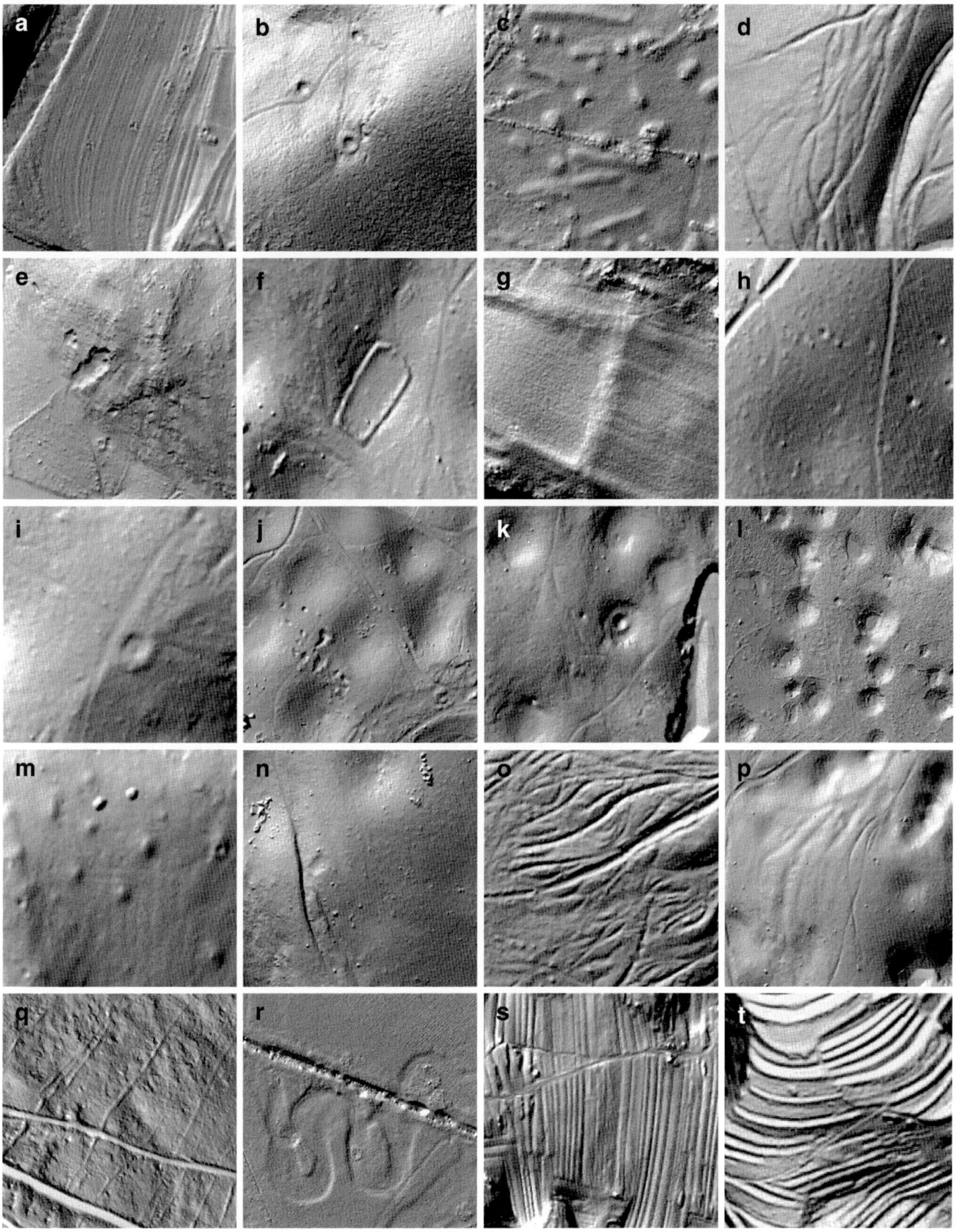

Fig. 6.2 Hill-shaded image of traces encountered in the high-resolution topographic data: ridge and furrow (**a**, **s**), limekilns (**b**, **k**), earthworks (**c**), holloways (**d**, **o**, **p**), quarries (**e**),enclosures (**f**), plowing headlands (**g**), tree throws (**h**), charcoal-burning platforms (**i**, **n**), extraction pits (**j**), dry stone walls (**l**), clearance mounds (**m**), wood removal chutes (**q**), paleochannels (**r**), and agricultural terraces (**t**)

The process of interpretation is thus not one-directional chaining of inscriptions but includes circulating along the chain of mediations and inscriptions and transformations, back to the point cloud, or ultimately back to the material landscape we study. Thus, for example, LiDAR survey of woods in Austria recorded several possible "mounds." After ground-truthing some mounds proved to be clearance piles. This made possible to adapt filtering algorithms to remove the woodland clearance piles from full waveform laser data (Doneus et al. 2008).

This is one reason why "automated feature extraction" (Cowley 2012) will never replace the skillful interpreter. It is up to interpreter to spot and understand and interpret the traces that might be important to understand what happened in a landscape. This is a skill that "needs to be learned, practices tested, and honed" (Palmer 2012, p. 88) through our own practices of manipulation and interpretation of high-resolution topographic data.

The question is where to start when we are faced with the vast amount of information contained in high-resolution topographical data? I would suggest *in medias res*, in the midst of the things, and start and reassemble from there:

> Learning about, and from, image interpretation never stops and that is one of the most interesting things about working with these data. To be able to start with an image—be it an old vertical photograph or last week's LiDAR data—and end with an analytical map provides a way into the past that few other archaeological activities can match. (Palmer 2012, 88)

Interpretation of high-resolution topographical data involves constant movement, zooming in the traces, and interaction with them through different visualizations, drawing information about them from different inscriptions, circulating along the chain of references, and then again zooming out, panning to another trace, and establishing connections with other traces and the wider landscape. This is what Rachel Opitz and Laure Nuninger (2010) call "contextual topography" and is a way of creating knowledge through practices of mapmaking, transformations and translations of maps, and juxtaposing different strands of evidence and between scales. In this way our own interpretation practices become interwoven with past practices that created material traces in the landscape. Through our encounter with traces, by moving between them, we reiterate connections between them (see Lucas 2001, p. 202).

And good practice of interpretation includes reflection on how knowledge is produced through our own practices of contextual topography (Mlekuž 2012; Halliday 2012).

6.6 Time and Palimpsests

The fundamental characteristic of material world is duration. Traces, by definition material, are durable remnants of past events. Traces of different periods can exist simultaneously enduring in a land for different lengths of time because there are variations in change or turnover (Lucas 2005). The landscape is therefore multi-temporal, made up of a series of past durations (Fig. 6.3).

The material world is composed of objects of differential duration. By definition, material objects, things, and traces have duration that extends their creation to the current moment of observation. Moments in time that leave no material traces are unknowable, at least from the archaeologist's perspective. Past is therefore incorporated and reworked into the present, as Oliver convincingly demonstrated describing his house:

> The house … was built towards the beginning of this century, in the courtyard of an ancient farm whose structure is still visible… I see an interweaving of houses and constructions, most of them dating back to the nineteenth century, sometimes including parts of earlier constructions from the eighteenth or seventeenth century. The twentieth century here looks so localized, so secondary: it is reduced to details, such as windows, doors or, within houses and flats, furniture … Right now, the present here is made up of a series of past durations that makes the present multitemporal. (Olivier 2001, p. 62)

On the other hand, LiDAR is an image-making technology which specializes in the freezing of time (Hauser 2007, p. 71).

What makes archaeology different from other disciplines is our concern with time depth of human engagement with the world and landscape. Landscape is continuously produced. Thus time is inscribed in its very constitution at multiple levels and scales. Landscape has a

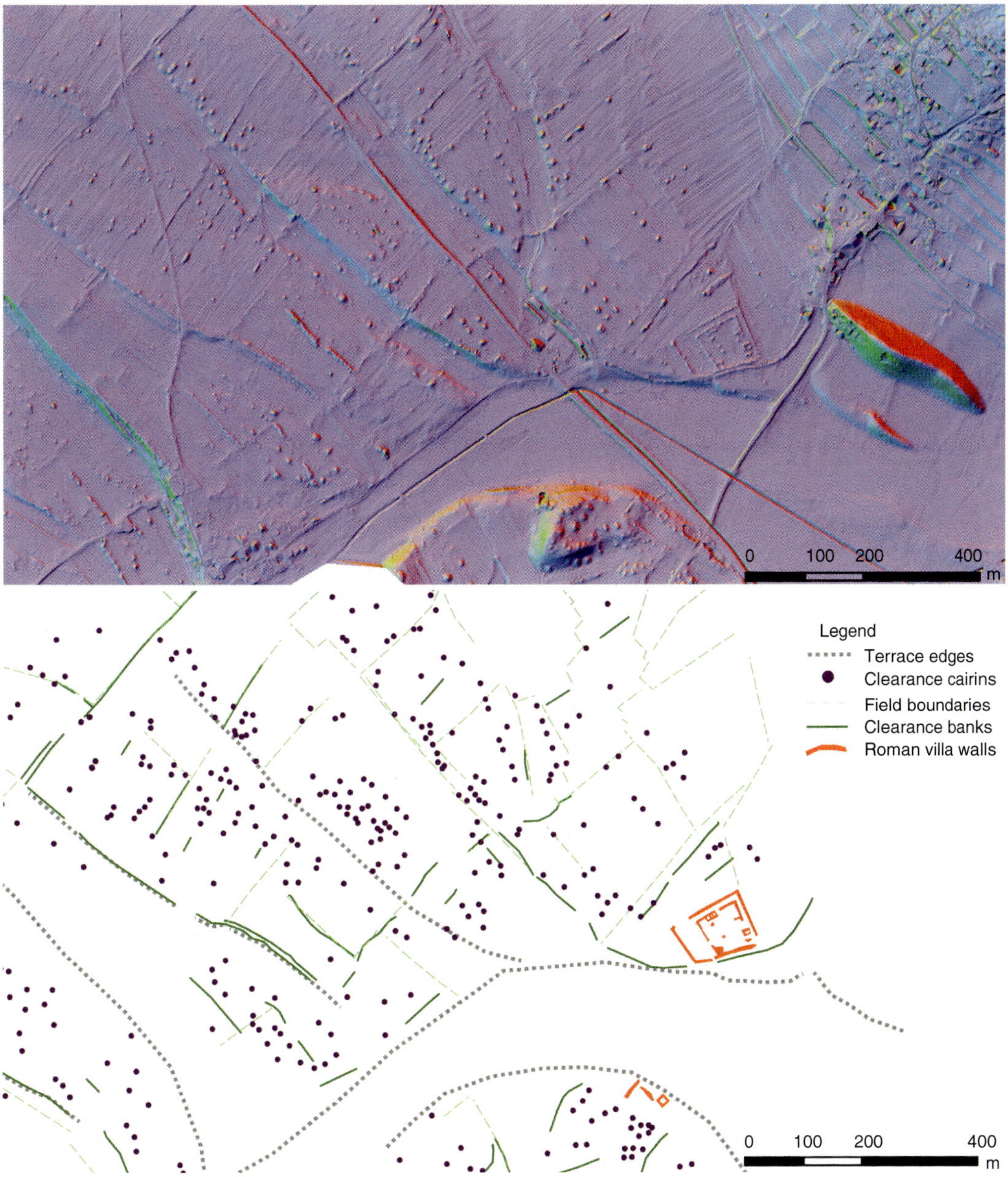

Fig. 6.3 Landscape of duration. Field clearance cairns cluster around Roman villa set in a moraine landscape. In some of the cairns Slavonic burials were placed. Clearance cairns became incorporated into medieval field system (Rodine, Northern Slovenia). Sky-view factor image combined RGB composite of first three principal components of hill-shaded images, illuminated from 16 different directions with 45° elevation of light source

temporal dimension. Traces are accumulated one upon another. As Barbara Bender (2002, p. 103) observes "Landscape is time materialized. Or, better, landscape is time materializing: landscapes, like time, never stand still. Landscapes are in a constant process of becoming."

The most often used metaphor to describe the building up of landscape is palimpsest. Palimpsest is a parchment on which earlier writing has been erased to make way for a new text. It refers to the traces of multiple, overlapping activities over variable periods of time and the variable erasing of earlier traces. It is a kind of flattened time. This passage from Crawford (1953, pp. 51–52) beautifully captures the idea of historical layering in the landscape:

> The surface of England is like a palimpsest, a document that has been written on and erased over and over again; and it is the business of the field archaeologist to decipher it. The features concerned are of course the roads and field boundaries, the woods, the farms and other habitations, and all the other products of human labor; these are the letters and words inscribed on the land. But it is not easy to read them because, whereas the vellum document was seldom wiped clean more than once or twice, the land has been subjected to continual change throughout the ages.

Although Crawford's use of palimpsest implied the possibility of separating discrete layers, palimpsest usually refers to a process of superimposition of successive activities, which partially destroyed or reworked earlier traces (see Lucas 2005, p. 37). Thus a palimpsest involves both the total removal of all information except the most recent as well as accumulation and transformation of successive and partially preserved activities, in such a way that the resulting totality is different from and greater than the sum of the individual constituents (Bailey 2007, p. 203).

Geoff Bailey (2007) places palimpsest between two extreme cases. True palimpsests are palimpsests in the strict sense of the term where all traces of earlier activity are removed except for the most recent, so they are impossible to distinguish from the single episode (Fig. 6.4). On the other hand, a cumulative palimpsest is one in which the successive episodes of deposition, or layers of activity, remain superimposed one upon the other without loss of evidence but are so reworked that it is difficult or impossible to separate them out into their original constituents (Bailey 2007, p. 20; Fig. 6.5). In this way, cumulative palimpsest is close to stratigraphy, where different traces and interfaces between them are interpreted in terms of different events that can be either successive or contemporaneous (Lucas 2012, p. 90).

Most palimpsests share elements of both and can be made both by mixing of material of different ages and destruction of material resulting from successive episodes of clearance and removal or progressive decay of material. The key trait they share in common is that both result from the repetition of activities and the deposition of material in the same location, or in similar locations with considerable overlap.

Thus instead of treating palimpsest as "flattened time," we should focus on the activities and events of erasure and inscription that produced them.

All these processes combine the products of different temporalities in different ways: destroy, blur, or sharpen their apparent boundaries. These effects are important, for they determine where we see traces of past human practices and what these look like. Some traces of human activities get worked into or get buried in the soil. Some episodes are buried or obscured from view, some are destroyed, some are disturbed, some retain high integrity and resolution of patterning, and some are accessible to archaeological discovery and analysis. Therefore, even if some episodes are completely destroyed, even if destruction starts history anew, it is only an episode within the wider palimpsest. Processes of destruction and erasure are just another trace, in spatial relation to other traces, within the wider pattern. In other words, we are still dealing with a palimpsest, except that we are dealing with a palimpsest at a larger spatial scale.

Due to the processes of reworking, mixing, and erasure, dating of individual episodes of palimpsest is difficult if not impossible. However, palimpsest still has considerable information potential because of their precise location in space and their duration.

Palimpsests are not anomalies that need to be untangled and separated into layers before they can be interpreted and understood but an inherent feature of the material world (Lucas 2012, pp. 115–123).

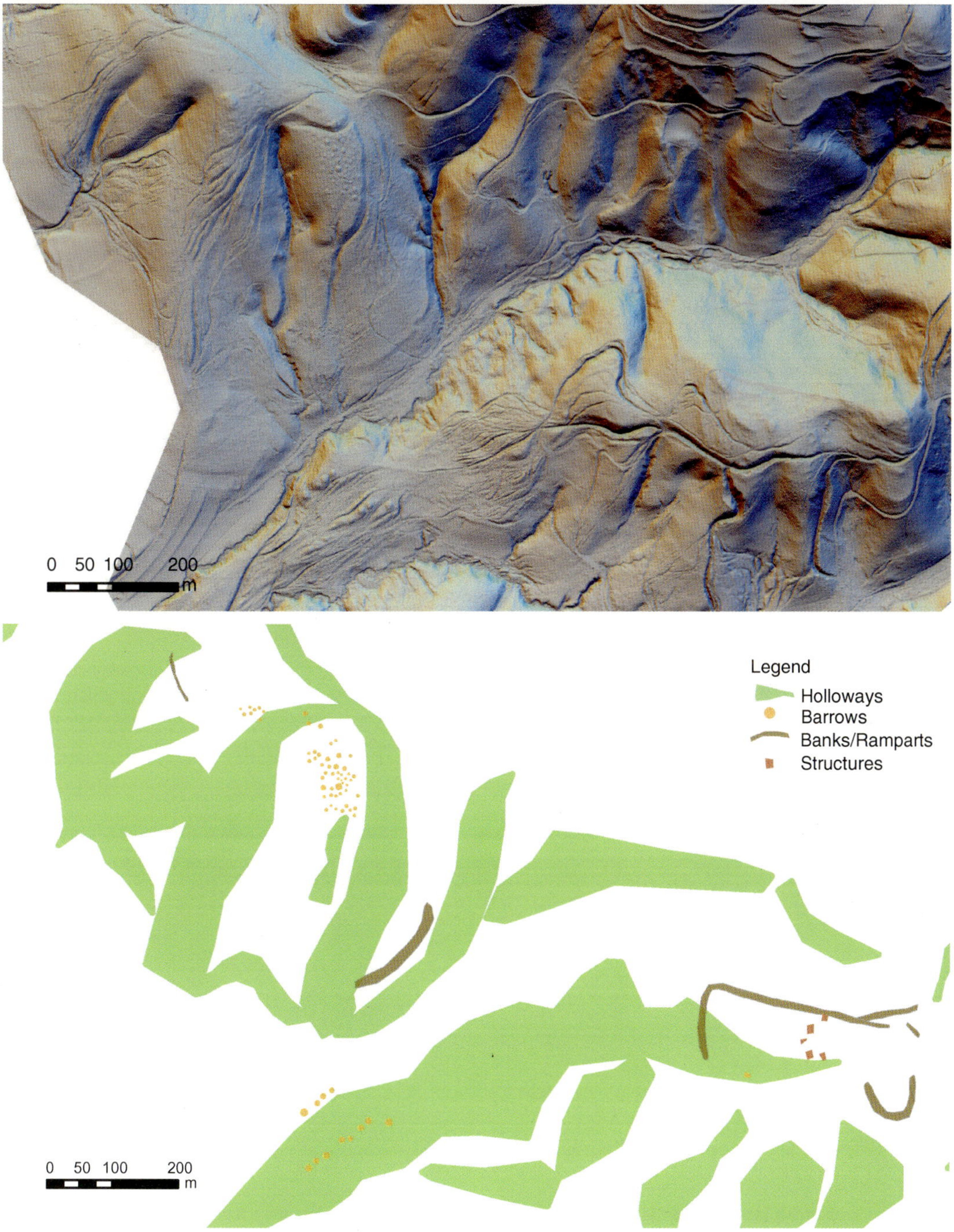

Fig. 6.4 True palimpsest. Holloways eroded by the flow of people, animals, and carts. Water erosion speeded the hollowing-out process and made some lanes muddy and impassable, leading to formation of river branching and converging pattern of paths. Holloways are associated with Iron Age barrows and hilltop settlement, but precise relation between them is impossible to establish (Tupaliče, Northern Slovenia)

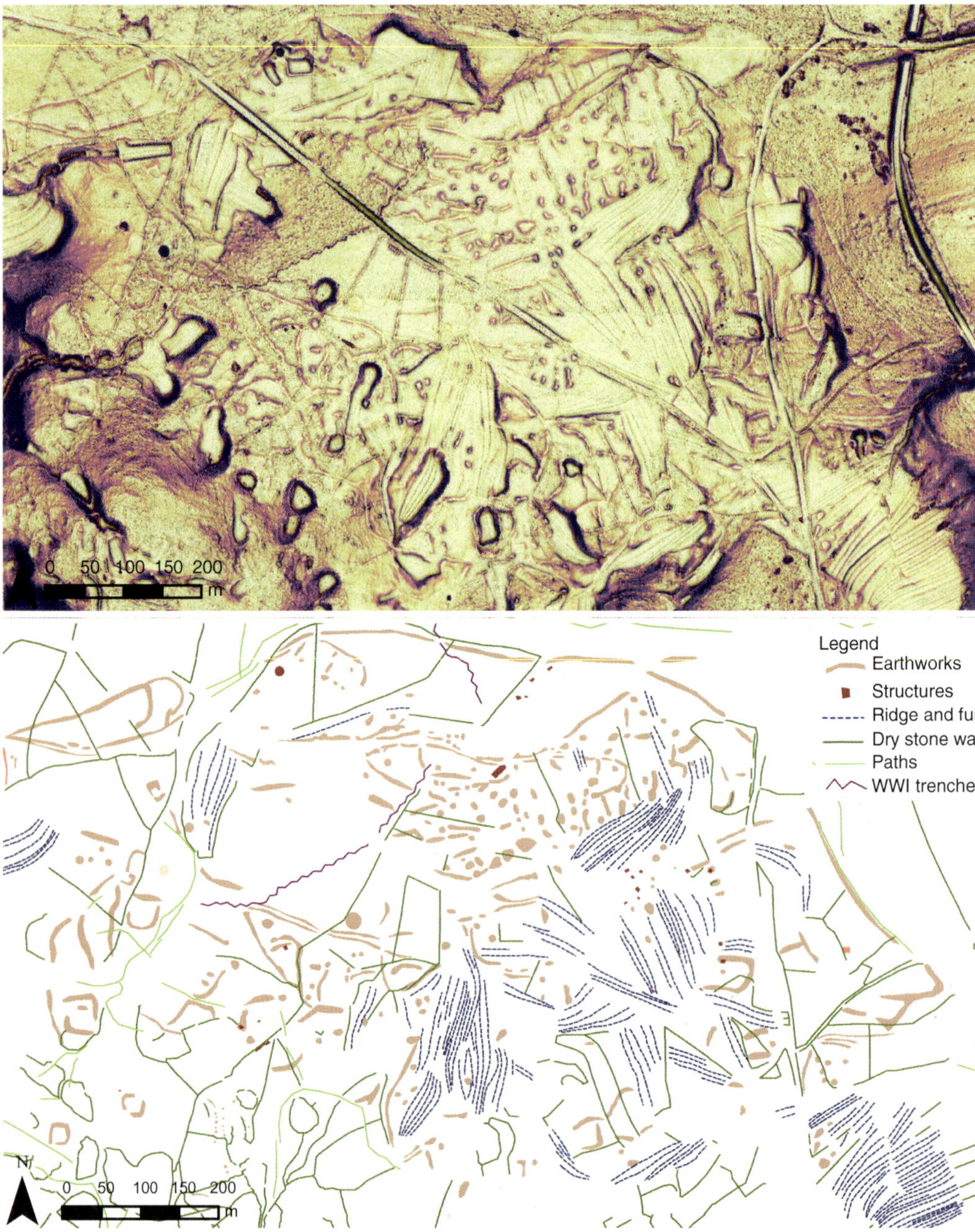

Fig. 6.5 Cumulative palimpsest. Prehistoric (Iron Age) landscape with irregular enclosed fields and settlement erased with medieval ridge and furrow fields in the areas with deeper soil, superimposed by medieval-modern dry stone walls marking outfield boundaries, cut across with WWI trenches protecting the road and railway (Goriče, Slovenian Karst). Slope map combined with the sky-view factor image

Conclusion

LiDAR is related to topographic survey and one of the oldest field techniques in the landscape archaeology's toolbox. Topographic survey means that location of surface traces, anomalies, and "humps and bumps" are recorded. But the sheer quantity of data that can be quickly and relatively cheaply collected with LiDAR has transformed it into new quality and new way of observing landscape. LiDAR does not separate between site and its environment or landscape, but treats them as the same.

LiDAR does not limit itself merely to "significant," isolated features of the landscape and does not separate them from the landscape as separate "sites." All locations are fully incorporated into the surrounding area; their form, dimension, context, and structure are the result of complex and lasting interactions with a changing landscape.

Thus LiDAR forces us to treat complex sites as integral part of landscape, osmotically connected to the wider landscape through practices of people who moved around.

It turns out that nowhere the landscape is empty; everywhere it is full of traces of practices and activities that have been materialized in the landscape. These scars and traces range from "ordinary" archaeological sites such as buildings, walls, roads, and burial mounds to traces of human activities such as sunken lanes, lynchets, clearance cairns, field boundaries, lime kilns, charcoal-burning platforms, quarries, ridge, and furrow, but also boar digs, animal trails, paleochannels, tree throws, landslides, etc.

LiDAR allows us to understand landscape, not as assemblage of sites, but as an assemblage of traces, produced by humans, animals, machines, and their various mixes and hybrids. We should be open to any entities that may participate in a given situation. We should be aware of imposing any arbitrary limits on the range of things, traces, and actors who participated in the landscape. Only when multiplicity of things is included, landscape becomes pluralistic, democratic, and relevant.

Complex sites are thus part of the lifeworld of people who inhabit it. People are not just situated on the sites; there is always a mutual relation where, on the one hand, the features constrain and enable social practices, and, on the other hand, the features are modified and rearranged by the inhabitants. The actions of people refer to these features, and the features structure the way people act. People inhabit the world shaped by their predecessors.

When put together, this way of thinking results in a messier world, and certainly a complex one, in continuous process of composition (Thrift 2003; Mlekuž 2011, 2012; Waterton 2012). As archaeologists, we need to focus on processes of inscription and erasure that continuously produce these messy palimpsests, rather than on the cumulative products of these events. We need to understand how the complexity of the material connections accumulated over time creates massive inertia in landscape and understands forces that reorganize it.

In this way we can treat complex sites as landscapes in themselves, created through numerous processes and practices that accumulated and inscribed new traces or erased old ones. These effects are important, for they determine where we see traces of past human practices and what they look like. Some traces may disappear as marks on the surface but can be retrieved by other ways of mapping, for example, by geophysics or artifact survey.

This complexity requires different sensibilities in interpreting remote-sensing data. The practice of landscape archaeology is thus essentially a "motley" of practices (Turnbull 2000, 39), instrumentations, theories, and people, more or less successfully brought together. Perhaps we should think more how we employ tools, knowledge, and skills when we map the landscape and produce interpretations of landscape.

Good practice is not about following fixed arbitrary recipes and rules. Instead, good practice in interpretation of LiDAR-derived high-resolution topography data constitutes fidelity to the richness and multiplicity of things of the past that left traces recorded by LiDAR.

References

Allen CD (2011) On actor-network theory and landscape. Area 43(3):274–280

Bailey G (2007) Time perspectives, palimpsests and the archaeology. J Anthropol Archaeol 26:198–223

Bender B (2002) Time and landscape. Curr Anthropol 43:103–112

Challis K, Howard AJ (2012) The role of lidar intensity data in interpreting environmental and cultural archaeological landscapes. In: Opitz RS, Cowley DC (eds) Interpreting archaeological topography: lasers, 3D data, observation, visualisation and applications. Oxbow, Oxford, pp 163–172

Challis K, Kincey M, Carey C, Howard AJ (2011) Airborne lidar intensity and geoarchaeological prospection in river valley floors. Archaeol Prospect 18:1–13

Cowley DC (2012) In with the new, out with the old? Auto-extraction for remote sensing archaeology. In: Bostater CR, Mertikas SP, Neyt X, Nichol C, Cowley D, Bruyant J-P (eds) Remote sensing of the ocean, sea ice, coastal waters, and large water regions, 2012. Proceedings of the SPIE, Volume 8532. SPIE, Edinburgh, article id. 853206

Crawford OGS (1953) Archaeology in the field. Phoenix House, London

Devereux BJ, Amable GS, Crow P (2008) Visualisation of LiDAR terrain models for archaeological feature detection. Antiquity 82(316):470–479

Doneus M, Briese C, Fera M, Janner M (2008) Archaeological prospection of forested areas using full-waveform airborne laser scanning. J Archaeol Sci 35:882–893

Halliday S (2012) I walked, I saw, I surveyed, but what did I see? ...and what did I survey? In: Opitz RS, Cowley DC (eds) Interpreting archaeological topography: lasers, 3D data, observation, visualisation and applications. Oxbow, Oxford, pp 67–78

Hauser K (2007) Shadow sites: photography, archaeology, and the British landscape 1927–1955. Oxford University Press, Oxford

Hesse R (2010) LiDAR-derived local relief models – a new tool for archaeological prospection. Archaeol Prospect 17(2):67–72

Hobbs KF (1999) An investigation of RGB multi-band shading for relief visualisation. Int J Appl Earth Observ Geoinf 1(3–4):181–186

Ingold T (2000) The perception of the environment: essays in livelihood, dwelling and skill. Routledge, London

Ingold T (2004) Culture on the ground: the world perceived through the feet. J Mater Cult 9(3): 315–340

Johnson M (2007) Ideas of landscape. Blackwell, Oxford

Knappett C (2005) Thinking through material culture: an interdisciplinary perspective. University of Pennsylvania Press, Philadelphia

Kokalj Z, Zakšek K, Oštir K (2012) Visualizations of lidar derived relief models. In: Opitz RS, Cowley DC (eds) Interpreting archaeological topography: lasers, 3D data, observation, visualisation and applications. Oxbow, Oxford, pp 102–116

Latour B (1986) Visualization and cognition: drawing things together. Knowl Soc Stud Sociol Cult Present 6:1–40

Latour B (1987) Science in action: how to follow scientists and engineers through society. Harvard University Press, Cambridge, MA

Latour B (1994) Pragmatogonies: a mythical account of how humans and nonhumans swap properties. Am Behav Sci 37(6):791–808

Latour B (1999) Pandora's hope: essays on the reality of science studies. Harvard University Press, Cambridge, MA

Latour B (2001) What is iconoclash ? Or is there a world beyond. In: Weibel P, Latour B (eds) Iconoclash, beyond the image-wars in science, religion and art. MIT Press, London, pp 14–37

Latour B (2004) How to talk about the body? The normative dimension of science studies. Body Soc 10(2–3): 205–229

Lucas G (2005) The archaeology of time. Routledge, London

Lucas G (2010) Triangulating absence: exploring the fault-lines between archaeology and anthropology. In: Garrow D, Yarrow T (eds) Archaeology and anthropology. Understanding similarity, exploring difference. Oxbow, Oxford, pp 28–39

Lucas G (2012) Understanding the archaeological record. Cambridge University Press, Cambridge

Meng X, Currit N, Zhao K (2010) Ground filtering algorithms for airborne LiDAR data: a review of critical issues. Remote Sens 2(3):833–860

Mlekuž D (2011) Messy landscapes manifesto. AARGnews 44:22–23

Mlekuž D (2012) Messy landscapes: lidar and practices of landscaping. In: Opitz RS, Cowley DC (eds) Interpreting archaeological topography: lasers, 3D data, observation, visualisation and applications. Oxbow, Oxford, pp 102–116

Olivier L (2001) Duration, memory and the nature of the archaeological record. In: Karlsson H (ed) It's about time. The concept of time in archaeology. Bricoleur Press, Göteborg, pp 61–70

Olsen B (2010) In defense of things: archaeology and the ontology of objects. AltaMira Press, Plymouth

Opitz R (2012) An overview of airborne and terrestrial laser scanning in archaeology. In: Opitz RS, Cowley DC (eds) Interpreting archaeological topography: lasers, 3D data, observation, visualisation and applications. Oxbow, Oxford, pp 13–31

Opitz R, Nuninger L (2010) Thinking through topography. Presentation at the Theoretical Roman archaeology conference TRAC, Oxford, 25–28 Mar 2010

Palmer R (2012) Reading aerial images. In: Opitz RS, Cowley DC (eds) Interpreting archaeological topography: lasers, 3D data, observation, visualisation and applications. Oxbow, Oxford, pp 79–89

Samberg A (2007) An implementation of the ASPRS LAS Standard. The Analyst XXXVI:363–372. http://www.isprs.org/proceedings/XXXVI/3-W52/final_papers/Samberg_2007.pdf

Thrift N (2003) Performance and.... Environ Plan A 35: 2019–2024

Turnbull D (2000) Mason, tricksters and cartographers: comparative studies in the sociology of scientific and indigenous knowledge. Harwood, Amsterdam

Waterton E (2012) Landscape and non-representational theories. In: Howard P, Thompson I, Waterton E (eds)

The Routledge companion to landscape studies. Routledge, London, pp 66–75
Wylie A (2002) Thinking from things: essays in the philosophy of archaeology. University of California Press, Berkeley
Wylie JW (2007) Landscape. Routledge, London/New York
Zakšek K, Oštir K, Kokalj Ž (2011) Sky-view factor as a relief visualization technique. Remote Sens 3: 398–415

Part II

Geophysics

Magnetic Exploration of Archaeological Sites

7

Bruce W. Bevan and Tatiana N. Smekalova

Contents

B.W. Bevan (✉)
Geosight, Weems, VA, USA
e-mail: geosight2@gmail.com

T.N. Smekalova
Århus University, Århus, Denmark
e-mail: klats@hum.au.dk

7.1 A Basic Overview of a Magnetic Survey

A magnetic survey is done by carrying a small instrument around an archaeological site and recording changes in the Earth's magnetic field at thousands of points; see Fig. 7.1. Buried iron-containing objects distort or warp the simple pattern of the Earth's field into small complex patterns called anomalies. These anomalies will be apparent when one creates a map of the measurements, after the field work of the survey is finished. Note two things: You must walk to every point where a measurement is needed; you cannot stand back and take a magnetic photograph of a site. Second, you cannot see the two-dimensional pattern of your measurements as you make them with any available magnetometer; you must wait until later to see the patterns that you have measured.

All magnetometers measure the strength or amplitude of the Earth's magnetic field, but not usually its direction, which is usually determined with a magnetic compass. Strength, rather than direction, is measured because this is more precise.

After the field work of a magnetic survey is done, one moves inside to study the patterns that have been recorded. The data are first displayed in a map; Fig. 7.2 shows an example. General practices for creating topographic maps are applied; high measurements are just considered to be high elevations. The next step is the most interesting one: Looking for patterns in the

C. Corsi et al. (eds.), *Good Practice in Archaeological Diagnostics*, Natural Science in Archaeology,
DOI 10.1007/978-3-319-01784-6_7,

Fig. 7.1 Doing a magnetic survey in the Fayoum Oasis with a single magnetic sensor (the *white cylinder*) that is held at a height of about 25 cm

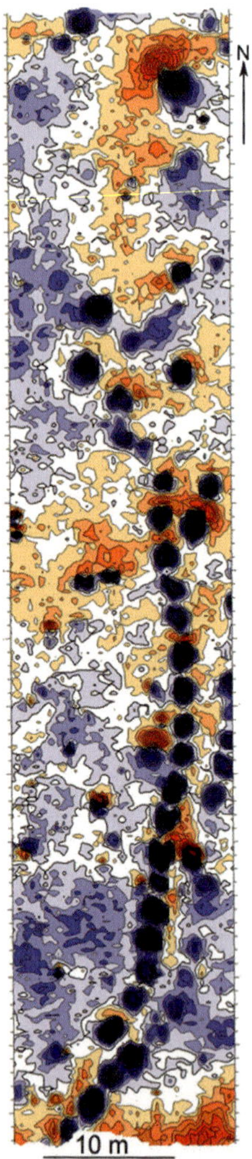

Fig. 7.2 Bronze-Age cooking pits in Denmark form lines of circular magnetic anomalies that have amplitudes of over 20 nT

magnetic map. Lines, particularly where they are perpendicular, and circles are excellent indicators of buried archaeological structures. However, many important features will be revealed by rather blobby patterns, without any clear shape. Concentrations of these patterns, and even individual anomalies, can provide guidance to the archaeological excavator and allow careful and economical excavations.

7.2 Is the Archaeological Site Suitable for a Magnetic Survey?

Most are, but some are more difficult than others. One must distinguish between the ease of doing a survey and the success of that survey. Many magnetic surveys in central Europe are done in farmed fields after the crop has been harvested; these are usually easy to do, although plowing may have removed much of the features that are sought, making the surveys less successful than wished. In the Mediterranean region, important areas for survey may frequently be in small fields, orchards, or on idle land; these sites have an intermediate difficulty. Some surveys are done in cities, or close to buildings; while these may be easy to do, they may be the least successful because of interference from metal and electric wires.

The two main difficulties with magnetic surveys are noise and access. If a site is dense with bushes, particularly thorny bushes, access to the area is a problem; forests are also difficult, and slow a survey, but do not stop it. However, because of the brush or woods, the preservation of archaeological features in these difficult areas may be better, and these sites may yield very valuable findings.

Noise can prevent a successful magnetic survey, and there are two types of noise: Spatial and temporal. Spatial noise is caused by magnetic stones or iron trash in the soil; so many unwanted anomalies can be detected that the desired patterns are hidden. Temporal noise is caused by passing cars or interference from electric wires; electrified railroads are a particular problem for magnetic surveys.

Many of these sources of trouble can be identified by viewing the site in Google Earth and by checking local maps of geology and surface sediments. Preliminary measurements on the site allow a better evaluation, but that adds to the cost and time of a survey.

It is generally more difficult to estimate whether the archaeological features of interest at the site will cause strong and distinctive anomalies. Experience is the best guide for this question. Earthen features will generally be invisible if they are deeper than their diameter; fired features or iron objects can be detected at a greater depth. An ample review of case-studies is in Aspinall et al. 2008.

7.3 Examples of Magnetic Surveys

A few illustrations will show some types of features that are excellent objects for magnetic surveys. These surveys have been done with an Overhauser magnetometer that measured the total flux density of the Earth's magnetic field; the height of the magnetic sensor was usually about 0.25 m, and the spacing between measurements and traverses was typically 0.5 m. For further examples of magnetic maps, see the book edited by Piro and Campana (2009).

In Denmark, long lines of cooking pits are readily traced with a magnetometer; see the examples in Figs. 7.2 and 7.3. These pits date to the Late Bronze Age (about 1000 BC); they are detectable because the fired stones within the pits create distinctive anomalies, often with a magnetic low north of the strong magnetic high. The intense anomalies at the upper edge of the magnetic map in Fig. 7.3 are caused by a steel pipe buried at the edge of the field; the broad anomaly certainly hides some pits even though the trench for the pipe has not disturbed them.

Igneous stones, even if they have not been refired by human activity, are usually very magnetic. Structural stones buried inside shallow earthen mounds in Denmark can be located with a magnetometer. The magnetic map of Fig. 7.4 locates a ring of stones around the perimeter of a mound, along with two lines of stones that lead into the central tomb. While one of these stone-lined entrances was known from an early excavation, the second entrance was first detected by this magnetic survey.

Stones that are not magnetic can also be detected with a magnetic survey, as long as the surrounding soil is rather magnetic. Foundations that were constructed of these nonmagnetic stones cause distinctive magnetic lows, as shown in Fig. 7.5. In the Mediterranean region, bedrock is commonly limestone, and this is essentially nonmagnetic. However, the soil in these areas is usually quite magnetic. Figure 7.6 illustrates the distinctive rectangular pattern of a limestone foundation.

Another example of the magnetic contrast between limestone and soil is shown in Fig. 7.7. The earthen barrow there is about 10 m tall and 60 m in diameter. The central tomb within this mound is constructed of limestone, as is a circular ring of stone buried below the perimeter of the mound. That ring creates a distinct magnetic low. However, the stone-lined tomb within this mound had collapsed and filled with soil, causing a magnetic high at that location. Without that collapse, a tomb constructed of limestone can be detected as a magnetic low.

Additional examples of the detection of buried nonmagnetic stones are shown in Fig. 7.8. Light tones in this magnetic map also show magnetic lows: the stone foundations of Iron Age buildings at B and the walls of a double-pen enclosure dating to the Bronze Age at E. However, this magnetic map also reveals soil contrasts. Farmed fields are apparent as pairs of bands at F. There is a magnetic low in the middle of each double band, and there are magnetic highs at the outer sides. This appears to be caused by asymmetrical plowing that has shifted soil from the middle of the field toward its edges. Finally, this very rich magnetic map also reveals a cluster of storage pits at P that date to the fourth century BC. This archaeological site is in western Crimea; while the field patterns can be visible in images from Google Earth, the other structures cannot be seen at the surface.

Kilns and furnaces are usually very magnetic and easy to detect. Thousands of iron-smelting furnaces (dating from the second to the sixth centuries AD) have been located in Denmark with

Fig. 7.3 This larger group of cooking pits was also found in Denmark

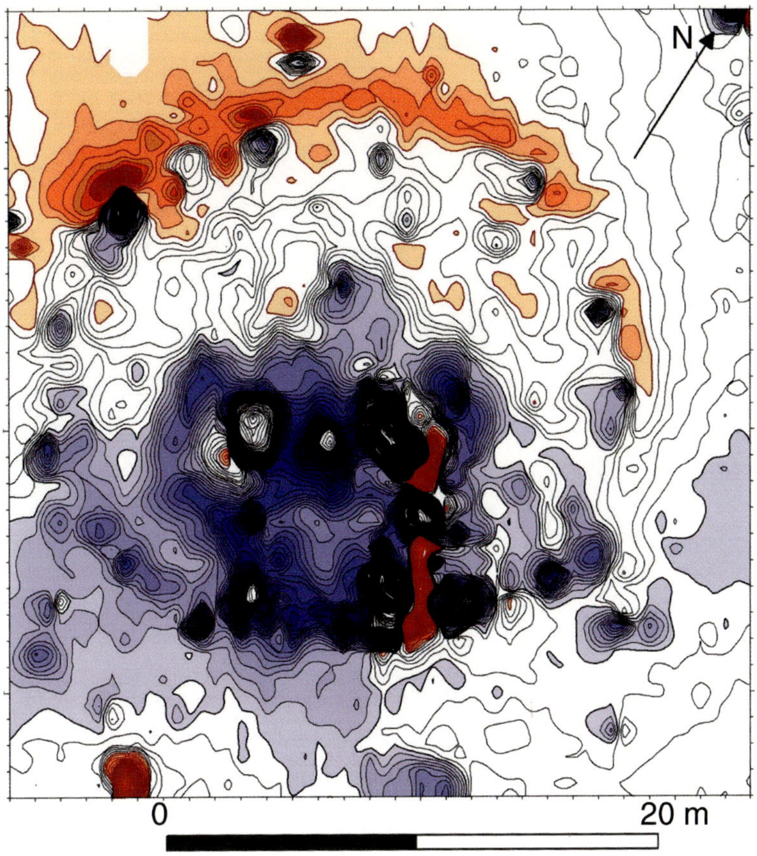

Fig. 7.4 Granite boulders within a megalithic grave in Denmark create magnetic anomalies of 50–200 nT

magnetic surveys. A block of iron-containing slag that typically weighs 200 kg is found at a depth of about 0.4 m below each furnace; the clay chimney that originally stood above each slag block was destroyed long ago. Figure 7.9 shows a line of slag blocks that were readily detected; the upper panel lists the amplitudes of the anomalies. The same map is drawn with contour lines at intervals of 10 nT in the middle panel. The lower panel locates where slag blocks were found in a later excavation; it appears that one or two slag blocks that caused weak anomalies were not identified by this survey.

Kilns for firing either ceramic or bricks can be large and are easy to find with a magnetic survey. Figure 7.10 shows both a magnetic map and also a drawing of the kiln that was excavated after that magnetic survey. A strong magnetic high was located over the kiln, while a magnetic low was found to the north. Magnetic objects create both highs and lows; however, the lows are usually weaker than the highs and these lows are not always visible in a magnetic map.

Metallic iron is readily detected with a magnetometer; much of the time this iron is modern trash and unwanted. However, important iron artifacts can also be found, and Fig. 7.11 shows an example. The sword that was excavated (on the left) was somewhat less than 1 m long.

7.4 Selecting a Magnetometer

Not only are there many manufacturers of magnetometers, there are also many different types of magnetometers. Almost all of these can be useful

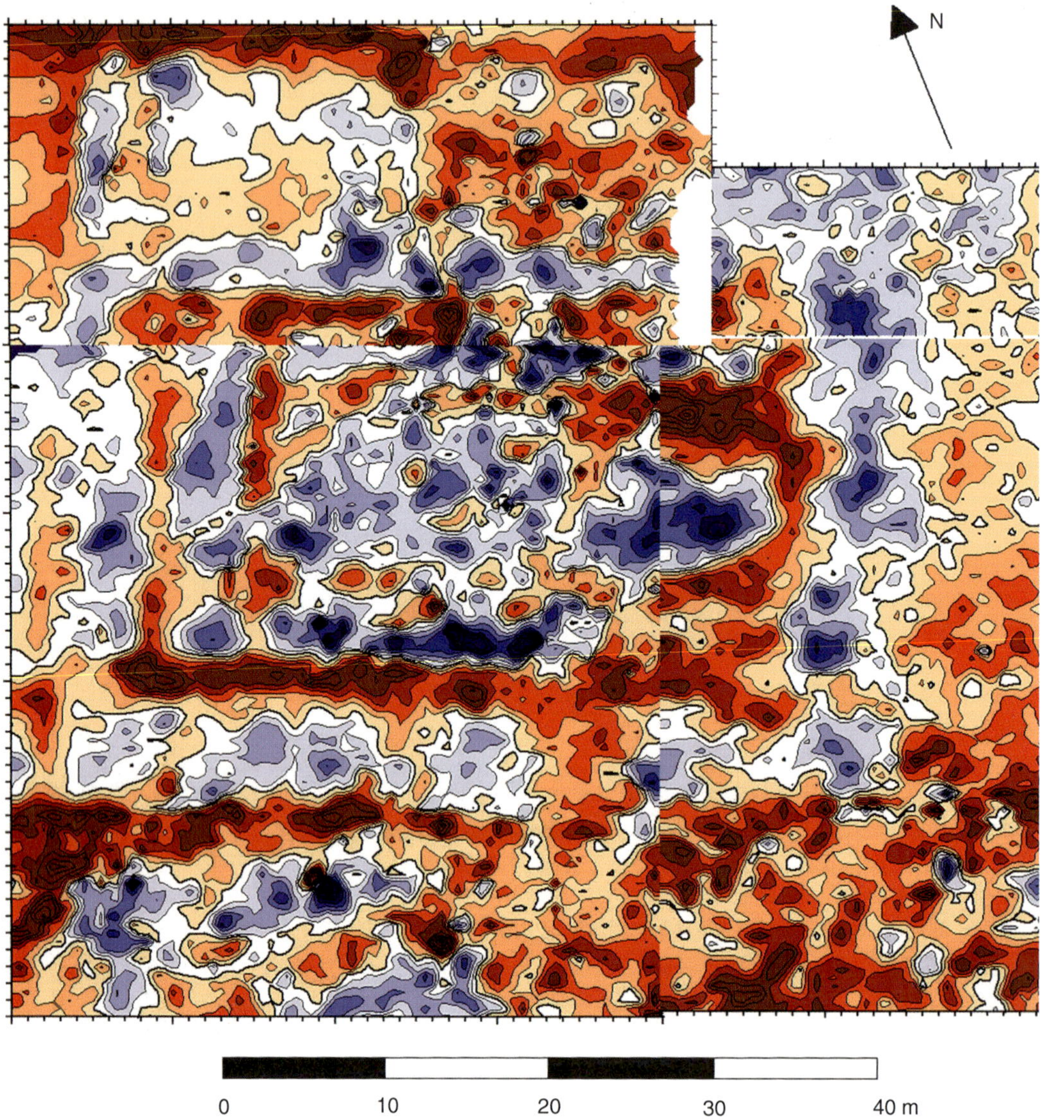

Fig. 7.5 Magnetic lows (about 30 nT) trace the rectangular foundation of a basilica in Antioch

for archaeological exploration; the most common types are fluxgate and perhaps cesium and Overhauser magnetometers. While these different magnetometers operate with very different physical principles, those principles will not be described here. However, the special characteristics of some types will be noted:

High speed and a large number of measurements are needed: Probably any instrument except for a proton magnetometer will be good.

Simplicity is most important: A proton magnetometer is needed (it may have only two buttons or controls), although fluxgate magnetometers are second best.

The cost of the equipment must be the lowest possible: This instrument will be a proton magnetometer, possibly bought used.

Extremely magnetic features must be mapped: Fluxgate or cesium magnetometers will be best, for these have a good tolerance of high magnetic gradients.

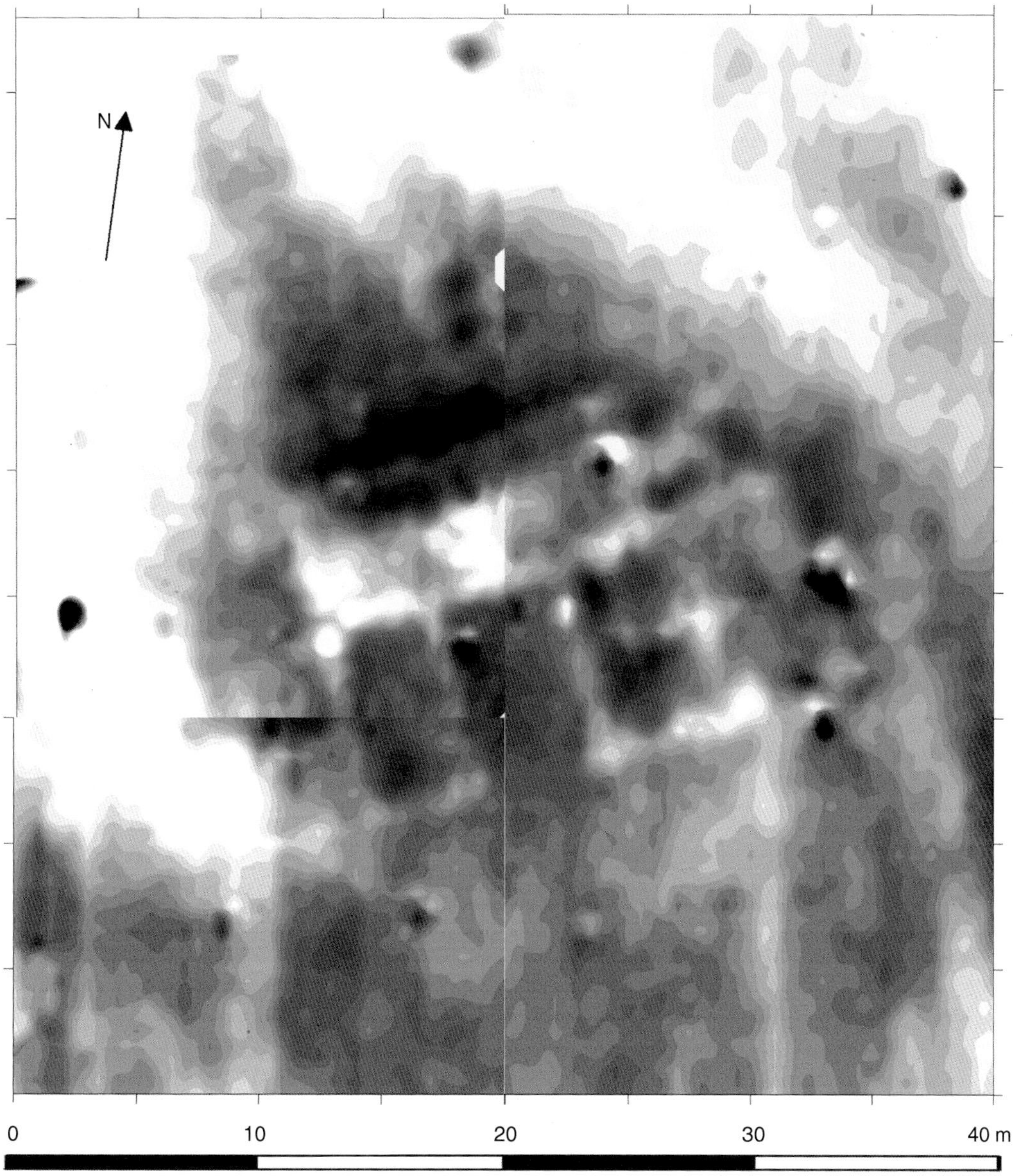

Fig. 7.6 The limestone walls of a Roman villa in Macedonia are detected as magnetic lows with an amplitude of 5–10 nT

The highest precision in the readings is needed: Overhauser or cesium magnetometers may allow faint anomalies to be mapped.

In addition to these fundamentally different types of magnetometers, there are different modes for using these instruments. The primary difference is between a total-field magnetometer (with one sensor) and a magnetic gradiometer (which has two sensors that are typically along a vertical line and spaced by 0.5–1 m). A gradiometer allows an easy and accurate correction for natural and daily changes in the Earth's magnetic field. However, a gradiometer is heavier than a magnetometer with a single sensor, and it will be more difficult to operate in brush. A gradiometer allows a greater spatial resolution than a total-field magnetometer

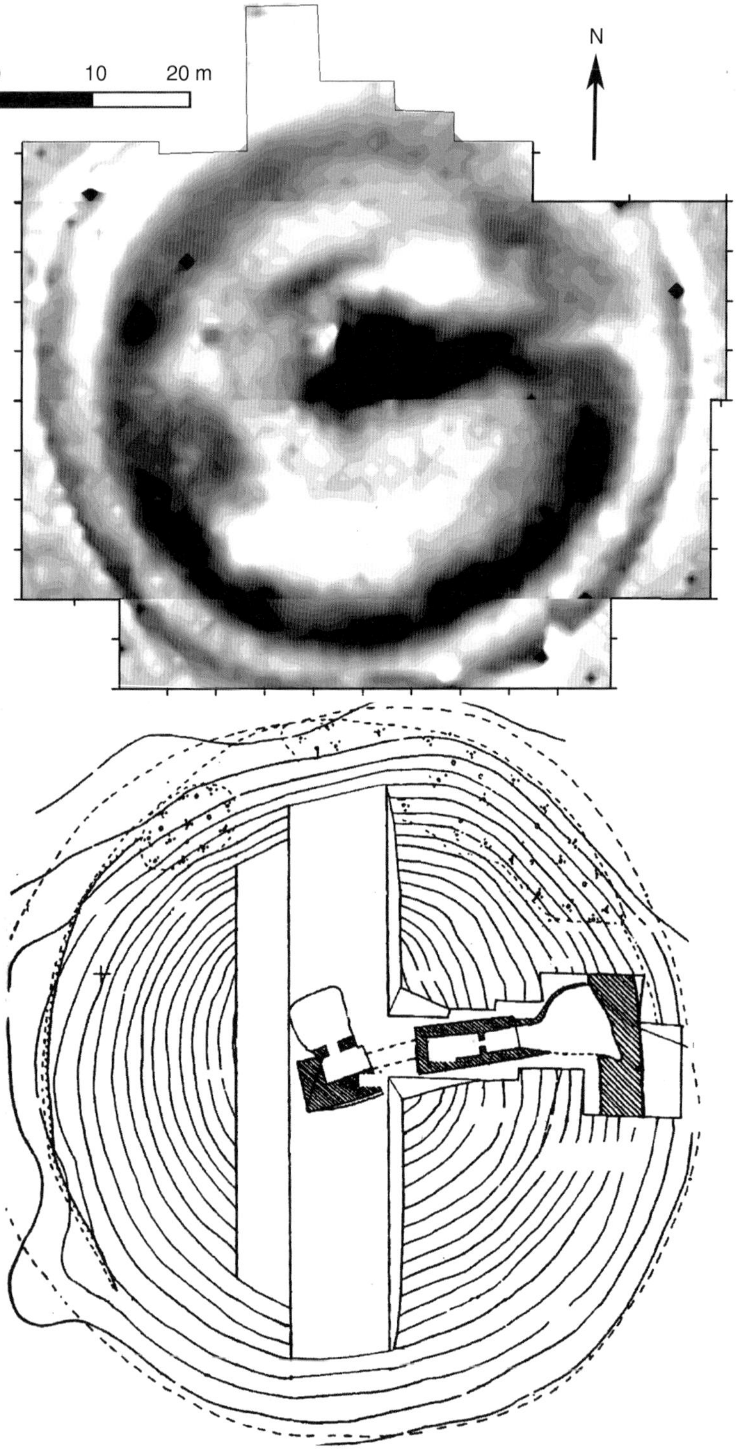

Fig. 7.7 Light tones indicate low readings in the magnetic map (*above*) of a Scythian barrow in Crimea that was later excavated (*below*)

(at the same sensor height); however, the gradiometer must then make its measurements at a closer interval to take advantage of that higher resolution. A gradiometer accentuates shallow features and this is often good, except that trash is shallow also. If it is important to detect features that are deeper than 2 m, a gradiometer should not be used. Fluxgate magnetometers are almost always

Fig. 7.8 The magnetic map in the center of this Google Earth image reveals ancient buildings (B), a double enclosure (E), farm fields (F), and a cluster of filled pits (P)

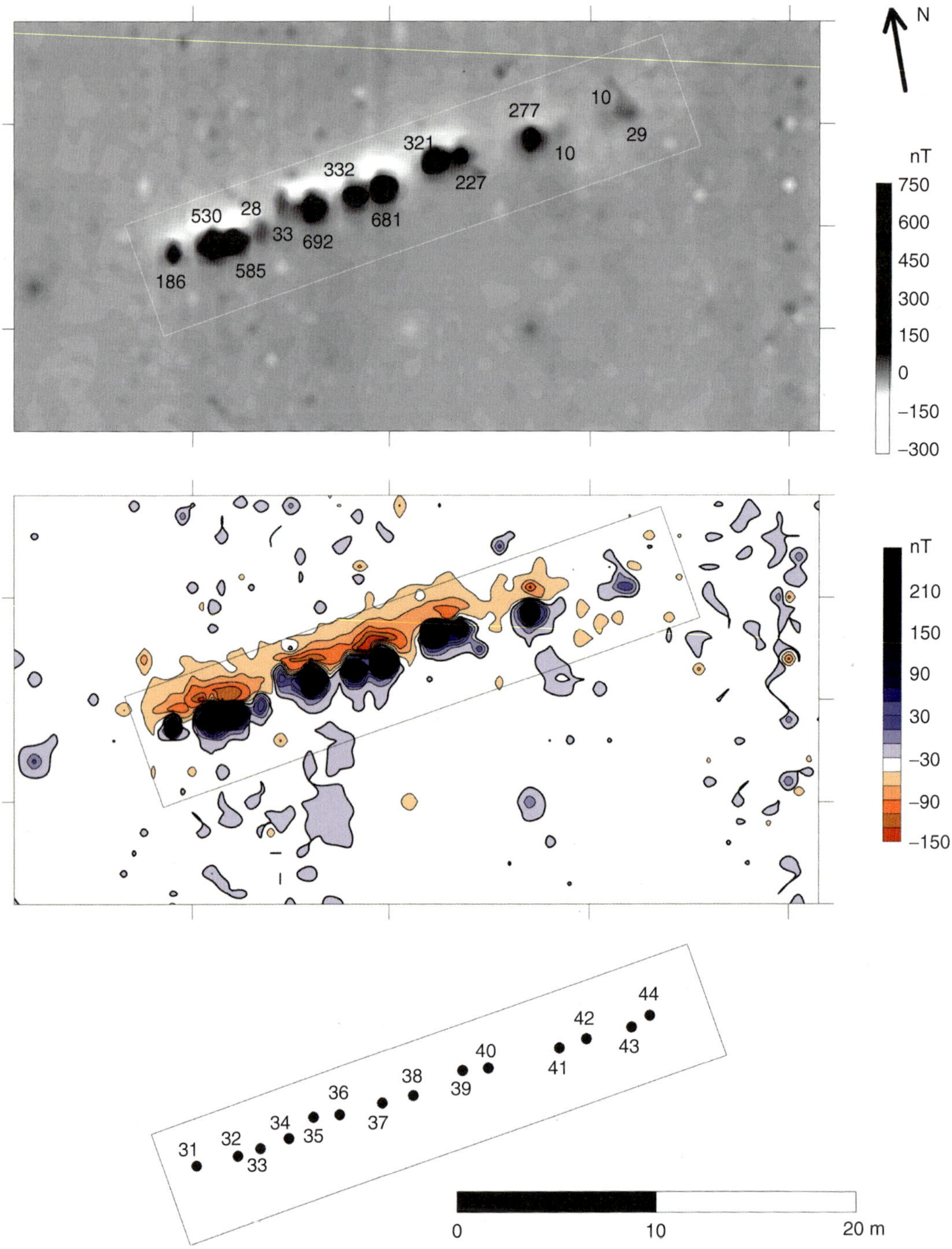

Fig. 7.9 Blocks of slag below former iron furnaces in Denmark create a line of magnetic anomalies with highs that are usually greater than 200 nT

gradiometers, while other types of magnetometers can be operated as either gradiometers or total-field magnetometers. Total-field magnetometers usually require a second magnetometer to monitor natural changes in the Earth's magnetic field, so they are not less expensive than a gradiometer.

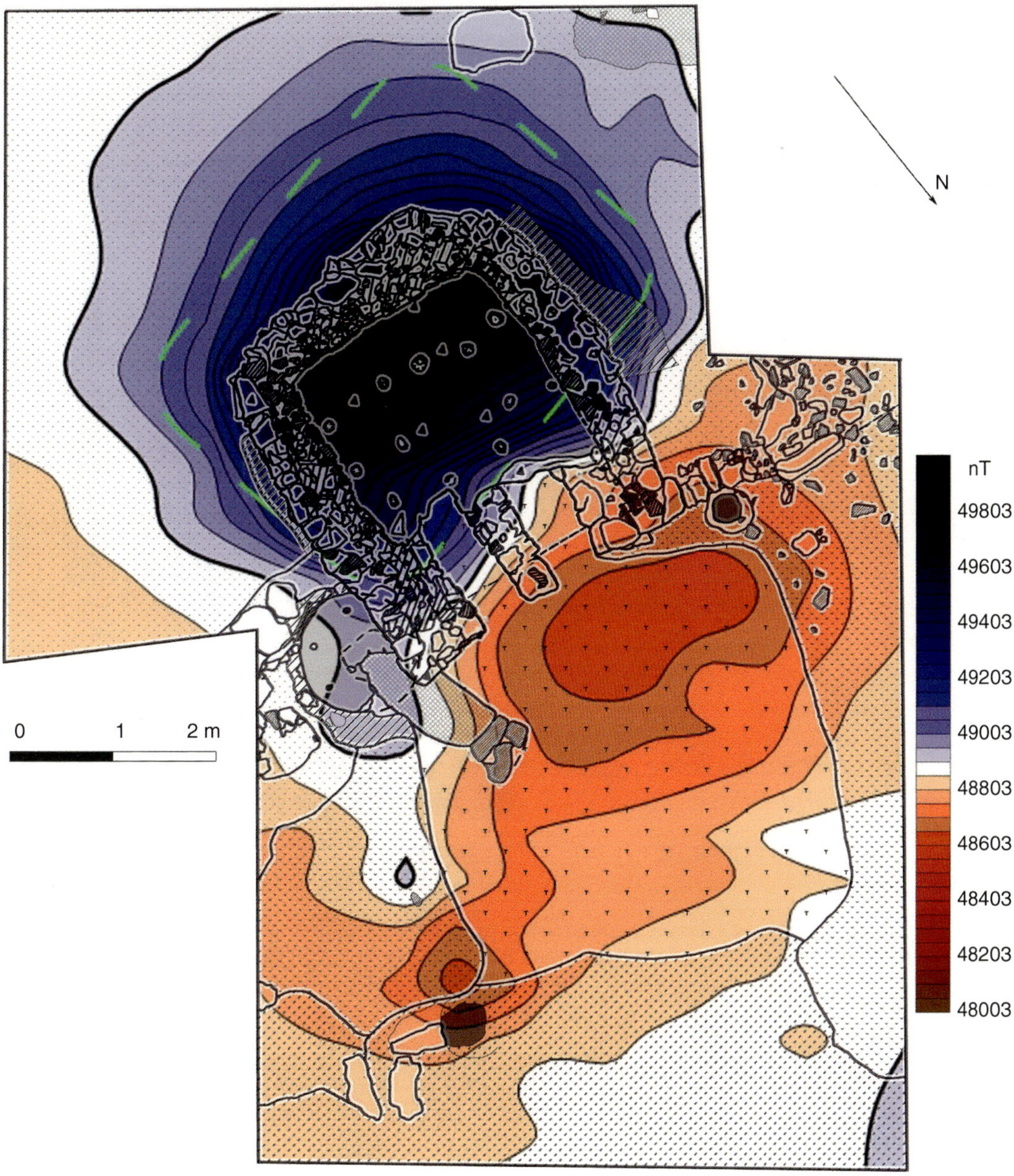

Fig. 7.10 A pottery kiln from the eighth to tenth century AD in Crimea was detected with an anomaly amplitude of more than 800 nT

Magnetometers also differ in the component of the Earth's magnetic field that is measured. Total-field magnetometers measure the magnitude of the magnetic field; the unit of measurement is the nanotesla, abbreviated nT. Fluxgate magnetometers usually measure the vertical component of the magnetic field; the unit is the same, and the patterns of the anomalies are almost the same as those with a total-field magnetometer (if the large vertical component is vectorially added to the smaller horizontal component, the result is the total field).

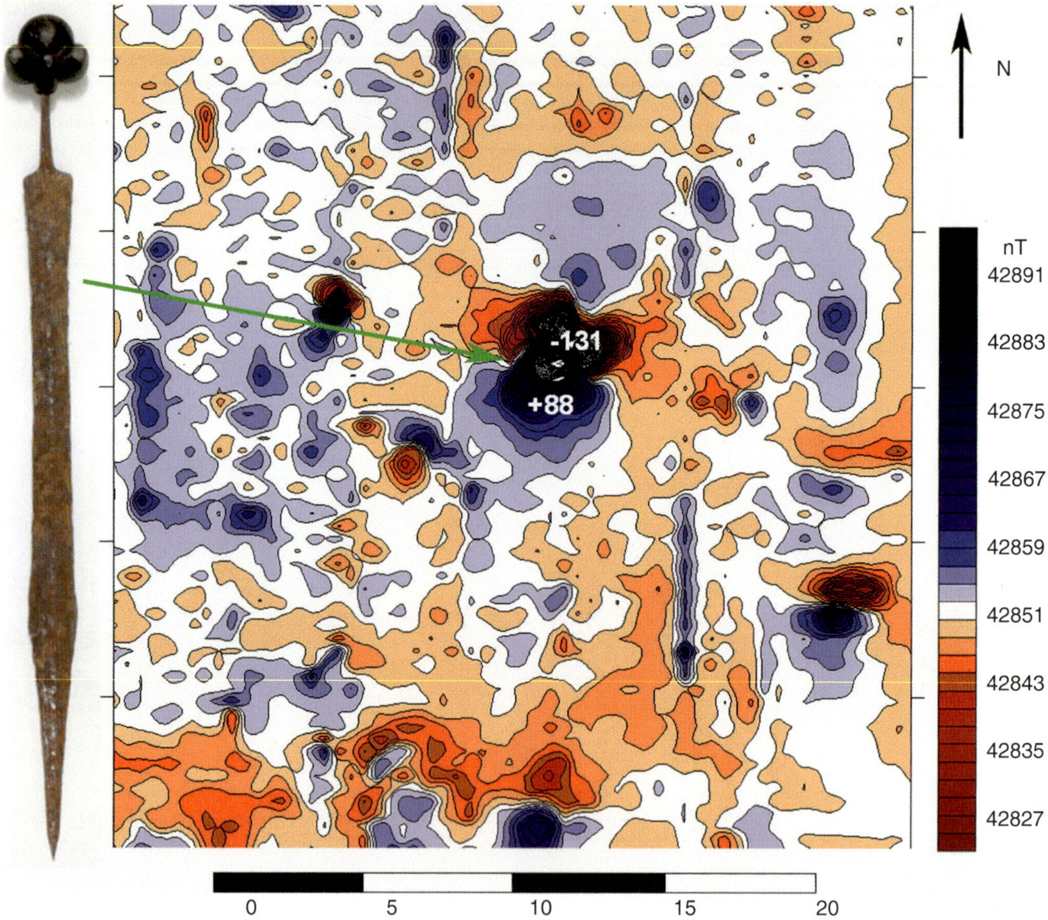

Fig. 7.11 In the Fayoum Oasis, a dipolar magnetic anomaly with an amplitude of about 100 nT located an iron sword (shown on the *left*) that dates to about the second century AD

7.5 Field Procedures

Before starting to measure a magnetic map, it can be helpful to do a quick reconnaissance of the area to be explored. This will reveal the range of the anomalies to be measured, and it may even show that all of the anomalies are found in a small area. Most magnetometers will allow readings to be noted on their digital or graphical displays while wandering in the area of interest. In rare cases, this reconnaissance survey may also achieve the goal of an exploration: Perhaps a few isolated and distinctive features are sought.

Select a procedure for locating the points of measurement. A basic decision to be made is whether instantaneous locations should be determined with a GPS receiver. This would have to be a precise receiver, allowing points to be measured to an accuracy of probably better than 0.5 m and preferably better than 0.1 m. These good receivers are costly, and they add to the complexity of a survey, but they also speed a survey. Do not use a GPS receiver for your first surveys; instead, plan to set up somewhat rectangular areas for your exploration. Normal procedures for spatial mapping can be applied, and the maps can be referenced to fixed points (such as buildings or roads) or wooden stakes for later relocation of the area.

Your area of survey may be filled with a set of rectangles or squares, and each one can be called a grid (or perhaps the entire area is called a grid).

Stretch a nonmetallic measuring tape along two opposite sides of a rectangle; then stretch a guide rope or cord between those two tapes. The guide rope may have colored bands spray painted at 1 m intervals. The length of the guide rope and rectangle can vary between about 10 and 100 m; longer spans are more time efficient, although smaller rectangles allow the area of survey to be modified more easily. You can probably estimate distances from the guide rope well enough that you do not need to place a rope on each line of measurement that you make.

The operator of the magnetometer must carefully check for iron objects that could be carried in clothing. Remove everything from your pockets; paper money (it is the ink) may be more magnetic than coins. Be particularly watchful for steel bars in the bottom of shoes and also zippers. Some magnetic objects can be detected by touching them with a small permanent magnet; invisible iron objects can be revealed by moving them toward and away from the sensor of your magnetometer as you watch the readings.

It is best that the magnetic sensor be as close to the soil as possible; this will provide the strongest anomalies and the greatest spatial resolution. If the shallow soil is cluttered with magnetic objects, and the features of interest are large, it may be better to raise the magnetic sensor. Vegetation will usually limit the lower elevation of the magnetic sensor. Also, some magnetic sensors allow only a single height, at least with comfortable operation. Most magnetic surveys are done with the magnetic sensor within a height range of 0.2–0.4 m.

The spacing between readings along lines of measurement should usually be less than the "depth" to the objects of interest; depth is the sum of the sensor height and the actual depth underground. If features are expected at a depth of 0.5 m, and the sensor height is 0.2 m, a measurement spacing of 0.5 m will probably be adequate; if time allows, this spacing can be halved. The spacing is somewhat dependent on noise; if there is spatial or electrical noise, a closer spacing may help. Some fluxgate magnetometers can be unavoidably noisier in their readings than Overhauser or cesium magnetometers, and a closer spacing between readings may then be necessary.

Fig. 7.12 Four magnetic sensors (the *white cylinders*) on a cart allow an exploration of a band that is 2 m wide

A reduction in the spacing between readings along lines of traverse may increase the time that it takes to measure each line only a little. The spacing between lines of measurement has a major effect on the time required to survey an area. It is seldom practical to measure with a line spacing of less than 0.5 m.

With most types of magnetometers, one can orient the lines of measurement or traverse in any direction; they do not need to be north–south. Beginners should probably try to make traverses that do not follow architectural directions; faults in the survey may appear to be buried walls. Measurement traverses may be unidirectional (just to the north, for example) or they may be bidirectional (perhaps alternately to the north and south). Beginners might start with unidirectional traverses; for bidirectional surveys add the complication of amplitude and locational shifts, which create striations in magnetic maps. These errors are caused by iron moving with the magnetic sensor (heading error), visual parallax, and recording delays.

A magnetic survey is speeded when several magnetic sensors are operated in parallel; two, four, eight, or more sensors can be moved together like a push broom to explore an area 1–4 m wide in a single traverse. These sensors are often mounted on a wheeled cart, for they can be too heavy or awkward to carry; see Fig. 7.12. Multisensor magnetometers are ideal for large farmed fields; they are less usable in the Mediterranean region because fields are often small. Since some of those fields may be ancient, there may be few archaeological features in the fields. A large proportion of archaeological sites in the Mediterranean can be on idle land or in orchards, where brush and trees would prevent the operation of a magnetic cart.

7.6 Data Processing and Display

Those errors that were mentioned above may partially be corrected later, by data processing. However, it is always best to process (modify) one's measurements the least amount that is possible. Still, data processing can salvage some information from poor data, and this is not particularly dishonorable. While it is difficult to write software that does a good job of data processing, several computer programs are available to aid this step, although there is a charge for most of the ones that work easily.

Magnetic maps are most commonly displayed as gray-scale images; dark gray and black will probably mark areas with high magnetic field, while light gray or white will show where the field is low. Colored maps will allow a greater dynamic range of distinguishable readings; in one convention, cool colors (blue and green) can mark magnetic highs, while warm colors (red and orange) reveal magnetic lows. Shaded relief maps that are colored can look spectacular, but may show little additional information. For a technical analysis of magnetic maps, they should be drawn with contour lines, for this allows one to see important lateral gradients. Unless a map shows simple patterns, it will probably not be improved with an oblique perspective, for this view can hide important anomalies; it will also make it difficult to locate anomalies.

7.7 Interpretation of Magnetic Maps

This is the step where one tries to understand the patterns in the magnetic map; it can result in a simplified map that summarizes the findings of the survey. There are several different types of interpretation that may be done: Pattern interpretation, technical interpretation, and archaeological interpretation. The final one is the best.

With pattern interpretation, lines are drawn to mark almost-linear or curvilinear anomalies. Other distinctive areas, such as dense concentrations of anomalies, are outlined. It is best to distinguish anomalies that are magnetic highs from magnetic lows. As an improvement, it would be valuable to note the amplitudes of the different anomalies that are summarized; this, also, will help to distinguish different features. While a magnetic high and a low are often associated with a single feature (these are called bipolar, or sometimes dipolar, anomalies) some anomalies may be seen as only one or the other polarity. Wall foundations that are constructed of igneous stone can be revealed as randomly high and low readings that may be difficult to see as lines.

Technical or parametric interpretation provides estimates of the quantity of magnetic material that is underground and also its depth, and perhaps even a better indication of the shape of buried features. While this is valuable information, a technical knowledge of magnetics is required. Since this type of interpretation is seldom done, further details are left to a later section of this writing.

An archaeological interpretation of a magnetic map provides reasonable archaeological descriptions of what has been discovered; it is much better to be able to say "a Bronze-Age fortress" than to say "several straight lines." This type of interpretation may follow one of the interpretations above, or it may be the only interpretation that is done.

7.8 The Geophysical Report

If the geophysical surveyor is not the archaeological excavator, the report on the magnetic survey must provide as much guidance to the archaeologist as possible. It is most important to include an interpretation map that classifies the findings of the survey, revealing where things are similar and where they are different. This is because the excavation will probably sample a variety of the different findings of the survey; the excavator may not wish to dig four walls and no kilns. If it is possible to estimate the reliability of the different findings, that might also aid the excavator.

In addition to a detailed interpretation map, the report should start with a good summary; a busy archaeologist may not have time for a thorough reading of much more. Even if they are not read immediately, many important points must be included in the report; these might aid a future evaluation of the survey. The area of survey must be described: Vegetation, soil, geology, topography, trash, and proximity to sources of noise.

The parameters of the magnetic survey should be listed. It is particularly important to say what actually was measured. If a gradiometer was used, are the units in nanotesla (as a difference between two sensors) or in nT/m (which includes the sensor spacing)? With a gradiometer, it is conventional to subtract the reading of the upper sensor from the lower sensor, but this should be specified. List the spacing between the sensors, and the height of the lower sensor (indicate the height of the active center of the sensor if an analysis of depth is done). The traversing parameters, such as the spacing between lines and the interval between readings along lines, should be stated.

The magnetic maps with the report should include an arrow for magnetic north and a horizontal scale. Amplitude distinctions must also be made, so that the viewer can determine both the polarity and the approximate magnitude of anomalies. It is probably not necessary to state what software program generated the maps in the report.

It is necessary to mention what data processing and filtering has been applied to the data. Unless it is an unimportant report, the processing must be described in detail; it is not usually adequate to just state that you pushed the button that applied filter X in program Y. For example, if a smoothing filter (more accurately called a blurring filter) was applied to the data, one could state the width of the smoothing window and its other parameters.

7.9 Excavations

Since you, the archaeologist, have done this magnetic survey, perhaps you will get to do some excavation tests; these will be very educational, and they may be necessary.

In some cases, the magnetic map may show the important features of the site so well (perhaps locating buildings and streets) that no excavation is needed. While publications emphasize this type of finding, actual practice is less favorable. Your magnetic map will allow you to sample what may be a variety of archaeological features and leave similar features untouched. If you are wise, you will also excavate at many locations where the magnetic survey revealed nothing or only ambiguous findings; these excavations may teach you humility.

The source of your magnetic anomaly may be very clear in your excavation, but not always. You can make some additional magnetic tests to become more confident that you have uncovered the entire source of your magnetic anomaly. Magnetic measurements could be made in the bottom or on the sides of the excavation; gradiometers are poor for this because their two sensors may detect, and therefore confuse, two different features on the side of the excavation. However, you can bring samples of soil, stone, or ceramic close to the lower sensor of your stationary gradiometer, then rotate the sample and change its distance while noting differences in the magnetic readings.

If you believe the soil may be unusually magnetic, pour some dry soil across the face of a strong permanent magnet and see how much sticks to it. If you think a stone in your excavation might be quite magnetic, dangle a pendulum magnet next to it, and see if it sticks as you pull the magnet away. Make a pendulum magnet by tying a thin string to the middle of a bar magnet that is about 20 mm long and 2 mm in diameter; this magnet can also help you to find iron in your shoes and zippers.

A magnetic compass might also help you to find magnetic materials in your excavation. If it is possible, set the compass on the ground and move a stone or artifact by it, very close but not touching, and watch for the needle to deflect. You can also slide the compass straight and steady along a wooden board that passes next to your feature.

If you can afford a small and portable magnetic susceptibility meter, this will be better than any of the simplifications above. A fluxgate magnetometer with a single, small sensor is also excellent for identifying features on the faces of excavations or at a shallow depth behind them. With this instrument, one can make thousands of measurements in a square meter in order to reveal small features and thin strata that may be invisible to the eye of the excavator.

7.10 The Technical Side of Geophysics

Valuable information can be acquired from a magnetic survey without having to understand any details of the physical principles that are involved. However, a greater amount of information can be revealed if one has a technical knowledge of geophysics. Technical studies of magnetic maps that have been measured at archaeological sites are seldom done, but this summary will outline what is possible.

The depths of features may be estimated from a magnetic map. These estimates generally determine the maximum depth to the middle of features; it is more difficult or impossible to determine the minimum depth. A simple approximation is called the half-width rule (Breiner 1999, p. 31): The diameter of a magnetic anomaly at half its peak amplitude is an estimate of the maximum depth to the middle of a compact object below the magnetic sensor. Figure 7.13 illustrates how this approximation is made along a profile that crosses a magnetic object (in this case, the curve is a calculation of a magnetic dipole). While this approximation is good for the analysis of total-field magnetic data, it can be adequate for gradiometer surveys also. Figure 7.13 also shows how one may next calculate an estimate of the buried mass if one can guess the composition of the object.

There are two general classes of computer programs that are applied to the analysis of magnetic maps: Modeling software and deconvolution programs. Modeling programs provide the most accurate interpretations, but they are slower than an analysis with the deconvolution approach.

With a modeling program, an initial estimate of the parameters of a body is made; these parameters are location, depth, shape, and magnetic properties. The magnetic field of that model is calculated and compared to the measurements. Then one or more of the parameters are changed, and the field is recalculated; if this yields a magnetic map that is even more similar to the measurements, the parameters continue to change as they were. This automatic iteration continues until the calculated map is most similar to the measured map. The final parameters will include the best estimate of depth; however, these programs also allow one to approximate the amount of magnetic material that is underground, and they can provide an approximation of the shape of the body.

An illustration of the final calculation of a modeling program is in Fig. 7.14. The plus symbols locate magnetic dipoles whose calculated field is plotted; the high magnetic values are at the upper left side, while areas with low field (on the lower right) have tick marks along their contour lines. The magnetic field of the five dipoles within the magnetic high approximates the measurements of the kiln shown in Fig. 7.10; the directions of magnetization of these five dipoles were all constrained to be the same. This direction, the locations of the dipoles, and their magnetic moments were iterated with a computer program. The final pattern of five dipoles suggests the location and shape of the kiln that was later excavated. The total magnetic moment of the dipoles was about 80 Am^2; this approximates the amount of magnetic material in the kiln, and this number agrees with the analyses of other kilns in Crimea.

A modeling program generally analyzes one anomaly at a time; a deconvolution program is like a filter that analyzes an entire map at once. It provides a faster, but less accurate, analysis; it also

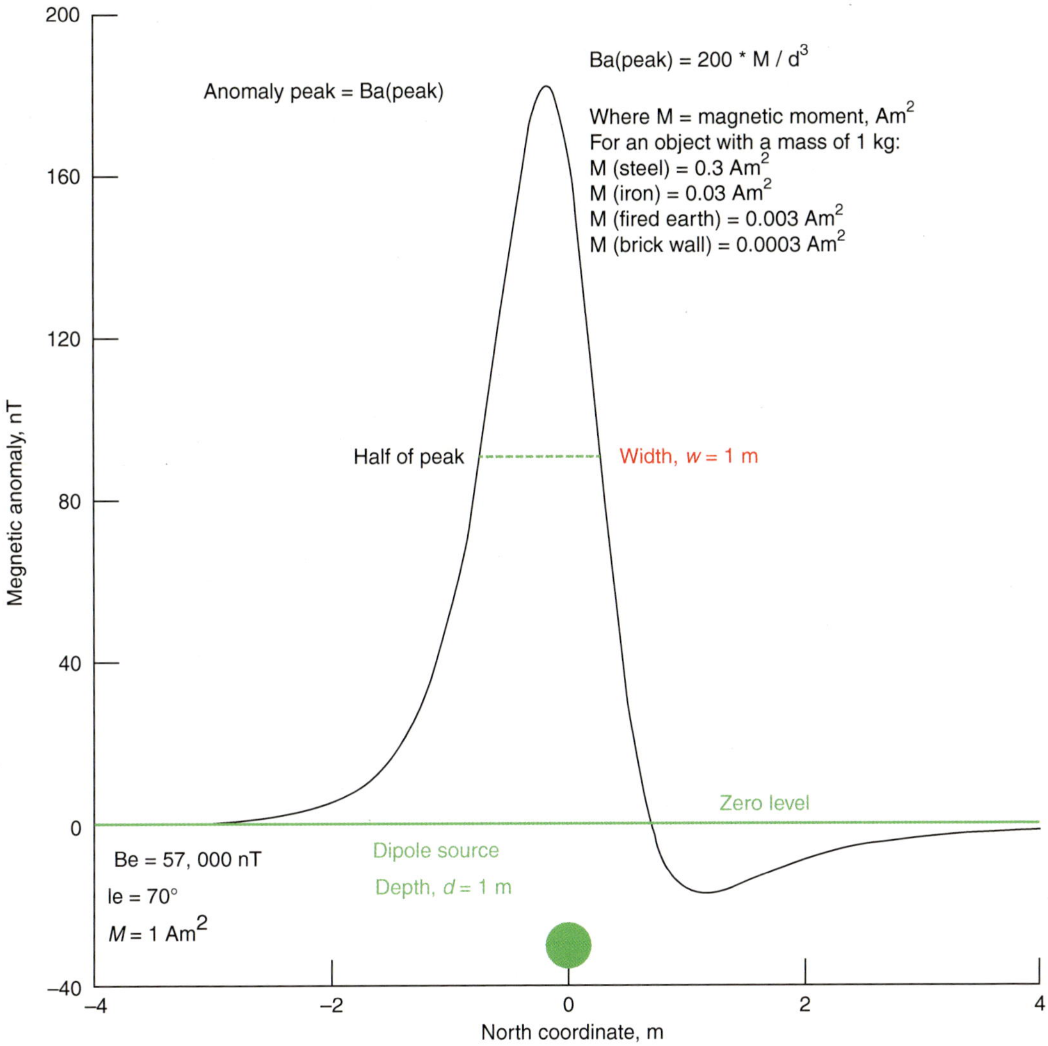

Fig. 7.13 The half-width rule approximates the depths of objects and estimates of mass are also possible

may not be possible to estimate the quantity of magnetic material with this approach. Euler deconvolution appears to be quite suitable for archaeological surveys, particularly where there are many anomalies to analyze (Desvignes et al. 1999).

Many computer programs are available for these technical analyses; both free programs and costly programs are listed below. None of these programs appear to have been designed for actual gradiometer data, although their analyses are still helpful.

7.11 Sources of Information

The names and Internet links here will lead to further and more detailed information about magnetic surveys. Search the Web for the following companies (and their products):

Manufacturers of magnetometers: Bartington Instruments (fluxgate), Gem Systems (Overhauser), Geometrics (cesium), Geoscan Research (fluxgate), Institut Dr. Foerster (fluxgate). Journals that include papers on magnetic

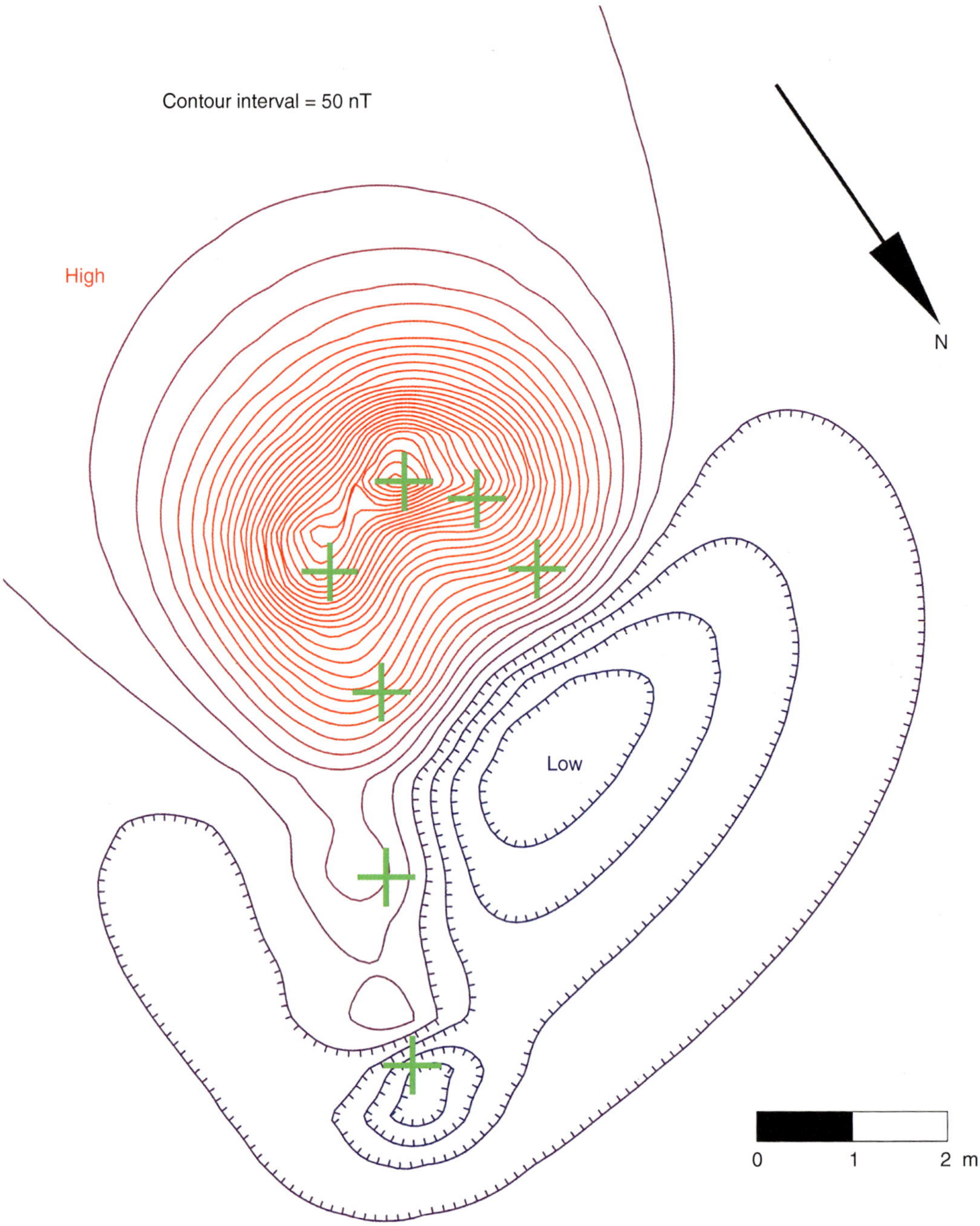

Fig. 7.14 The calculated magnetic field of a modeling program approximates the measurements that were made on the underground kiln in Fig. 7.10

surveys at archaeological sites: Archaeological Prospection, and Geoarchaeology. Software for plotting magnetic maps and doing some data processing: DW Consulting (ArcheoSurveyor) and Golden Software (Surfer); some manufacturers supply software that has been optimized for their instruments: Geometrics (included) and Geoscan Research (added cost). Developers of technical software for the analysis of magnetic maps: Geophysical Software Solutions (Potent), Geosoft

(Oasis montaj), PetRos Eikon (Emigma), Pitney Bowes Software (Encom ModelVision), and Zond Software (Mag3D).

The following books, newsletters, courses, and software are free on the Internet; where long URLs are listed, it may be easier to search by titles. These web sites were available in September 2012; if the sites have moved, search for some of the text below:

An excellent monograph titled "Geophysical Survey in Archaeological Field Evaluation" is available from English Heritage. This publication has a thorough discussion of many topics about archaeological surveys in a relatively small number of pages. It can be found at http://www.english-heritage.org.uk/content/publications/publicationsNew/guidelines-standards/geophysical-survey-in-archaeological-field-evaluation/geophysics-guidelines.pdf.

A valuable, although old, monograph by Sheldon Breiner at Geometrics describes many important ideas about magnetic surveys and anomalies. This report is titled "Applications manual for portable magnetometers," and it is available from this manufacturer of magnetometers and other geophysical instruments at ftp://geom.geometrics.com/pub/mag/Literature/AMPM-OPT.pdf.

A collection of many magnetic maps of archaeological sites in Europe was prepared by Tatiana Smekalova and her colleagues; while many sites are in Scandinavia and Crimea, several sites are in Greece, Syria, and Egypt. This report is titled "Magnetic surveying in archaeology," and it is available from another manufacturer of magnetometers, Gem Systems: http://www.gemsys.ca/pdf/10_Years_of_Overhauser_for_Archaeology.pdf.

A report titled "Geophysical exploration for archaeology" was prepared in 1998 by Bruce Bevan; the discussion there on magnetic surveys is now getting a little old. However, one volume of this digital report is still available from the Midwest Archeological Center of the National Park Service in the USA at http://www.nps.gov/mwac/publications/pdf/spec1.pdf.

Very old books can also be very educational. One thorough book (1,000 pages) is titled simply "Geophysical exploration," and it was written by C. A. Heiland in 1946. While the equipment is entirely outdated, the mathematics is still good even if some unit names have changed. This book is available as a scanned copy from the Internet Archive at http://www.archive.org/details/geophysicalexplo00heil.

A free newsletter of geophysics, called FastTimes, is published by the Environmental and Engineering Geophysical Society. While only a few articles on archaeology are included, many near-surface geophysical topics are discussed: http://www.eegs.org/PublicationsMerchandise/FASTTIMES/LatestIssue.aspx

An excellent course on magnetic surveys (along with other geophysical courses) was written by Laurent Marescot while he was a lecturer at several universities in Europe. These are now available at his Tomoquest web site: http://www.tomoquest.com/Lectures_in_Geophysics.php.

One of these courses is on archaeological exploration (in French), but that course does not include magnetic surveys.

Lisa Tauxe, who teaches at the University of California at San Diego, has a superb and thorough course titled "Magnetic techniques in geology and archaeology". This course includes detailed texts and notes and also audio recordings of the lectures. It is available at http://magician.ucsd.edu/Essentials/index.html

Steven D. Sheriff, at the University of Montana, has a course titled "Geoscience 436 – Subsurface imaging for archaeology"; he applies several free geophysical programs and his notes for this and other courses have helpful tips about these programs: http://www.cas.umt.edu/geosciences/people/facultyDetails.php?ID=622.

Another complete course, in French, can be found at the École Polytechnique de Montréal Génies C. G. M. at http://geo.polymtl.ca/.

Courses on many topics of geophysical exploration are available in Greek from the School of Geology at the University of Thessaloniki (since these courses include so many illustrations, language causes little difficulty): http://www.geo.auth.gr/en_e-teach.htm.

Many useful programs for the analysis of magnetic maps are freely available on the

Internet; most of these are only for individuals who do a technical analysis of magnetic maps. One simple and very helpful program called Pdyke is from Geophysical Software Solutions. This program calculates the magnetic field of a long magnetic prism; this is a basic and adequate model for many anomalies: http://www.geoss.com.au/downloads.html.

A program from Geometrics allows one to approximate small-area magnetic anomalies with dipoles, and this is suitable for the analysis of many archaeological maps. This program is called MagPick; and it is found at ftp://geom.geometrics.com/pub/mag/Software/magpick-latest.exe. A manual for this program is also there.

Gordon R. J. Cooper is a geophysicist in the School of Earth Sciences at the University of Witwatersrand. His web site includes many computer programs that he has written for geophysical data processing and analysis, including his program Mag2dc that does forward modeling and inversion of two-dimensional magnetic anomalies. His software and many of his publications are found at http://www.wits.ac.za/academic/science/geosciences/research/geophysics/gordoncooper/6511/software.html

The most modern and comprehensive set of software for magnetic analysis is from the US Geological Survey. These routines use a free program from Geosoft, the Oasis montaj viewer, for displaying maps and controlling the routines. That viewer can be downloaded from http://www.geosoft.com/support/downloads/viewers/oasis-montaj-viewer.

The geophysical programs themselves are called Geosoft eXecutables (GX); as examples, routines are available for upwards continuation and Euler deconvolution. These programs, along with a manual for their operation (USGS Open-File Report 2007–1355), can be found at http://pubs.usgs.gov/of/2007/1355/

An earlier suite of programs was, like the above, prepared by Jeff Phillips at the USGS; while these earlier programs were written for MS-DOS, they will still operate with Windows. The GX routines include only a few of the valuable programs that are in this early collection: http://pubs.usgs.gov/fs/fs-0076-95/FS076-95.html or http://pubs.usgs.gov/of/1997/ofr-97-0725/

Markku Pirttijärvi, a geophysicist at the University of Oulu, in Finland, has a good collection of his computer software on his web site; a program of his called MagPrism calculates the magnetic field along a single line over or near a three-dimensional, rotated prism: https://wiki.oulu.fi/pages/viewpage.action?pageId=20677906

Acknowledgment We thank Cristina Corsi (Universidade de Évora) and Frank Vermeulen (Universiteit Gent) for originating and coordinating the Radio-Past project and for inviting us to their 2011 workshop. We also appreciate the help that was given us by many Portuguese archaeologists during our stay at *Ammaia*.

References

Aspinall A, Gaffney C, Schmidt A (2008) Magnetometry for archaeologists. Altamira Press, Lanham

Breiner S (1999) Applications manual for portable magnetometers. Geometrics, San Jose

Desvignes G, Tabbagh A, Benech C (1999) The determination of the depth of magnetic anomaly sources. Archaeol Prospect 6:85–105. doi:10.1002/(SICI)1099-0763(199906)6:2<85::AID-ARP119>3.0.CO;2-I

Piro S, Campana S (eds) (2009) Seeing the unseen: geophysics and landscape archaeology. CRC Press, London

Earth Resistance Survey: A Mature Archaeological Geophysics Method for Archaeology

8

César Carreras Monfort

Contents

8.1 Introduction

In recent years, there have been important developments of technological applications in the field of geophysics applied to archaeology. New electromagnetic and magnetic equipments currently produce high-quality data in a very short time with a quite accurate geolocalisation thanks to the use of GPS. This recent evolution of geophysics in archaeology is complemented by an improvement in software for data interpretation and public dissemination.

Although this current trend is welcomed in the field of archaeology, people tend to forget that all the geophysical methods provide complementary data, and therefore, all the techniques are equally important. In the last few years, resistance surveys have been less fashionable because they are slow to complete and expensive in terms of time and effort compared to other newer methods. Nevertheless, they are still useful in many special conditions such as suburban areas with a lot of metal interferences and as complementary information to other methods.

8.2 Principles of Earth Resistance Surveys

Earth resistance was one of the first geophysical methods applied to archaeology, as early as 1946. It seems that it was used simultaneously in Mexico and England in that year, though more details are recorded from the latter application.

C. Carreras Monfort
Department of Classics,
Universitat Autònoma de Barcelona,
Barcelona, Spain
e-mail: cesar.carreras@uab.cat

C. Corsi et al. (eds.), *Good Practice in Archaeological Diagnostics*, Natural Science in Archaeology,
DOI 10.1007/978-3-319-01784-6_8,

The development of aerial photography after World War II meant that landmarks recorded from the sky should be contrasted by ground fieldwork (see Chap. 2 by Ceraudo, this volume). Apart from recording any material or structure on the surface, Atkinson (1946) proposed the use of earth resistance in order to identify potential archaeological works. He applied earth resistance in Dorchester-on-Thames, which was an Iron Age hill-fort with little evidence from the ground. A Megger earth tester was used with a Wenner array configuration providing an image of resistance values from which ditches and the hill-fort layout could be inferred. It was a rough picture of possible archaeological features, but it was a promising initial result for the application of geophysics to archaeology.

The method consists of feeding electric current into the ground and taking resistance measures at particular points (Scollar 1990; Clark 1996). Such resistance is highly dependent on the distribution of moisture in soils affected by drainage, the presence of structures and soil porosity. Clay and soil normally have resistances around 1–10 Ω-m, porous rocks 100–1,000 Ω-m and non-porous rocks may reach 103–106 Ω-m (Gaffney and Gater 2003, p. 112).

Voltage is applied to one end of an electric conductor (i.e. wire), so the size of the current depends on conductor resistivity. This is expressed with the formula:

$$I = V / R$$

where I = current, R = resistance (ohms – Ω) and V = potential difference (volts). Soils become natural conductors with rainfall, since rainfall contains carbon dioxide and carbonic acid that reacts with natural soils, making them electric conductors. Factors that influence material resistivity are the degree of what is called "compactivity", which is affected by depth, porosity and permeability. Normally, resistance can be detected only at a depth of 0.75–1.5 m depending on the equipment employed. With regard to porosity, the more porous a soil or material is, the less resistance is recorded. Finally, permeability favours moisture, therefore also favouring conductivity and less resistance.

Resistance in archaeology is measured with mobile equipment that requires current to be fed into the soil and resistance to be recorded at particular points. Since areas covered by geophysical surveys are large, a pair of electrodes is required to undertake fieldwork. Within the pair of electrodes, current passes through electrodes C (C_1 and C_2) and resistance is measured by electrodes P (P_1 and P_2). There are different electrodes configurations and different distances between them, which provide significant differences in results.

The earliest configuration (Wenner array) established the same distance between electrodes, where electrodes C were at the extremes, so only large structures were recorded (Clark 1996). The second configuration in time was called Schlumberger's array and defines 30 times the distance between electrodes C and P, which improved the fieldwork conditions. Nowadays, both arrays are only used in archaeology for large electric profiles (tomography) with multiple electrodes in order to recognise large earthworks, bedrocks, mines, aquifers, etc. (Szymanski and Tsourlos 1993). They are relatively slow but they record very deep profiles.

Modern arrays (double dipole and twin-probes) combine C and P electrodes as either remote or mobile electrodes, which make them quite handy and less time-consuming. In particular, the twin-probes array is designed for archaeological surveys, with the only disadvantage that it does not discriminate between geological and archaeological anomalies (Gaffney and Gater 2003, pp. 29–31). In the twin-probes array, the pairs of electrodes are C_1 combined with P_1 and C_2 with P_2, and in addition the distance between remote and mobile probes should be at least 30 times the distance between the 2 mobile probes.

The distance between mobile probes affects the readings depth as well as the distance of the remote probes. For instance, the Geoscan RM-15D illustrated in Fig. 8.1 shows a 0.5 distance m between the two mobile probes (Walker 1991). This configuration allows readings at a maximum of 0.75–1 m depth, whereas a configuration with 1 m distance reaches 1.5–1.75 m depth. Therefore, the method only recognises

Fig. 8.1 Resistance survey at the Roman villa of *C. Iulius Rufion* (Giano dell'Umbria, ICAC)

anomalies in the near surface. There is an adapted resistivity metre called a multiplexer that performs 3 simultaneous readings at 3 different depths with 6 mobile probes at different distances (0.5, 1 and 1.5 m) (Walker 2000). Although this may seem a great advantage, in fact the multiplexer is not practical in normal conditions, since there are no ideal flat ground surfaces and it becomes complicated to position the 6 mobile probes at the same time.

Compared to other geophysical methods, earth resistance is a relatively slow method since surveyors should push mobile probes into the ground every metre, which takes some time. An average grid of 30×30 m may take between 40 min and 1 h, according to different conditions. That is why there have been some experiments, for instance at Wroxeter, employing wheeled vehicles to record quickly many readings (Dabas et al. 2000). Nevertheless, mobile probes seem to deteriorate over time, so data quality is also affected.

The most suitable conditions for earth resistance surveys are the presence of sedimentary soils, lack of rock outcrops on the surface, limited short vegetation and a well-drained season such as spring or autumn in Mediterranean regions. As said, earth resistance surveys require soil moisture, which becomes common only in Mediterranean latitudes during the rainy months in spring and autumn.

With regard to equipment, one of the most commonly used is the Geoscan RM-15D with data logging for more than 30,000 readings and a robust structure that allows all kinds of arrays, as well as adapting the multiplexer (Walker 1991). An alternative equipment is an M.M instrument with a resistivity metre that only allows Dipole and Werner configurations. Nowadays, many geophysical surveyors create their own resistivity equipment's from well-known guidelines.

As happens in other geophysical surveys, a good earth resistance survey requires an initial visit to the potential site in order to evaluate whether conditions (i.e. natural, present uses) may affect results. Once the survey is planned, the whole area to be surveyed should be divided into grids, normally of 30×30 m, defined by tapes and plastic ropes. Surveyors normally work in pairs, taking readings every metre in a traverse mode (zigzag) that reduces movement of equipment. While one surveyor takes readings with the resistivity metre, the other carries electric cable and moves plastic ropes (Schmidt 2002).

Earth resistance surveys record all types of anomalies, whether due to archaeological features or to other natural or human causes. One of the most important tasks of the archaeologist is to discriminate these anomalies, sometimes due to rock outcrops, tree roots, climate conditions (dryness, coldness) or modern disturbances (such as rubbish disposal and pathways). Good survey conditions, such as those enjoyed by the survey of the Roman city of *Italica* in the 1990s carried out by Terra Nova Ltd. (Rodríguez Hidalgo et al. 1999) (see Fig. 8.2), may generate clear images after downloading data logging and displaying the results. Such images may be enhanced by employing suitable software with built-in filter options.

Archaeological anomalies are difficult to explain individually (Clark 1996), so they acquire meaning only when combined together and can then create architectural and urban shapes. Therefore, archaeologists should include any visible and excavated structures from a site that may help to interpret any potential archaeological anomalies. Besides, long occupied sites such as *Italica* may present different urban layouts corresponding to different periods. In the case of *Italica*, one can observe a Late Roman wall in the

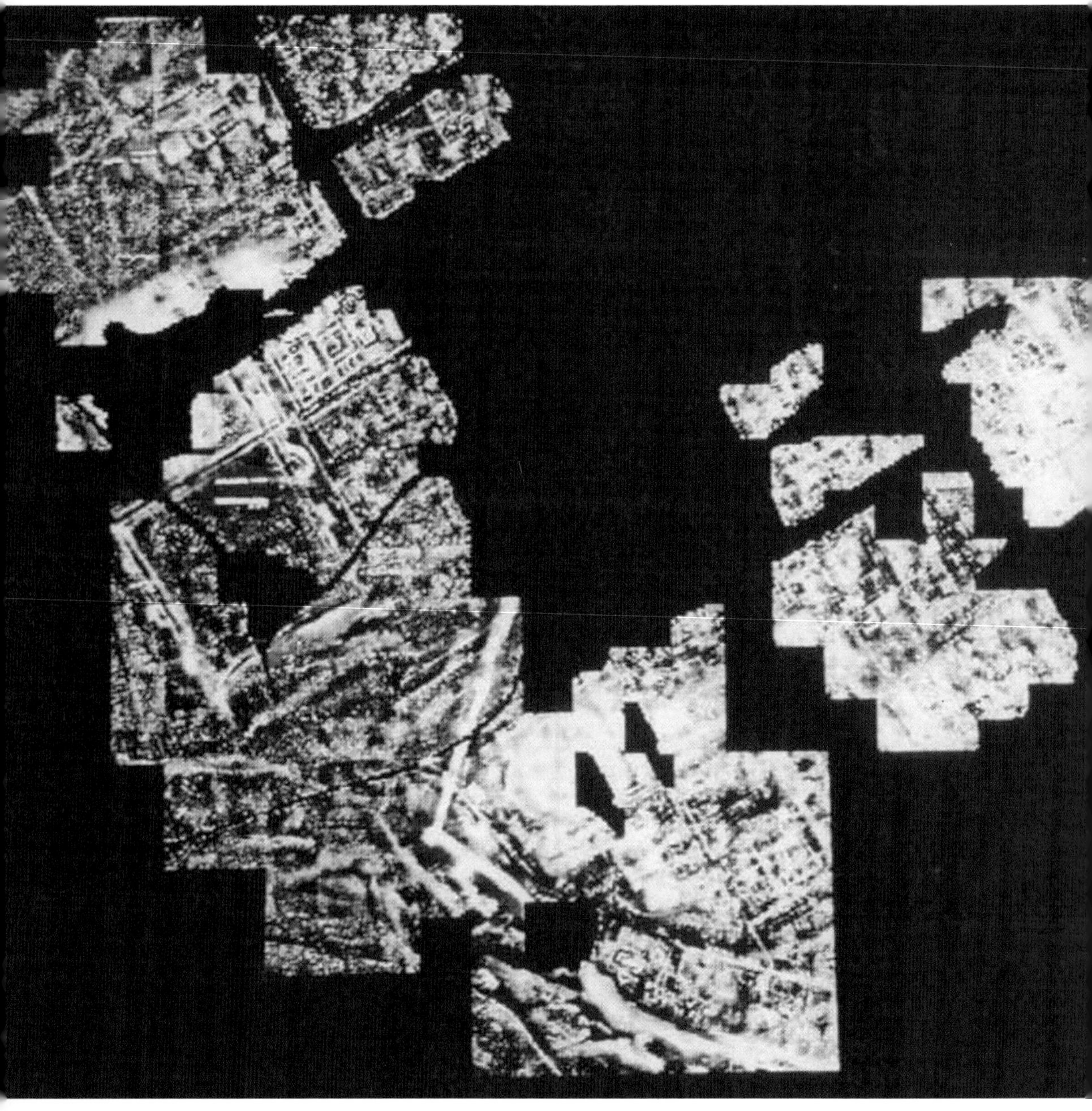

Fig. 8.2 Resistance survey at *Italica* (Terra Nova Ltd., Rodríguez Hidalgo et al. 1999)

south surrounding only a part of the city, whose layout is dated to the early second-century AD.

Apart from shapes, most common archaeological anomalies are recognised for their resistivity response (Clark 1996; Gaffney and Gater 2003). High resistances are recorded in materials such as stone from city walls or private residences, even brick walls. Moreover, high responses are documented for dumping, rubble and filling areas and floors (e.g. pavements, mosaics). In rural areas, high linear resistances of constant width normally refer to roads or pathways, either paved or simply pressed soils. Finally, high resistances in a necropolis area may identify sarcophagi, cists (i.e. tombs surrounded by stones) or burial constructions.

By contrast, low resistance densities may be due to the presence of ditches and pits filled by less compacted earth, or any kind of water infrastructures for human consumption or disposal, such as gullies, slots and drains (Clark 1996; Gaffney and Gater 2003). Other low resistance densities may be recorded with metal pipes or graves.

Urban sites are one of the most appropriate archaeological venues to undertake earth resistance surveys. High resistance densities appear in most house walls and lower densities may turn up in all the drains that removed dirty water from individual and public buildings. However, resistances are never constant and depend strongly on the contrast with the natural background. The survey of the Iberian site of Puente Tablas (Fig. 8.3), located on the top of a hill, reveals clearly the urban layout, and in fact two orientations because it was adapted to the local topography. Besides, there are other anomalies with high resistance due to rock outcrops close to the surface which alter the regular picture of the orthogonal urbanism of Puente Tablas.

This example shows also that a site cannot always be completely surveyed due to areas that have already been excavated or have impossible surveying conditions. Filling these unsurveyed areas is another point to take into account when interpreting earth resistance results.

8.3 Typical Resistance Features in Classical Archaeology

Part of the interpretation process relies on practical experience and knowledge of types of anomalies of particular archaeological cultures. This proper archaeological knowledge is essential to understand the shapes of anomalies and material "signatures", or in other words the expected results from a typical site and typical construction materials, whose remains should be partially documented on surface materials.

One of the typical geophysical features in Classical archaeology detected by earth resistance is a *road*, *pathway* or *street*. Such features are long linear high resistance anomalies between 2 and 5 m in width produced by pressed or paved soil and sometimes with two lateral ditches defined by low resistance values (Gaffney and Gater 2003, pp. 142–143). There are minor differences in the standards of Roman roads, which may vary between provinces, according to traffic needs and local geological and geo-morphological conditions, and special problems are presented by medieval infrastructures. In many Western European regions, medieval roads are re-formed Roman ones with changes due to new political boundaries.

Resistance anomalies related to the presence of roads may be rather slight in contrast to the geological background. Sometimes, the central part of the road becomes indistinct, whereas the two parallel ditches define the road path. Earth resistance surveys searching for roads may be the result of following up evidence of excavated roads, entrances to towns and cities or testing linear features from aerial photography. Of course, the depths at which the remains of a road are buried affect the use of earth resistance, so other geophysical methods should be used instead.

Related to ancient roads, there are the so-called *field systems*, a series of anomalies that define agricultural properties and soil uses. Some of these shapes of ancient landscape are clearly visible from aerial photography but produce very slight anomalies in earth resistance. In general, it is very difficult to assign a particular historical period. In Classical times, the foundation of colonies involved a surrounding land division to assign each settler a plot of land. Between each settler's property there were boundaries made of stone, pathways and even public roads. Nowadays, some field surveys that document potential land division (i.e. *Ager Tarraconensis*) require geophysics such as earth resistance to recognise these potential limits of properties.

Those limits may supply high resistance anomalies when they are built (i.e. stone) or consist of trackways (i.e. beaten soils) and negative ones when they are defined by ditches. Sometimes earth resistance may provide negative evidence of modern agricultural changes in landscape that have altered ancient field systems (i.e. modern terraces). The features of field

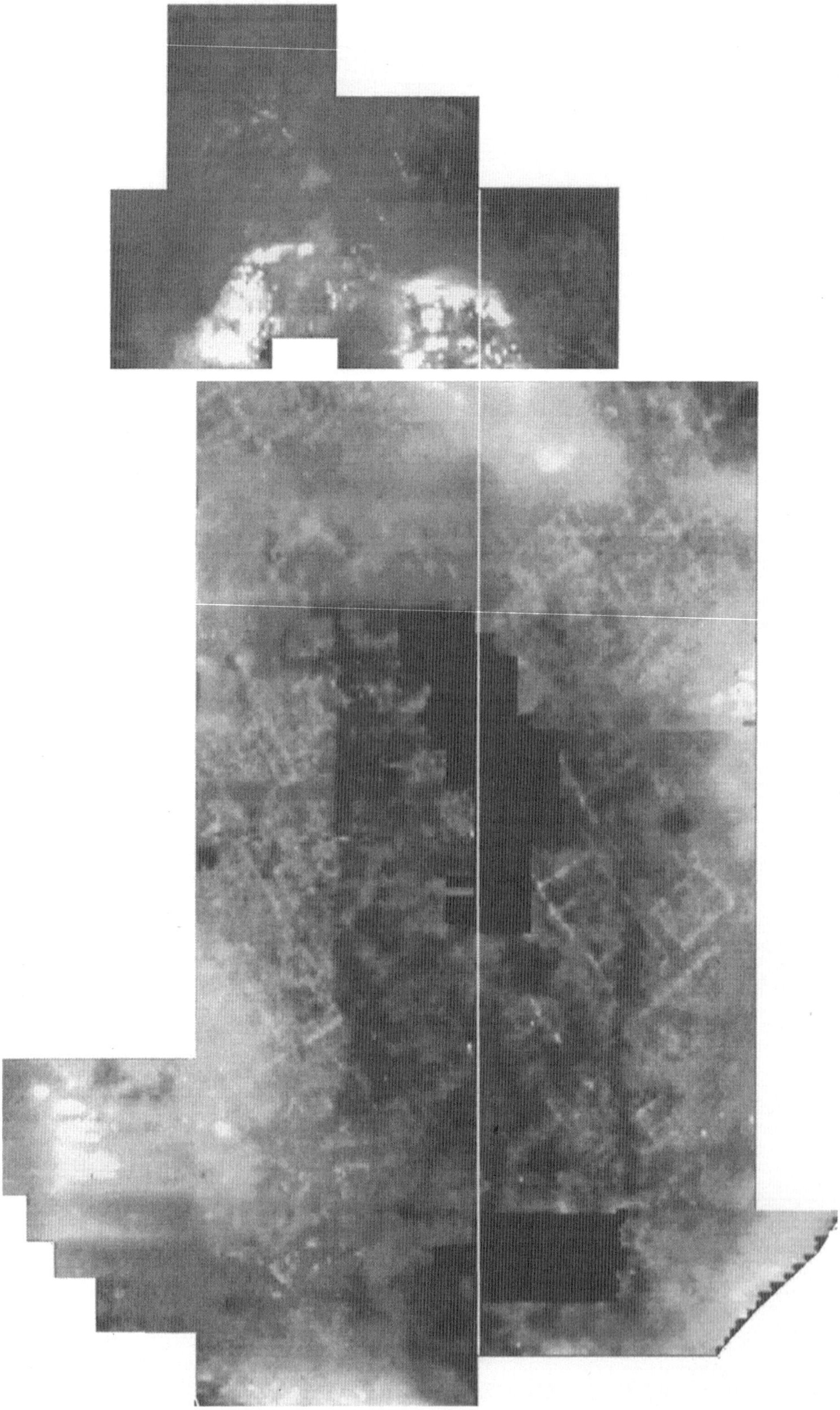

Fig. 8.3 Resistance survey at Puente Tablas (Terra Nova Ltd.)

systems are quite complex to detect and interpret, and not many case studies have been recorded so far. There is still a great potential to discover the traits of these historical landscapes thanks to geophysics and particularly earth resistance. Actually, it is a potential complement to give meaning to ceramic scatters documented in field surveys.

Normally, these field surveys only recognise large disturbances in rural areas identified as *farms* or *villae*, a Roman term to define a special intensive agricultural exploitation. These large disturbances, which may also be visible from aerial photography, are the only ones that have attracted geophysical attention. Due to their dimensions, it becomes relatively easy to distinguish by earth resistance the high anomalies of walls and floors (i.e. mosaics), as well as complex architectural shapes. Earth resistance is not the only geophysical method applied to identify those buildings, but it produces excellent results in calcareous soils where other methods fail (e.g. *Ager Tarraconensis*) (Prevosti et al. 2010).

Earth resistance and magnetometry were applied in the *Ager Tarraconensis* survey to identify the shapes of large scatters of pottery shards combined with visible wall structures (Strutt et al. 2011) which were described as Roman *villae*. More than ten rural sites were geophysically surveyed, providing a wide variety of buildings that do not really fit the traditional definition of Roman *villae*. Though more examples are required, geophysics and particularly earth resistivity may help to give sense and shape to ground pottery evidence recovered on the surface. For instance, an earth resistance survey of one of the sites called Les Bassases detected a large building (see Fig. 8.5) of 20×10 m with enormous walls, which was later excavated and revealed a possible large water basin. Many structures that were in the past considered to be farms or *villae* are in fact all types of industrial and agricultural buildings, whose typologies are still to be defined. Some of those potential rural sites are well-known from excavations and ancient literature, such as pottery workshops, wine and olive-oil production centres and garum pools, but little is known in geophysical results.

However, the quality of earth resistance survey results depends strongly on the size of these buildings, since they look for regular patterns that are difficult to interpret in small areas. Nevertheless, it is a quite promising field in archaeological geophysics, since it may give shape to structures detected by ceramic scatters on the ground.

As said before, earth resistance and other geophysical methods obtain excellent results in *urban sites* since they are large in size, making it easier to discover regular patterns of streets or house walls. There is a minor difference between sites that are fortified or not or are simply military settlements. Fortified sites are normally surrounded by large walls of wide dimensions (2–7 m) combined with a large ditch and bank enclosure surrounding the whole city wall as well. Such large anomalies may be repeated around the site more than once, as happens in Iron Age sites and some Roman military camps. In terms of earth resistance, they produce high resistance anomalies in the case of a large wall and lower ones for a ditch and bank enclosure. Although many geophysical methods may provide good results to discover such structures (Gaffney and Gater 2003, pp. 155–156), earth resistance is a quite reliable one.

Inside urban settings, more Classical cities keep an identical street grid to that laid out at their foundation. Therefore, once the grid module is identified, geophysical results are easy to interpret. In other words, such modularity means that areas with slight anomalies or modern interference may be easily reconstructed on the basis of neighbouring blocks. A special case arises when two possible foundations or layouts are expected, such as the case of *Complutum* (Spain), where there is supposed to be an early Augustean foundation in the western part and a later Claudian one with different *insulae* modules.

An earth resistance survey at the *region* IV of *Complutum* has revealed a series of anomalies of potential streets and houses walls that do not completely fit the ideal Claudian *insulae* layout recorded in other parts of the city (see Fig. 8.4).

It is still difficult to recognise whether these linear anomalies belong to streets or dividing

Fig. 8.4 Resistance survey at *Complutum* region IV (ICAC)

walls between houses. Besides, disturbances from nineteenth-century excavations are still providing some extra difficulties in making a direct interpretation. Classical urban sites are the places where more earth resistance surveys are performed and provide more outstanding results. Projects such as the cities of Roman *Baetica* or the Tiber Valley survey run by the University of Southampton are good examples of how geophysics (earth resistance and magnetometry) may enhance our knowledge of urbanism in a territory without any excavations.

As a follow-up to the Tiber Valley survey, the project continued in the area of *Portus* (Trajan port) close to Fiumicino (Keay et al. 2005), where all the port installations such as warehouses, moles and channels were documented. Again *ports*, either urban or not, are other types of site that are suitable for earth resistance surveys when there is no modern occupation of the site.

8.4 Filters in Earth Resistance Interpretation

Apart from the fieldwork and basic knowledge of archaeological structures one is looking for, as well as the resistance response of particular materials and geological backgrounds, geophysical operators should take full advantage of computer applications developed for enhancing survey images. Every equipment firm provides its own software to analyse and interpret geophysical data, though they are normally restricted to limited functions and filters.

For instance, the Geoscan RM-15D provides its own software called Geoplot, which downloads data from a resistivimeter and supplies standard functions to get rid of undesired data as well as filters for improving image resolution. It is important to bear in mind that earth resistance anomalies should be visible from the initial downloaded image and that filter application should only be used to enhance the quality of data. Applying filters without care may generate artificial anomalies that do not represent any actual buried structure or geological feature.

Normally, geophysical surveyors apply a *despike filter* to get rid of extreme values that result from errors in pushing mobile electrodes into the ground while field surveying. Another filter that should be initially used to remove defects is *edge matching*. The third recommended filter is *high pass, which* improves the contrast with the background geology (Gaffney and Gater 2003: 104). Besides, there are other filters that help to enhance images, such as low pass, smooth or interpolate.

The geophysics survey team from Southampton, who undertook large earth resistance surveys in the Tiber Valley and *Portus* projects, basically employed *despike* and *edge matching* filters to remove defects and, afterwards, the high-pass filter to improve the resulting images (Keay et al. 2005). However, they added data from a topographical survey and aerial photographs within a GIS in order to generate more comprehensive images for interpretation. Likewise, the *Ager Tarraconensis* geophysical surveys have used the same combination of filters to obtain high-resolution images (Prevosti et al. 2010).

By contrast, surveyors may create their own built-in programmes in Visual Basic to download data from resistivimeters and analyse data in standard GIS systems, other packages such as SURFER or may create their own filter programmes. For example, SNUFFLER (http://www.sussexarch.org.uk/geophys/snuffler.html) is a freeware geophysics software that includes all sorts of filters for any earth resistance equipment (Fig. 8.4 was created with this software), whereas ArchaeoSurveyor (http://www.dwconsulting.nl/archeosurveyor.html) is a commercial-friendly software with more potential filters and functions. There is still a wide scope for research in filters that are especially suited for archaeological features by taking into account the response of building materials as well as possible external "noises".

Brito-Schimmel (Brito-Schimmel and Carreras 2010) has created her own filters to discriminate potential individual noises in archaeological geophysical surveys. At the start, she generated synthetic models of the most typical noises that may turn up in a geophysical survey from modern disturbances (i.e. modern uses, infrastructure), types of soil (i.e. different kinds of geology), high density of archaeological material (e.g. rubble, pottery, stones) and tree roots. The idea is that these filters may make it possible to survey poor condition sites and obtain more data from already surveyed locations. The first tests of these filters with real data from earth resistance surveys have generated quite promising results. For instance, they were applied to some of the earth resistance results from rural sites in the *Ager Tarraconensis* project (Prevosti et al. 2010). Although standard Geoscan filters were employed to interpret data in places like Les Bassases (Fig. 8.5), later on a series of built-in filters were applied to remove modern disturbances and geology.

The resulting images reveal a series of negative anomalies in the eastern side, which were not recognised at first sight. As was said, Les Bassases has been partially excavated and documents a large basin or pool in an agricultural estate. Negative anomalies may be potential drains and channels related to this structure. Future excavations will probably confirm or reject this interpretation, but at least these built-in filters have provided new fresh information, previously masked by "noise".

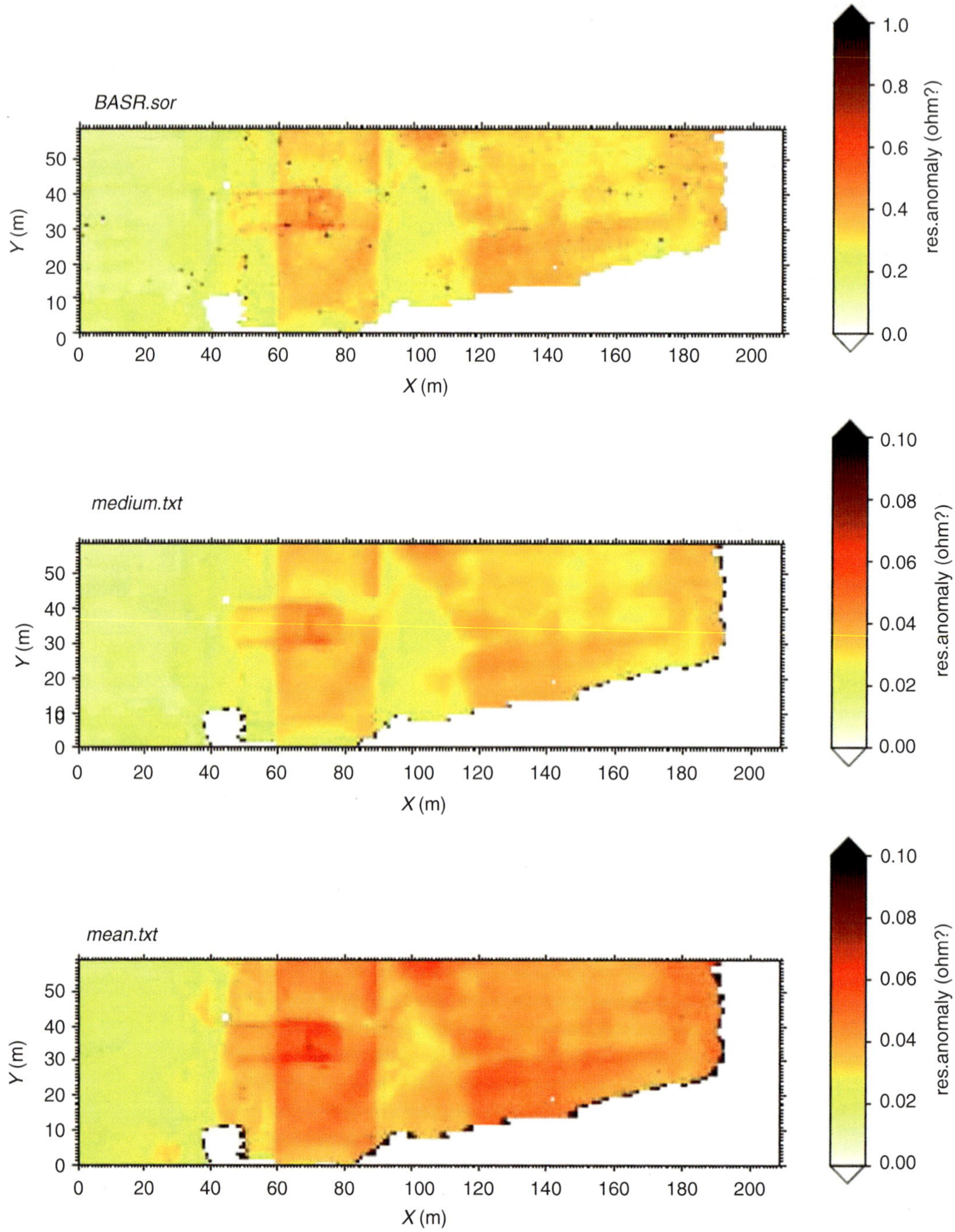

Fig. 8.5 Resistance survey at Les Bassases (Southampton University – ICAC)

8.5 Future Developments of Earth Resistance Surveys

Earth resistance surveys are still useful in archaeology. Despite the fact that other methods are currently generating great volumes of data faster, deeper and in difficult conditions, earth resistance still has a role to play in archaeology. First of all, earth resistance is very suitable for special soil conditions such as calcareous geological backgrounds, where both GPS and magnetometry yield very poor results. However, in some cases, the scant results obtained from those methods have complemented earth resistance interpretation.

Other examples of suitable applications of earth resistivity are those where modern disturbances make it useless to employ GPR or magnetometry (e.g. *Complutum*). The presence of numerous remains of metal in the ground as well as a series of modern structures generate so much noise that the resulting images are impossible to interpret. Only earth resistance can provide clear data from archaeological urban structures.

Nevertheless, if site conditions are normal, earth resistance may fulfil a special role as a source of complementary data to other methods. Since the principles of earth resistance are different from other techniques, RES shows structures or materials that may be less clearly recorded by other geophysical methods. Earth resistance can either be employed to survey the whole research area or simply used to check places that have already been surveyed and are rich in archaeology. Due to its slowness, it is recommended to employ magnetometry or GPR for the whole extension, and later earth resistance in selected areas with anomalies.

Combining results maps from different geophysical methods is relatively easy in any GIS environment by means of overlays, and the final image may contain complementary features that may be invisible in any individual image.

The required investment of time in the survey is one of the drawbacks of earth resistance. Although some experiments have been carried out in the past (Dabas et al. 2000; Walker 2000), the results cannot compete with the speed of other methods. Therefore, there is still new ground for research in improvements in the speed and depth for obtaining resistance data.

The other field in which one expects new technical improvements is computing applications to discriminate archaeological anomalies. Although these developments in software focussed on filtering and enhancing image results can apply to any geophysical technique, earth resistance may benefit from them mainly for low-value anomalies.

References

Atkinson RJC (1946) Field archaeology. Methuen, London

Brito-Schimmel P, Carreras C (2010) Metodología para la prospección geofísica en arqueología: apuntes a partir de los trabajos de Iesso, Can Tacó, Molins Nous y el Goleró. Lucentum XXIX:9–22

Clark AJ (1996) Seeing beneath the soil: prospecting methods in archaeology, 2nd edn. Batsford, London

Dabas M, Hesse A, Tabbagh J (2000) Experimental resistivity survey at Wroxeter archaeological site with a fast and light recording device. Archaeol Prospect 7(2):107–118

Gaffney C, Gater J (2003) Revealing the buried past. Geophysics for archaeologists. Tempus, Stroud

Hidalgo R, Manuel J, Keay S et al (1999) La Italica de Adriano. Resultado de las prospecciones geofísicas de 1991 y 1993. Arch Esp Arqueol 72:73–78

Keay S, Martin M, Lidia P, Kristian S (2005) Portus: an archaeological survey of the port of imperial Rome, vol 15, Archaeological monographs. The British School at Rome, London

Prevosti M, Strutt K, Carreras C (2010) The *Ager Tarraconensis* Project (right side of river Francoli) (PAT): the application of geophysical survey to identify rural Roman settlements typologies. In: Corsi C, Vermeulen F (eds) Changing landscapes. The impact of roman towns in the Western Mediterranean. AnteQuem, Bologna, pp 205–216

Schmidt A (2002) Geophysical data in archaeology: a guide to good practice.AHDS guides to good practice. Oxbow Books, Oxford. http://ads.ahds.ac.uk/project/goodguides/geophys/

Scollar I (1990) Archaeological prospection and remote sensing. Cambridge University Press, Cambridge

Strutt K., Fry R, Prevosti M, Carreras C (2011) Memòria de les prospeccions geofísiques realitzades a l'Ager Tarraconensis. In: Marta P, Josep G (eds) Ager Tarraconensis 2. El poblament. Institut d'Estudis Catalans & ICAC, col. Documenta, Tarragona, pp 223–249

Szymanski JE, Tsourlos P (1993) The resistivity tomography technique for archaeology: introduction and review. Archaeol Pol 31:5–32

Walker AR (1991) Resistance meter RM15. Manual version 1.2. Geoscan Research, Bradford

Walker AR (2000) Multiplexed resistivity survey at the Roman town of Wroxeter. Archaeol Prospect 7(2): 119–132

9 Ground-Penetrating Radar (GPR)

Alexandre Novo

Contents

A. Novo
Department of GeoRadar Division,
GeoSystems Business Unit, IDS North America Ltd.,
Montreal, Canada
e-mail: a.novo@idscorporation.com,
alexnovo@gpr-slice.es

9.1 Brief Historical Review

Ground-penetrating radar (GPR) has been one of the most utilized tools for archaeological prospection in the past two decades (Conyers and Leckebusch 2010) due to its high-resolution data and three-dimensional visualization capabilities. In particular, GPR has a successful history in discovering Roman villas and towns (Neubauer et al. 2002; Leckebusch 2003; Piro et al. 2003; Linford 2004; Berard and Maillol 2008).

According to Conyers (2004) and Gaffney (2008), it was in the mid-1970s that GPR appeared with archaeological purposes. It is possible to find the works of Bevan and Kenyon (1975), Vickers and Dolphin (1975), Vickers et al. (1976) and Kenyon and Bevan (1977) in the international bibliography. These pioneers investigated the use of subsurface radar to obtain field-readable cross sections of shallow anomalies in archaeological and historical sites in the USA (New Mexico and Philadelphia).

It was in the 1980s, with the consolidation of the first commercial equipment's from Geophysical Survey Systems Inc. (OYO Corporation), when GPR becomes to be used worldwide as non-destructive technique (NDT) in archaeological site investigations: Japan (Imai et al. 1987), UK (Stove and Addyman 1989), Sweden (Ulriksen 1982), Egypt (El-Baz 1989), Germany (Forkmann and Petzold 1989), China (Qiang and Chang 1985) and Canada (Vaughan 1986). Moreover in 1986 the first international conference on GPR takes place in Georgia (USA).

C. Corsi et al. (eds.), *Good Practice in Archaeological Diagnostics*, Natural Science in Archaeology,
DOI 10.1007/978-3-319-01784-6_9,

But it was in the 1990s when the technology was consolidated by the appearance of a new generation of commercial digital systems from different manufacturers: GSSI, Sensors & Software, MALÅ Geoscience, IDS Ingegneria Dei Sistemi SpA and ERA Technology. The increasing use of GPR in the archaeological field is found in many contributions to international conferences, specially the biannual International Conference on Ground Penetrating Radar (2nd in Florida (USA) in 1988, 3rd in Lakewood (USA) in 1990, 4th in Rovaniemi (Finland) in 1992, 5th in Ontario (Canada) in 1994, 6th in Sendai (Japan) in 1996, 7th in Lawrence (USA) in 1998). The contributions in the field of archaeology started with only 4–5 communications in the conferences of 1990 and 1992 – all of them showed results in only two dimensions yet. Nonetheless, it was in this decade when the first contributions related to 3D GPR imaging appear (van Deen and de Feitjer 1992), being the most significant those of Grasmueck (1996), Grasmueck and Green (1996) and Grandjean and Gourry (1996). In archaeological site investigation the works of Conyers and Goodman (1997); Goodman (1994, 1996), Goodman et al. (1995) must be highlighted.

Alternately to the International Conference on Ground Penetrating Radar, the second most important international conference related with GPR is the IWAGPR (International Workshop on Advanced Ground Penetrating Radar). Of great importance too, the biennial conferences on Archaeological Prospection have contributed to spread worldwide GPR amongst the archaeological community.

In addition, the journal Archaeological Prospection (http://onlinelibrary.wiley.com/journal/10.1002/(ISSN)1099-0763), initiated in 1994, has become the main vehicle for publication of relevant research and case studies. As well, an International Society for Archaeological Prospection (ISAP) was initiated in 2003 (http://www.archprospection.org) (English Heritage 2008).

During the last decade, some good references related to the development of 3D GPR systems and software are the University of Miami (Grasmueck et al. 2004a, b, 2005 and Grasmueck and Viggiano 2006, 2007), Politecnico Milano (Lualdi and Zanzi 2003; Lualdi et al. 2003, 2006) and Universität Postdam (Böniger and Tronicke 2010a, b, 2012). And with special attention to the archaeological field: Leckebusch 2000; Leckebusch and Peikert 2001; Leckebusch 2003, 2005a, b, 2007; Leckebusch and Rychener 2007; Leckebusch et al. 2008; The Vienna Institute for Archaeological Science (University of Vienna) together with Archeo Prospections (Central Institute for Meteorology and Geodynamics, Department for Geophysics): Neubauer et al. 2002 and recently the Ludwig Boltzmann Institute for Archaeological Prospection and Virtual Archaeology (Trinks et al. 2012a, b) together with Archaeological Sciences, School of Applied Sciences, University of Bradford: Gaffney et al. (2012); The University of Salento: Leucci 2002; Nuzzo et al. 2002; Leucci and Negri 2006; Negri and Leucci 2006 and Negri et al. 2008; The Geophysical Archaeometry Laboratory together with Nara National Cultural Properties Research Institute and Istituto per le Tecnologie Applicate ai Beni Culturali (ITABC-CNR): Goodman et al. 2002, 2006a, b, 2007; Piro et al. 2003; The Geophysical Team, English Heritage: Linford 2004; Linford et al. 2010; The University of Denver: Conyers and Goodman 1997; Conyers 2004, 2006; The university of Vigo: Lorenzo et al. 2002a, b; Lorenzo and Arias 2005; Novo et al. 2008, 2010, 2012a, b; Solla et al. 2010; Geostudi Astier: together with SOT Prospection (Novo et al. 2011), together with UC San Diego (Lin et al. 2011a, b) and together with Geocarta (Novo et al. 2012a, b).

Recent and upcoming books of interest are Conyers (2012) and Goodman and Piro (2013).

9.2 Brief Note on Principle and Systems

GPR transmits a very short pulse of electromagnetic energy and measures a reflected signal that is dependent upon the dielectric properties of the subsurface material (Fig. 9.1). The time delay of the reflected signal is used to estimate the depth

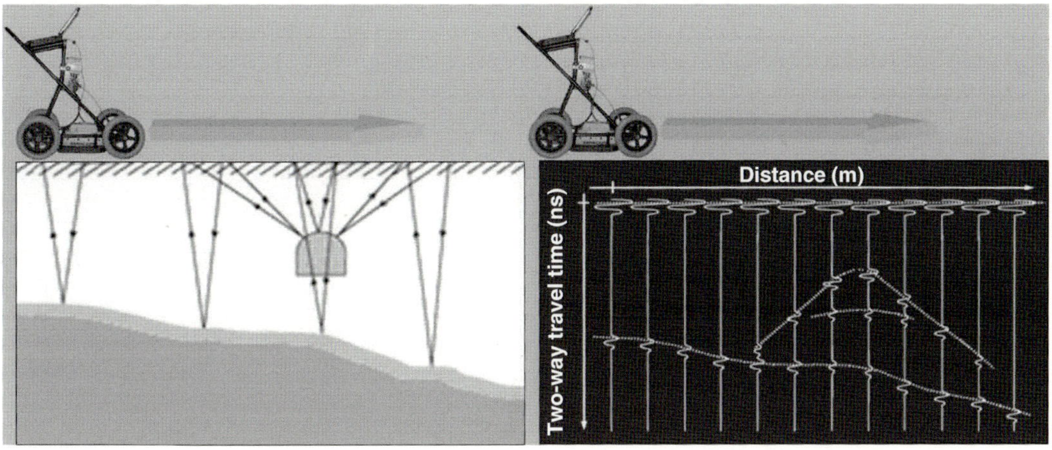

Fig. 9.1 The GPR cart is moved along a certain distance in order to equidistantly acquire reflected energy traces

range of the target. The latter and the resolution are related to the radar frequency, the dielectric properties of the host material and the shape and dimensions of the target.

Annan (2003) and Daniels (2005), offer a thorough technical description of GPR technology and applications which is beyond the scope of the present document.

There are many different configurations and designs of GPR systems. The ones used for archaeology are usually composed by a control unit, the antenna box with one or multiple channels and single or multiple frequencies, an encoder (often attached to a wheel) and a laptop adapted for field surveys (Fig. 9.2). Generally, the antenna box contains both transmitter and receiver antennas. The control unit is the main part of the system as it controls the data acquisition, generates and synchronizes the pulses and all the needed control signals. It is connected to the laptop and ensures a high data transmission rate. Finally, the odometer wheel is often used for measuring the distance of the profiles acquired. Its advantages are low-cost, easy-to-use device and its rather high reliability (1–10 cm). The transmitted pulse is often triggered at constant distance intervals (few centimetres). Therefore, calibration of the wheel for each terrain is strongly recommended for mitigating factors which affect the accuracy of radar trace location such as microtopography and wet or snowy terrains.

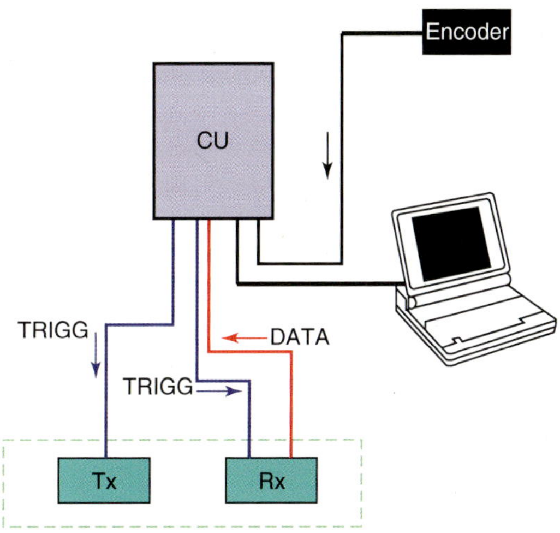

Fig. 9.2 Basic scheme of a GPR system. *Tx* transmitter, *Rx* receiver, *CU* control unit, *TRIGG* trigger

Positioning devices as GPS or total stations are sometimes added for very accurate data positioning.

Figure 9.3 shows two different approaches for different needs on archaeological prospection.

9.3 Basic 3D Imaging

GPR profiles generate two-dimensional data that are usually visualized as B-scans or radargrams (Fig. 9.4).

Fig. 9.3 *Left* single-antenna configuration for small areas. *Right* multi-antenna array system for large areas

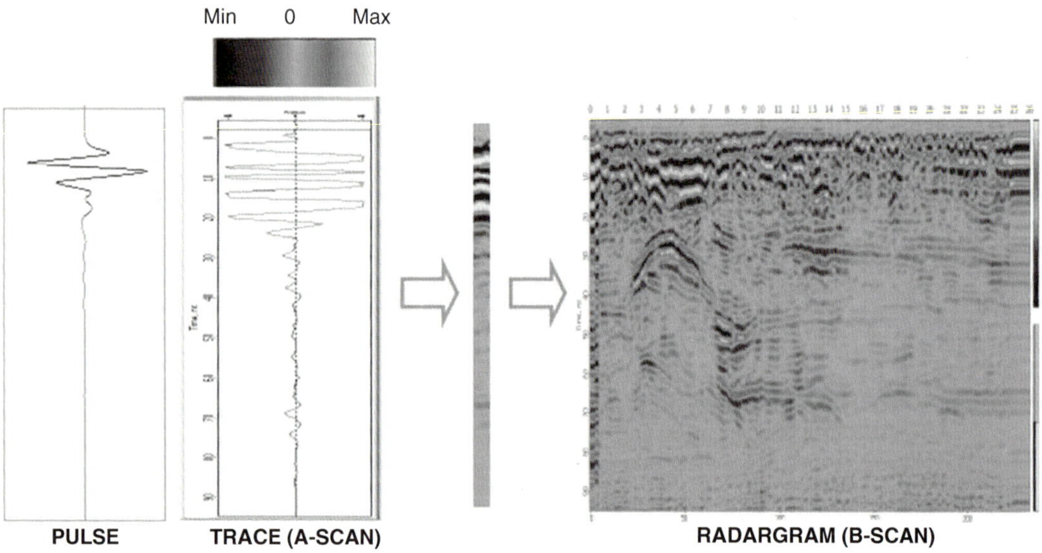

Fig. 9.4 From *left* to *right*: *Pulse* transmitted by the transmitter. *Trace (A-scan)* recorded by the receiver. *Radargram (B-scan)* representing a sequence of A-scans equidistantly spaced along the *X* axis

For obtaining 3D images, data are usually collected in parallel lines. Then, the resulting B-scans are assembled in order to generate a three-dimensional data volume (Fig. 9.5). Then, the resulted data cube can be sliced in any desirable direction (plane) resulting in images called slices, cuts or C-scans.

A standard way of analysing and presenting 3D GPR data is through horizontal time slices. They represent maps of the amplitudes of the recorded reflections across the survey area at a specified time (Fig. 9.6). Time to depth conversion is possible after estimation of the velocity of propagation of the radar wave in the host material.

These horizontal slice maps are often the most used because they can effectively show the size, shape, location and depth of buried targets. Subtle features that are difficult to understand on radargrams can be imaged and interpreted on

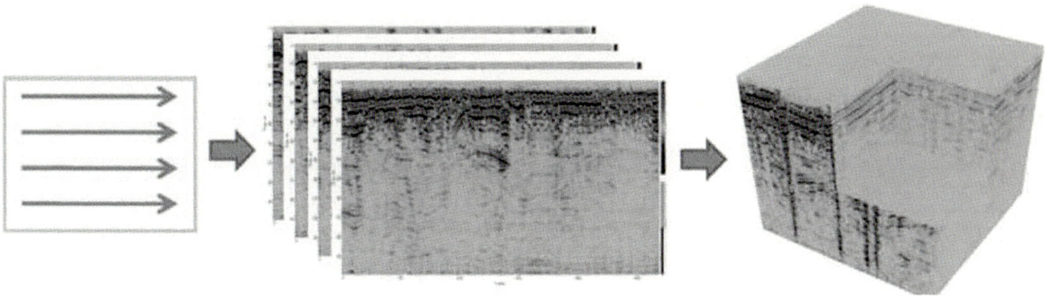

Fig. 9.5 From *left* to *right*: Grid plot of data collection shows four parallel profiles. The raw data acquired are four B-scans. During post-processing the B-scans can be compiled into a 3D volume (or C-scan)

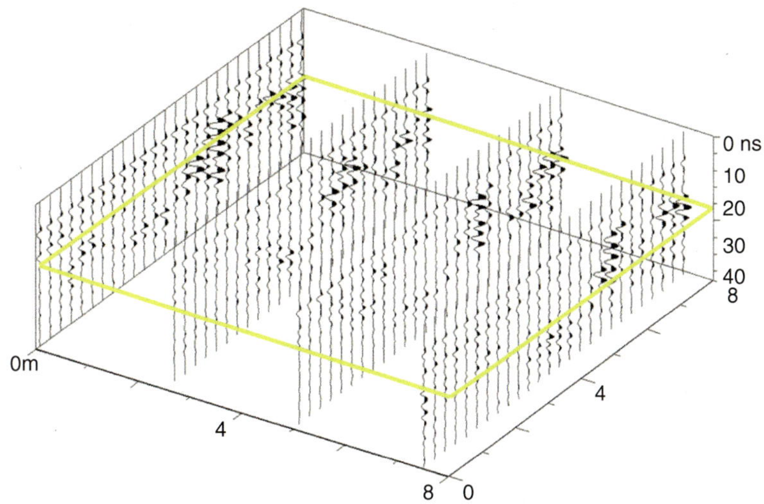

Fig. 9.6 Sketch of time slice formation (Figure adapted from Goodman 2012)

horizontal slice maps extracted from 3D volume datasets.

Horizontal slices can be generated by spatially averaging the squared wave amplitudes or by using the amplitude value of radar reflections. Data are usually gridded using different types of algorithms for data interpolation such as linear weight, square weight, inverse distance and kriging. On the other hand, very densely collected data do not need interpolation algorithms and final images are obtained faster by regularizing the slices in cells which dimensions are equal to trace spacing.

Images and animations of horizontal slices are usually generated as the final step for basic as well as advanced 3D data visualization (Watters 2006; Lin et al. 2011b; Trinks et al. 2012a).

9.4 3D GPR Methodologies for Archaeological Prospection

9.4.1 Standard 3D Surveying or Pseudo 3D

Present standards of 3D GPR in archaeological prospection are based on pseudo 3D methodologies which are characterized by a cross-line spacing which ranges from 0.25 to 1 m (being 0.5 m the most common), the use of 250–500 MHz antennas and vast interpolation to fill in the data gaps. Such methodologies along with powerful 3D visualization techniques are widely applied in GPR surveys with archaeological purpose (Leckebusch 2003; Goodman et al. 2006a, b). These surveys are usually image sites containing

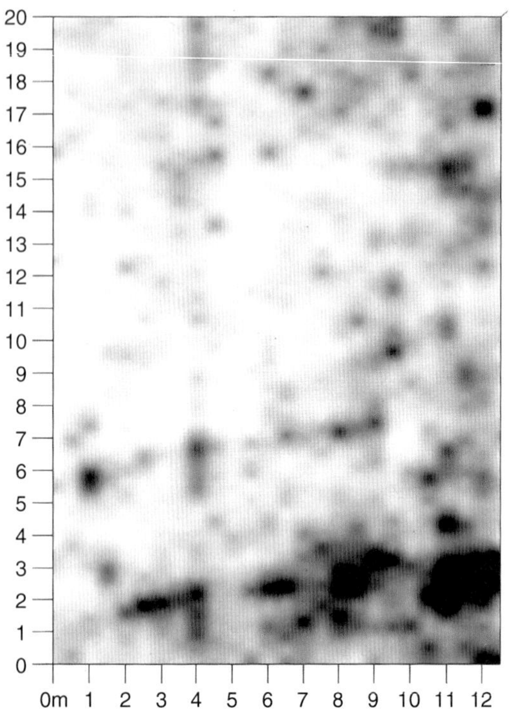

Fig. 9.7 Horizontal depth slice at 52 cm depth generated from a standard pseudo 3D surveying and processing. Some linear targets (*black colour*) are hardly visible

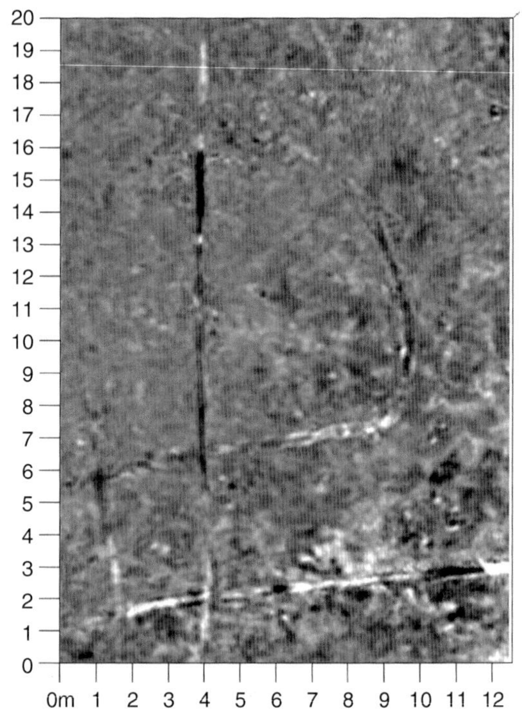

Fig. 9.8 Horizontal depth slice at 52 cm depth generated from full-resolution data acquisition and processing. Linear anomalies represent as small objects as 5 cm diameter PVC pipes and irrigation lines

continuous linear features extending over several meters length such as pipes, foundations, ditches, walls and roads (Fig. 9.7). However, archaeologists sometimes need to locate isolated features smaller than walls such as objects, pits, postholes, burials or cisterns (Gaffney 2008). Hence, the following questions emerge: What is being lost by decimating data acquisition and applying data interpolation? For example, could we obtain a much higher-resolution image of Fig. 9.7?

9.4.2 Full-Resolution or True 3D or Ultradense 3D Surveying

However, most archaeological surveyors have not yet pushed GPR to its full potential and experienced the benefits of maximum resolution achieved with very dense data acquisition and processing. Full-resolution GPR honouring spatial Nyquist sampling theorem have already been successfully utilized to obtain unaliased 3D images of heterogeneous subsurface geometries such as dune stratigraphy, tree roots and rock fractures (Grasmueck et al. 2005). High-resolution 3D images of the subsurface can be obtained if the space among traces is reduced to a quarter of the wavelength in the host material in all directions. In addition, a highly precise positioning of the GPR antenna during data acquisition is crucial, as it has been pointed out by other authors (Groenenboom et al. 2001) and (Lualdi et al. 2006). Otherwise post-processing corrections should be taken into account (Leckebusch 2003).

Figure 9.8 shows a depth slice over the same area as Fig. 9.7. The new full-resolution image clearly depicts linear targets as well as small objects. Novo et al. (2008) fully described the methods and results of this test.

The currently prevailing paradigm that archaeological GPR datasets are already being gathered with enough density seems to be the main reason why the applications of ultradense 3D GPR methodologies are still limited in archaeological exploration. Besides, other geophysical techniques (such as magnetometry (Zöllner et al. 2011) or electrical resistivity (Dabas 2008)), in adequate subsurface environments, can produce the same pseudo 3D image quality than GPR and resolve the main archaeological features in much less time (several hectares per day).

9.4.3 Other Methodologies

Another interesting approach is multi-offset GPR imaging. This methodology has already shown its capability of producing very high-resolution images (Pipan et al. 1999; Booth et al. 2008, 2010; Berard and Maillol 2008) although its current drawback is the extremely slowness 3D data acquisition which makes the methodology rarely utilized. However, new multichannel systems may speed up this type of data acquisition in the near future. As well, multipolarized data acquisition has been also approached in order to increase target detection sensitivity (Lualdi et al. 2006).

9.4.4 Best Practices

Independently of the data acquisition strategy selected, there are some common steps for best practices in 3D GPR for archaeological prospection. First of all, before selecting the most suitable strategy, a deep knowledge of the equipment in use is necessary. It will vary depending on several factors, above all, the size of the area, density of data acquisition and GPR system limitations.

On the other hand, accurate data positioning is pivotal in order to achieve good 3D imaging. Still, the most used positioning system is based on laying out grids using measurement tapes and the strings or spray marks. Even quite rudimentary, this method may be accurate if it is properly performed. On the other hand, a most advanced way of data positioning involves the combination of GPR with GPS which provides accurate trace positioning in a global reference coordinate system. Therefore, several authors have reported works related to this topic in archaeological prospection: Urbini et al. (2007) and Leckebusch (2005a, b). In places where GPS coverage is poor (i.e. urban environment, indoor areas), robotic total stations are used instead (Lehmann and Green 1999; Leckebusch 2005a, b; Gustafsson and Alkarp 2007; Tronicke and Böniger 2008). Other systems as rotary lasers have been used with the same purpose (Grasmueck and Viggiano 2007).

9.5 Current State-of-the-Art Technology: Multichannel Array Systems

Although full-resolution images can be generated with single-channel systems (Grasmueck et al. 2005; Novo et al. 2008) and can benefit the interpretation of subtle features over complex archaeological sites (Novo et al. 2012a, b), the extra time needed for collecting this type of data has minimized its use. Therefore, as mentioned in Sect. 9.4, the standard methodology is still based on coarse profile spacing data collection and subsequent interpolation to fill in the data gaps during three-dimensional processing (Novo et al. 2010).

Small array systems using two antennae in parallel have shown a potential simple solution (Leckebusch 2005a, b). In fact, Grasmueck et al. (2010) were able of halve the time needed for obtaining full-resolution images with a self-developed system (Grasmueck and Viggiano 2007). However, the use of such modular array systems for full-resolution surveying in large-scale archaeological prospection still presents some limitations (Verdonck and Vermeulen 2011).

The new generation of multichannel GPR instruments try to solve the spatial sampling bias while speeding up data acquisition. A paramount advantage of multichannel systems is that the full-resolution of GPR recording on the ground can be handled adequately by systems in which the antenna channel separation approaches distances close to a quarter of the wavelength of the transmitted wavelength of the central antenna frequency. These design characteristics have

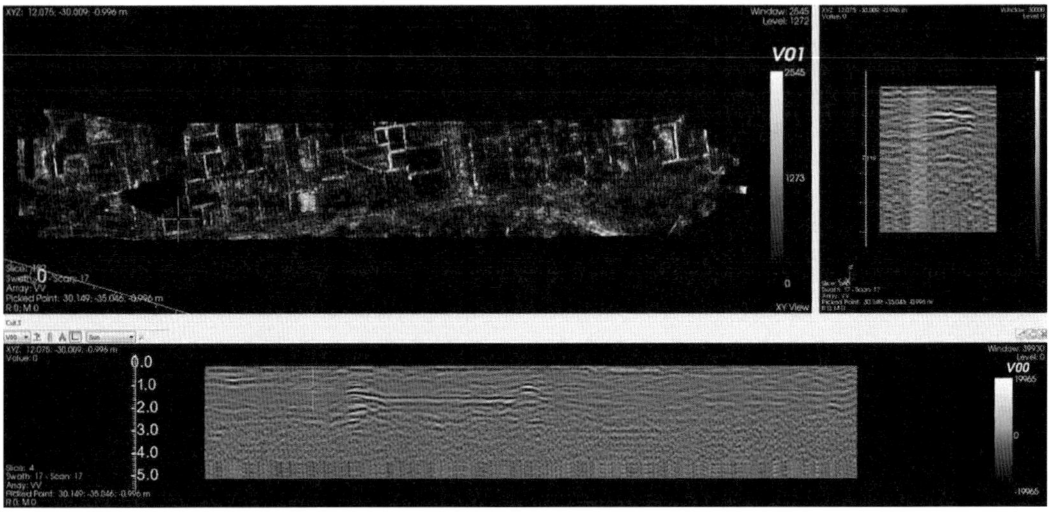

Fig. 9.9 Interactive view of depth slice at 0.996 m depth (*top left*), perpendicular section (*top right*) and longitudinal section (*bottom*). The area is approximately 1 ha and was collected in 2 h at 6 cm × 12 cm resolution

been accomplished by many of the current GPR manufacturers of multichannel systems.

Even though the first introduction of multichannel GPR systems dates back more than 15 years (Warhus et al. 1993), the complete acceptance of multichannel recording was limited by the quality of the data and complex data processing (Francese et al. 2009). Wildly different frequency responses of the multichannel antenna prevented useful amalgamation of the individual profiles into useful images (Novo et al. 2013). However, in the past few years, most of the multichannel manufacturers have provided GPR systems where the antenna responses of the individual elements are much closer (Linford et al. 2010; Trinks et al. 2010; Novo et al. 2011, 2012a, b). Figure 9.9 shows an example of data collected with the Stream X multichannel system in Empuries, Spain (Novo et al. 2012a, b). The horizontal depth slice at approximately 1 m shows with great detail (in white colour) many square and rectangular buildings and roads belonging to the urban subdivision of part of this Roman city.

Conclusions

3D GPR imaging is an excellent tool for non-invasively detecting and interpreting archaeological remains. Intuitive comprehension of geophysical data by archaeologists or historians can be very useful as well as cost-saving for planning most archaeological excavations, and this purpose is achieved if data are accurately positioned during sufficiently dense acquisition and then properly processed and displayed in three dimensions.

How dense 3D archaeological GPR surveys should be recorded is still open to debate. Results during the last two decades show no standard density scheme. Dense grid spacing is suggested but may be relaxed depending on the characteristics of the target (size, depth, orientation…), the scope of the survey (full-resolution or pseudo 3D imaging), the characteristics of the survey area and the equipment in use. If the GPR survey is done as a previous phase followed by immediate archaeological excavations of small areas, ultradense data acquisition may be too costly because subtle information can be already extracted by archaeologists during digging. However, in case excavations are not planned, archaeologists themselves ask for very detailed information of the subsurface which usually is only possible after acquiring data using full-resolution data acquisition. In fact, in the last 3–4 years, massive antenna array systems have been developed for making high detailed subsurface maps of large sites. The high density of antennas allows recording full wave-field data in a fraction of time needed previously. The capabilities of modern software tools permit an automated data processing producing final georeferenced results shortly after finishing the field survey. Data processing takes no more several times longer than the survey itself but is accomplished in a fraction of time spent in the field. Therefore, these systems show the underground targets very clearly and seem to be very adapted for mapping of large archaeological sites with high resolution in the shortest possible time.

For good 3D imaging, radar signal data processing must be careful from the very first filter applied. Nowadays, commercial software packages usually provide macros of filters to automatically produce a final image from raw data. The quality of the results depends upon proper data acquisition and favourable soil and topographic conditions.

How to visualize the final images is crucial for understanding and therefore final interpreting the data. Geological strata evolution or archaeological remains distribution at different depths are complex subsurface environments which can only be fully explore with 3D full-resolution images and special software tools for their visualization and interpretation. Animations of time slices and real-time visualization of 3D data cubes help to better understand GPR data. Unfortunately, this is impossible to display on printed documents.

Some of the ongoing and future research lines involve different aspects:

- *Data acquisition*: The use of multichannel systems permits combining different frequencies, orientations, transmitters and receivers and opens new ways of acquiring data.
- *Data positioning*: Integration of different systems for centimetre-precise positioning of continuous full-resolution mapping without present limitations regarding satellite coverage or range.
- *Software development and visualization*: Real-time 3D processing and visualization during massive data acquisition. Once generation of the best possible imaging is accomplished, the next goal should be to obtain more qualitative information about the underground materials.
- *Hardware development*: Robotization of the current systems for unmanned full-resolution data acquisition. New GPR antennas capable of sending out a beam and actually collecting a beam of information. Current GPR technology sends a beam (3D) and records a trace (1D), as explained in point 3 of this document.
- *Integration* with other geophysical (i.e. Electromagnetic) and non-geophysical sensors (i.e. Laser scanner).

References

Annan AP (2003) Ground penetrating radar. Principles, procedures and applications. Sensors & Software, Inc., Mississauga

Berard BA, Maillol J-M (2008) Common- and multi-offset ground-penetrating radar multi-offset study of a Roman villa, Tourega, Portugal. Archaeol Prospect 15:32–46

Bevan BW, Kenyon J (1975) Ground-penetrating radar for historical archaeology. MASCA Newslett 11:2–7

Böniger U, Tronicke J (2010a) Improving the interpretability of 3D GPR data using target–specific attributes: application to tomb detection. J Archaeol Sci 37(4):672–679

Böniger U, Tronicke J (2010b) Integrated data analysis at an archaeological site: A case study using 3D GPR, magnetic, and high-resolution topographic data. Geophysics 75(4):B169–B176

Böniger U, Tronicke J (2012) Subsurface utility extraction and characterization: combining GPR symmetry and polarization attributes. IEEE Trans Geosci Remote Sens 50(3):736–746

Booth AD, Linford NT, Clark RA, Murray T (2008) Three-dimensional, multi-offset ground-penetrating radar imaging of archaeological targets. Archaeol Prospect 15:93–112

Booth AD, Clark RA, Hamilton K, Murray T (2010) Multi-offset ground penetrating radar methods to image buried foundations of a medieval town wall, Great Yarmouth, UK. Archaeol Prospect 17(2): 103–116

Conyers LB (2004) Ground-penetrating radar for archaeology. AltaMira Press, Walnut Creek

Conyers LB (2006) Innovative ground-penetrating radar methods for archaeological mapping. Archaeol Prospect 13(2):139–141

Conyers LB (2012) Interpreting ground-penetrating radar for archaeology. Left Coast Press, Walnut Creek

Conyers LB, Goodman D (1997) Ground penetrating radar: an introduction for archaeologists. AltaMira Press, Walnut Creek

Conyers LB, Leckebusch J (2010) Geophysical archaeology research agendas for the future: some ground-penetrating radar examples. Archaeol Prospect 17(2): 117–123

Dabas M (2008) Theory and practice of the new fast electrical imaging system ARP. In: Campana S, Piro S (eds) Seeing the unseen: geophysics and landscape archaeology. CRC Press/Taylor & Francis Group, London, pp 105–126

Daniels DJ (2005) Ground penetrating radar. In: Kai Chang (ed) Encyclopedia of RF and microwave engineering. Wiley, New York

El-Baz F (1989) Remote sensing in archaeology: a case study. Sensors 6(10):33–38

Forkmann B, Petzold H (1989) Principle and application of the ground penetrating radar for researching of the near surface range. Freiberger Forschungshefte C(432). Deutscher Verlag für Grundstoffindustrie, Leipzig

Francese RG, Finzi E, Morelli G (2009) 3-D high-resolution multi-channel radar investigation of a Roman village in Northern Italy. J Appl Geophys 67(1):44–51

Gaffney C (2008) Detecting trends in the prediction of the buried past: a review of geophysical techniques in archaeology. Archaeometry 50(2):313–336

Gaffney C, Gaffney V et al (2012) The Stonehenge hidden landscapes project. Archaeol Prospect 19(2):147–155

Goodman D (1994) Ground-penetrating radar simulation in engineering and archaeology. Geophysics 59(2): 224–232

Goodman D (1996) Comparison of GPR time slices and archaeological excavations. In: Proceedings of the 6th international conference on ground penetrating radar (Sendai, Japan, September 30-October 3, 1996). Department of Geoscience and Technology, Tohuku University, Sendai, pp 77–78

Goodman D (2012). GPR-SLICE v7.0 manual. Geophysical Archaeometry Laboratory Inc. http://www.gpr-survey.com

Goodman D, Piro S (2013) GPR remote sensing in archaeology, vol 9, Geotechnologies and the environment. Springer, New York

Goodman D, Nishimura Y, Daniel Rogers J (1995) GPR time slices in archaeological prospection. Archaeol Prospect 2(2):85–89

Goodman D, Piro S, Nishimura Y (2002) GPR time slice images of the villa of emperor Trajanus, Arcinazzo, Italy (AD 52–117). In: Koppenjan S, Hua Lee (eds) Proceedings of the 9th international conference on ground penetrating radar, vol 4758, SPIE, pp 268–272

Goodman D, Nishimura Y, Hongo H, Higashi N (2006a) Correcting for topography and the tilt of ground-penetrating radar antennae. Archaeol Prospect 13(2):157–161

Goodman D, Steinberg J, Damiata B, Nishimura Y, Schneider K, Hongo H, Higashi N (2006b) GPR overlay analysis for archaeological prospection. In: Daniel JJ (ed) Proceedings of the 11th international conference on ground penetrating radar (Columbus, Ohio, June 19–21, 2006). Ohio State University, Columbus, CD-ROM

Goodman D, Hongo H et al (2007) GPR surveying over burial mounds: correcting for topography and the tilt of the GPR antenna. Near Surf Geophys 5(6): 383–388

Grandjean G, Gourry J-C (1996) GPR data processing for 3D fracture mapping in a marble quarry (Thassos, Greece). J Appl Geophys 36(1):19–30

Grasmueck M (1996) 3-D ground-penetrating radar applied to fracture imaging in gneiss. Geophysics 61(4):1050–1064

Grasmueck M, Green AG (1996) 3-D georadar mapping: looking into the subsurface. Environ Eng Geosci 2(2):195–200

Grasmueck M, Viggiano DA (2006) 3D/4D GPR toolbox and data acquisition strategy for high-resolution imaging of field sites. In: Daniel JJ (ed) Proceedings of the 11th international conference on ground penetrating radar (Columbus, Ohio, June 19–21, 2006). Ohio State University, Columbus, CD-ROM

Grasmueck M, Viggiano DA (2007) Integration of ground-penetrating radar and laser positioning sensors for real-time 3-D data fusion. IEEE Trans Geosci Remote Sens 45(1):130–137

Grasmueck M, Weger R, Horstmeyer H (2004a) Full-resolution 3D imaging for geoscience and archeology. In: Tenth international conference on ground penetrating radar (Delft, The Netherlands, 21–24 June 2004), vol 1, pp 329–332

Grasmueck M, Weger R, Horstmeyer H (2004b) Three-dimensional ground-penetrating radar imaging of sedimentary structures, fractures, and archaeological features at submeter resolution. Geology 32(11):933–936

Grasmueck M, Weger R, Horstmeyer H (2005) Full-resolution 3D GPR imaging. Geophysics 70(1): K12–K19

Grasmueck M, Marchesini P, Eberli GP, Zeller M, van Dam RL (2010) 4D GPR tracking of water infiltration in fractured high-porosity limestone. In: Proceedings of 13th international conference on ground penetrating radar (Lecce, Italy, 19–22 June 2010), pp 1–6

Groenenboom J, van der Kruk J, Harry Zeeman J (2001) 3D GPR data acquisition and the influence of positioning errors on image quality. In: 63rd EAGE conference and technical exhibition (Amsterdam, 11–15 June 2001)

Gustafsson J, Alkarp M (2007) Array GPR investigation of the cathedral of Uppsala. Near Surf Geophys 5(3):203–207

Imai T, Sakayama T, Kanemori T (1987) Use of ground-probing radar and resistivity survey for archaeological investigations. Geophysics 52(2):137–150

Kenyon JL, Bevan B (1977) Ground-penetrating radar and its application to a historical archaeological site. Hist Archaeol 11:48–55

Leckebusch J (2000) Two- and three-dimensional ground-penetrating radar surveys across a medieval choir: a case study in archaeology. Archaeol Prospect 7: 189–200

Leckebusch J (2003) Ground-penetrating radar: a modern three-dimensional prospection method. Archaeol Prospect 10:213–240

Leckebusch J (2005a) Precision real-time positioning for fast geophysical prospection. Archaeol Prospect 12(3):199–202

Leckebusch J (2005b) Use of antenna arrays for GPR surveying in archaeology. Near Surf Geophys 3(2):107–111

Leckebusch J (2007) Pull-up/pull-down corrections for ground-penetrating radar data. Archaeol Prospect 14(2): 142–145

Leckebusch J, Peikert R (2001) Investigating the true resolution and three-dimensional capabilities of ground-penetrating radar data in archaeological surveys: measurements in a sand box. Archaeol Prospect 8(1):29–40

Leckebusch J, Rychener J (2007) Verification and topographic correction of GPR data in three dimensions. Near Surf Geophys 5(6):395–403

Leckebusch J, Weibel A, Bühler F (2008) Semi-automatic feature extraction from GPR data for archaeology. Near Surf Geophys 6:75–84

Lehmann F, Green AG (1999) Semiautomated georadar data acquisition in three dimensions. Geophysics 64(3):719–731

Leucci G (2002) Ground-penetrating radar survey to map the location of buried structures under two churches. Archaeol Prospect 9(4):217–228

Leucci G, Negri S (2006) Use of ground penetrating radar to map subsurface archaeological features in an urban area. J Archaeol Sci 33:502–512

Lin AY-M, Novo A, Har-Noy S, Ricklin ND, Stamatiou K (2011a) Combining GeoEye-1 satellite remote sensing, UAV aerial imaging, and geophysical surveys in anomaly detection applied to archaeology. IEEE J Sel Top Appl Earth Observations and Remote Sens 4(4):870–876

Lin AY-M, Novo A, Weber PP, Morelli G, Goodman D, Schulze JP (2011b) A virtual excavation: combining 3D immersive virtual reality and geophysical surveying. In: Boyle R, Parvin B, Koracin D, Paragios N, Tanveer S-M, Tao Ju, Zicheng Liu, Coquillart S, Cruz-Neira C, Möller T, Malzbender T (eds) Advances in visual computing, vol 6939, Lecture notes in computer science. Springer, New York/ Heidelberg, pp 229–238

Linford N (2004) From hypocaust to hyperbola: ground-penetrating radar surveys over mainly Roman remains in the UK. Archaeol Prospect 11(4):237–246

Linford N, Linford P, Martin L, Payne A (2010) Stepped frequency ground-penetrating radar survey with a multi-element array antenna: results from field application on archaeological sites. Archaeol Prospect 17(3):187–198

Lorenzo H, Arias P (2005) A methodology for rapid archaeological site documentation using ground-penetrating radar and terrestrial photogrammetry. Geoarchaeology 20(5):521–535

Lorenzo H, del Carmen Hernández M, Cuellar V (2002a) Selected radar images of man-made underground galleries. Archaeol Prospect 9(1):1–7

Lorenzo H, Arias P et al (2002b) Ground-based radar, close-range photogrammetry and digital terrain data applied together to archaeological heritage documentation. In: Koppenjan S, Hua Lee (eds) Proceedings of the ninth international conference on ground penetrating radar, vol 4758, SPIE, pp 527–532

Lualdi M, Zanzi L (2003) 3D GPR investigations on building elements using the PSG. In: Proceedings of the 16th EEGS symposium on the application of geophysics to engineering and environmental problems (San Antonio, Texas, 6–10 April 2003), pp 131–152

Lualdi M, Zanzi L, Binda L (2003) Acquisition and processing requirements for high quality 3D reconstructions from GPR investigation. In: Proceedings of the international symposium on non-destructive testing in civil engineering (Berlin, September 16–19, 2003)

Lualdi M, Zanzi L, Sosio G (2006) A 3D GPR survey methodology for archaeological applications. In: Daniel JJ (ed) In proceedings of the 11th international conference on ground penetrating radar (Columbus, Ohio, June 19–21, 2006). Ohio State University, Columbus, CD-ROM

Negri S, Leucci G (2006) Geophysical investigation of the Temple of Apollo (Hierapolis, Turkey). J Archaeol Sci 33(11):1505–1513

Negri S, Leucci G, Mazzone F (2008) High resolution 3D ERT to help GPR data interpretation for researching archaeological items in a geologically complex subsurface. J Appl Geophys 65(3–4):111–120

Neubauer W, Eder-Hinterleitner A, Seren S, Melichar P (2002) Georadar in the Roman civil town Carnuntum, Austria: an approach for archaeological interpretation of GPR data. Archaeol Prospect 9(3):135–156

Novo A, Grasmueck M, Viggiano DA, Lorenzo H (2008) 3D GPR in archaeology: what can be gained from dense data acquisition and processing? In: Proceedings of 12th international conference on ground penetrating radar (Birmingham, UK, 16–19 June 2008), CD-ROM

Novo A, Lorenzo H, Rial FI, Solla M (2010) Three-dimensional ground-penetrating radar strategies over an indoor archaeological site: convent of Santo Domingo (Lugo, Spain). Archaeol Prospect 17(4):213–222

Novo A, Sala R, Morelli G, Leckebusch J, Trilla JT (2011) Full wave-field recording: STREAM X at Empuries (Girona, Spain). In: Drahor MG, Berge MA (eds) 9th International conference on archaeological prospection (Izmir, Turkey, September 19–24, 2011), pp 213–217

Novo A, Lorenzo H, Rial FI, Solla M (2012a) From pseudo-3D to full-resolution GPR imaging of a complex Roman site. Near Surf Geophys 10:11–15

Novo A, Dabas M, Morelli G (2012b) The STREAM X multichannel GPR system first test at Vieil-Evreux (France) and comparison with other geophysical data. Archaeol Prospect 19:179–189

Novo A, Leckebusch J, Goodman D, Morelli G, Piro S, Catanzariti G (2013) Advances in GPR imaging with multi-channel radar systems. J Surv Mapp Eng 1(1):7–14

Nuzzo L, Leucci G, Negri S, Carrozzo MT, Quarta T (2002) Application of 3D visualization techniques in the analysis of GPR data for archaeology. Ann Geophys 45(2):321–338

Pipan M, Baradello L, Forte E, Prizzon A, Finetti I (1999) 2-D and 3-D processing and interpretation of multifold ground penetrating radar: a case history from an archaeological site. J Appl Geophys 41(2):271–292

Piro S, Goodman D, Nishimura Y (2003) The study and characterization of Emperor Traiano's Villa (Altopiani di Arcinazzo, Roma) using high-resolution integrated geophysical surveys. Archaeol Prospect 10(1): 1–25

Qiang B, Chang Y (1985) Experimental underground radar for historical archaeology. In: IEEE international on symposium on electromagnetic compatibility. Institute of Electrical and Electronics Engineers, New York, pp 359–361

Solla M, Lorenzo H, Novo A, Rial FI (2010) Ground-penetrating radar assessment of the medieval arch bridge of San Anton, Galicia, Spain. Archaeol Prospect 17:223–232

Stove GC, Addyman PV (1989) Ground probing impulse radar: an experiment in archaeological remote sensing at York. Antiquity 63(239):337–342

Trinks I, Johansson B, Gustafsson J, Emilsson J, Friborg J, Gustafsson C, Nissen J, Hinterleitner A (2010) Efficient, large-scale archaeological prospection using a true three-dimensional ground-penetrating radar array system. Archaeol Prospect 17(3):175–186

Trinks I, Kucera M, Hinterleitner A, Löcker K, Nau E, Neubauer W, Zitz T (2012a) Large-scale, high-definition ground penetrating radar prospection in archaeology. In: EGU general assembly conference abstracts, vol 14, EGU2012-13447. EGU General Assembly, Vienna

Trinks I, Neubauer W, Doneus M (2012b) Prospecting archaeological landscapes. In: Ioannides M et al (eds) Progress in cultural heritage preservation. Proceedings of the 4th international conference, EuroMed 2012 (Limassol, Cyprus, October 29 – November 3, 2012). Springer, Berlin/Heidelberg, pp 21–29

Tronicke J, Böniger U (2008) Efficient and flexible TPS based 3-D GPR surveying: An archaeological case study. Proceedings of 12th International Conference on Ground Penetrating Radar, June 16-19, Birmingham, UK. CD-ROM

Ulriksen CPF (1982) Application of impulse radar to civil engineering. Department of Engineering Geology, University of Technology, Lund

Urbini S, Cafarella L, Marchetti M, Chiarucci P, Bonini D (2007) Fast geophysical prospecting applied to archaeology: results at "Villa ai Cavallacci" (Albano Laziale, Rome) site. Ann Geophys 50(3):291–299

Van Deen JK, de Feijter JW (1992) Three-dimensional ground probing radar. Geolog Surv Finland, Special Paper 16:35–40

Vaughan CJ (1986) Ground penetrating radar surveys used in archaeological investigations. Geophysics 51(3):595–604

Verdonck L, Vermeulen F (2011) 3-D GPR survey with a modular system: reducing positioning inaccuracies and linear noise. In: Drahor MG, Berge MA (eds) 9th international conference on archaeological prospection (Izmir, Turkey, September 19–24, 2011), pp 204–212

Vickers RS, Dolphin LT (1975) A communication on an archaeological radar experiment at Chaco Canyon, New Mexico. MASCA Newslett 11(1):6–8

Vickers RS, Dolphin LT, Johnson D (1976) Archaeological investigations at Chaco Canyon using subsurface radar. In: Lyons TR (ed) Remote sensing experiments in cultural resource studies. U.S. Department of the Interior/National Park Service and University of New Mexico, Albuquerque, pp 81–101

Warhus JP, Mast JE, Nelson SE, Johansson EM (1993) Ground-penetrating imaging radar development for bridge deck and road bed inspection. Technical report for the US Department of Energy. Lawrence Livermore National Laboratory, Livermore

Watters M (2006) Geovisualization: an example from the Catholme Ceremonial Complex. Archaeol Prospect 13(4):282–290

Zöllner H, Kniess R, Meyer C, Trinks I (2011) Efficient large-scale magnetic prospection using multi-channel fluxgate arrays and the new digitizer LEA D2. In: Drahor MG, Berge MA (eds) 9th international conference on archaeological prospection (Izmir, Turkey, September 19–24, 2011), pp 201–203

Interpretation and Guidelines for Reporting

10

Cornelius Meyer

Contents

10.1 Introduction

In the transaction of science and humanities, inaccurate acting in the field of archaeological interpretation of geophysical data causes hardly reparable damages to mutual trust, an indispensable precondition in interdisciplinary research. A proper interpretation of data not only requires being native in one's own special field, in this case in the field of applied geophysics, but it also challenges qualities of a versed translator. Since geophysical works in archaeology may be related to manifold kind of archaeological sites starting from Palaeolithic times up to contemporary history, a geophysicist is neither supposed to be familiar with all archaeological subdisciplines nor it is expected that geophysicists develop comprehensive visions of archaeological sites or landscapes based only on the geophysical data.

The required additional armamentarium for a serious archaeological interpretation of geophysical data has to encompass a general interest in the physical expressions of human activity, an understanding of natural and anthropogenic origins of shaping landscapes and an overall view on archaeological categories and terms. Thus, an interpretation that can be used by archaeologists should entangle from the concept of geophysical anomalies and has to be proved by embedding it into the context of a man-made landscape. Mutual trust and transparency between geophysicists and archaeologists are the best concepts for an optimal flow of information benefiting for both disciplines. If all the mentioned requirements

C. Meyer
Eastern Atlas GmbH & Co. KG, Berlin, Germany
e-mail: cornelius@eastern-atlas.com

C. Corsi et al. (eds.), *Good Practice in Archaeological Diagnostics*, Natural Science in Archaeology,
DOI 10.1007/978-3-319-01784-6_10,

can be matched, the geophysical prospection in archaeological research becomes a real interdisciplinary approach, mainly in Anglo-American literature called "archaeogeophysics" (Kvamme 2003).

Besides the interpretation, the reporting on field measurements, data processing and interpretation are self-evident necessities, like in any other kind of scientific work. To avoid an artificial ranking, the minimum standards of an archaeogeophysical report are not depending on the entity carrying out the work, if a private contractor or a research institute is involved. The quality of reporting and archiving the data is crucial for a long-term usability of the data and the derived knowledge. The inescapable evanescence of archaeological remains caused by intensive agriculture, mining and human construction activity will make many datasets of today to priceless evidence of human history in the future (García Sanjuán and Wheatley 2002).

10.2 Interpretation of Geophysical Data

10.2.1 Basic Principles

Interpretation of geophysical data of archaeological surveys must be undertaken by qualified scientists familiar with applied geophysics as well as the geomorphological and archaeological conditions of the investigated site. Consultation of other specialists like landscape archaeologists, geologists or geographers is highly recommended.

From interpretation maps and textual descriptions, the degree of determination must emerge, e.g. a distinction of scientifically demonstrable results and assumptions based on scientific knowledge. The lack of geophysical anomalies does not necessarily mean negative evidence or a lack of archaeological features. The possible reasons for missing geophysical anomalies must be explained and other methods have to be considered to clear up the archaeological situation.

General guidelines for the integration of geophysical prospection techniques, from survey layout to reporting and archiving the data, were developed mainly in Great Britain and Central Europe. Their most comprehensive presentation was published by the English Heritage (Jones 2008).

Table 10.1 Sources of a priori information in archaeological surveys

Natural factors	Anthropogenic factors
Geology of the basement	Landscape history
Soil types	Archaeological data (surface data or excavations, historical sources)
Geomorphology	Agricultural practice (ploughing)
Topography	Crop plants
Surface conditions	Sources of modern interferences
Natural vegetation	Survey parameters
Weather conditions	Data treatment

10.2.2 A Priori Information

Starting point of all archaeological interpretation work is relating the geophysical data with all relevant a priori information. In general, natural and anthropogenic (artificial) factors can be differentiated. Table 10.1 shows a list of parameters to be considered without claiming to be exhaustive.

Any additional information coming from aerial photographs, current and historical maps and historical tradition can be taken into consideration with remarks related to the verification of their sources.

The outcome of the interpretation work can be presented in many ways. Basic requirements are maps in a suited scale (usually 1:500 to 1:2,000) with colour-coded drawings of interpreted structures and a scripted explanation of the results. Maps should show the limits of the surveyed areas and include a key of the applied colour and texture scheme. The drawings should clearly indicate the differences between explicit information directly derived from the geophysical data and assumptions deduced from the context of the surveyed site or landscape.

In case different geophysical methods were applied on a site, it is expedient to present individual interpretation maps for all datasets as well as a summary of the most important findings. Realisation and design of interpretation maps depend on the complexity of the

Fig. 10.1 Magnetic data and interpretation in GIS, *Ammaia*, Portugal (Meyer et al. 2012)

datasets and the dimensions of the surveyed areas. If there is a high density of both anthropogenic and natural structural elements, individual classified maps are recommended. Large areas of dozens of hectares (more and more common in landscape archaeology) should be divided in comprehensible sectors, accompanied by a small-scale overview map.

The elaboration of the presentation of both data and interpretation has to be made using preferentially vector maps (Fig. 10.1). When using GIS, the vital advantage is that attributions of interpreted structures can be added, discarded or changed easily if further data (from direct archaeological works or other survey methods) is added to the survey project. Open source desktop GIS like QuantumGIS provide efficient tools for the treatment of any georeferenced data and can already be applied in the field for merging and presentation of relevant data (Robinson et al. 2011). Crucial in the application of any vector-based interpretation software is the availability of reliable transformation tools for coordinate systems and their projections.

10.2.3 Magnetic Data

The archaeological interpretation of magnetic mapping data typically includes several intertwining steps (Figs. 10.2 and 10.3). In most cases of magnetic surveys at archaeological sites and

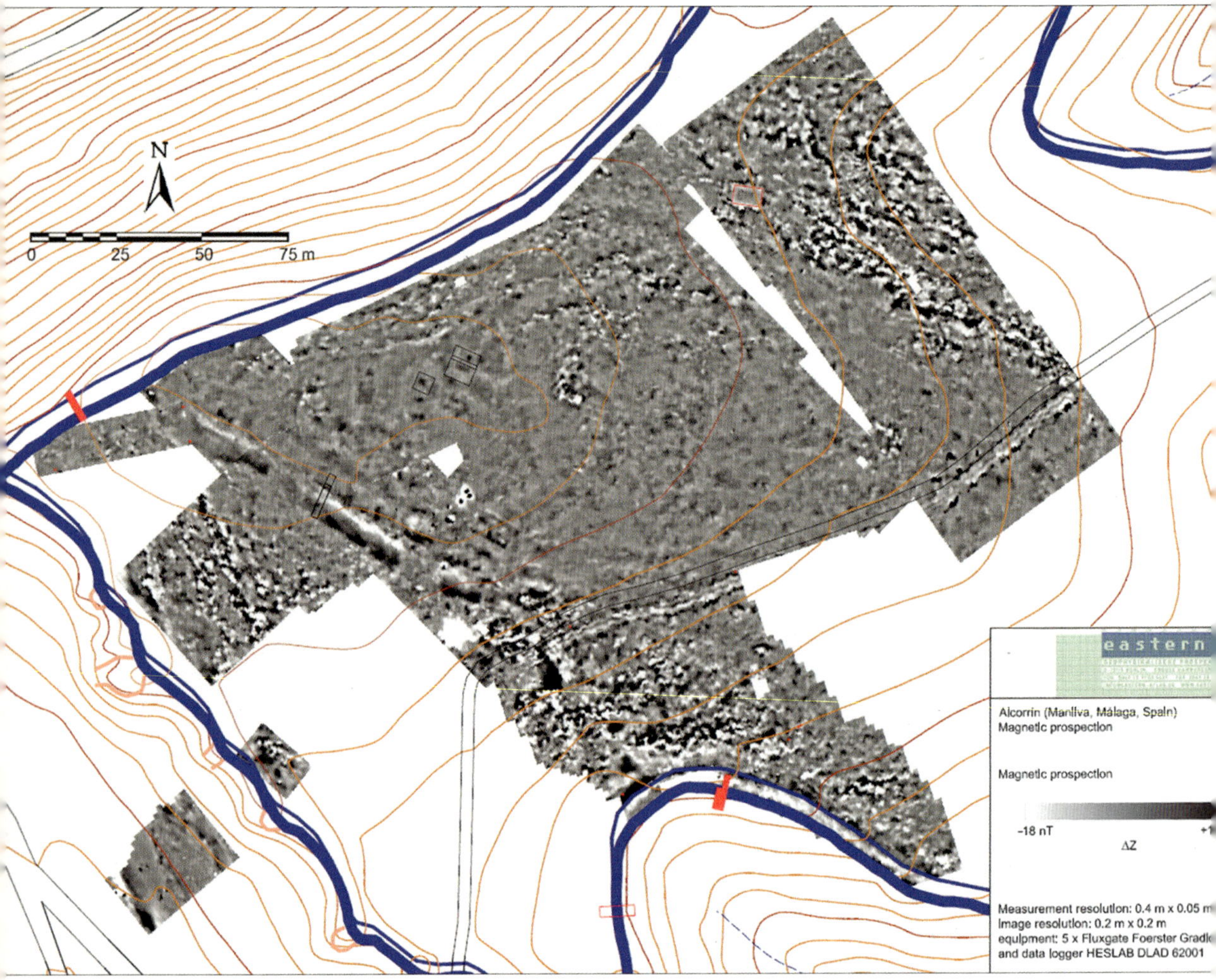

Fig. 10.2 Topographic map and magnetic data, Alcorrín, Spain (Marzoli et al. 2010)

landscapes, there is a certain part of anomalies caused by the modern overprint of the terrain (pipelines, scrap metal, agricultural impact). For that reason the first stage of interpretation is the separation of all magnetic anomalies of undoubtedly modern origin. If this strategy is adopted, it is most helpful to record all sources of magnetic interferences visible at the surface already while working in the field (fences, large metal parts, gullies, traces of excavations, etc.). Another effective possibility for a clear separation of these anomalies is a metal detector survey previous to the actual magnetic survey.

The classification of anomalies continues by separating magnetic anomalies referring to the geological underground. The visibility of deeper underground structures depends on the methodology. While gradient data accentuate near-surface structures, total field measurements are more sensitive in relation to deeper geological features. Studying the geological and geomorphological features is essential to assess their archaeological relevance.

Regarding the typically small-scale anomalies of actual archaeological origin, it is important to use data images of variable greyscale dynamics in order to verify the amplitude range of the features, primarily recognised by their geometry. In order to avoid the emergence of wishful patterns, one strategy is the subdivision of large areas in quadrants of reasonable size, whereby the features can be drawn for each quadrant

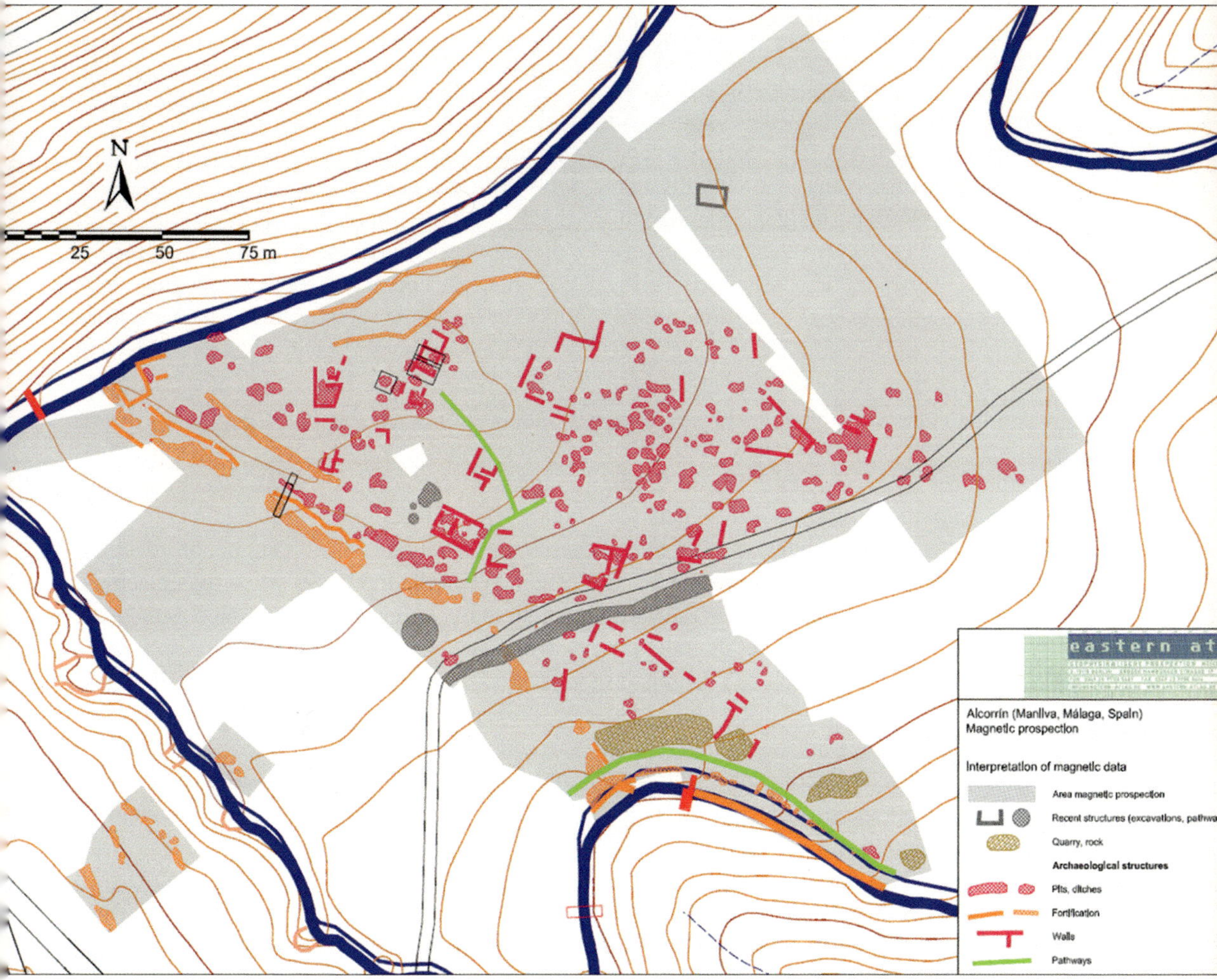

Fig. 10.3 Topographic map and interpretation of magnetic data, Alcorrín, Spain (Marzoli et al. 2010)

separately. In the work flow the plausibility of the interpretation has to be proven by reviewing the overall interpretation map. Especially at surveys of complex archaeological sites with an abundance of magnetic anomalies, this method is highly recommended.

It is important to add that not only the individual anomaly has to be considered: the observation of both distribution and density of magnetic anomalies can also contribute to the archaeological interpretation (Gaffney et al. 2000).

The imaging of archaeological features in magnetic data can be manifold. Several detailed publications are available to learn about the magnetic behaviour of soils, rocks and anthropogenic materials (Faßbinder 1994; Clark 1996; Neubauer 2001; Gaffney and Gater 2003; Aspinall et al. 2008).

10.2.4 GPR Data

The interpretation of GPR data requires not only experience in archaeological data interpretation but also a profound knowledge on theory and practice of geophysical wave-based methods. Instructive articles, monographs and guidelines have been published in the last decade (Leckebusch 2001; Neubauer et al. 2002; Linford et al. 2010; Conyers 2012).

The standard image for GPR data is a two-dimensional vertical profile, displaying depth on the vertical axis and distance along the ground on the horizontal axis. The analysis of single vertical profiles allows the interpretation of reflective structures, however, the spatial context of archaeological features is hardly recognisable.

More comprehensive analysis of GPR data offers the calculation and display of horizontal amplitude slices (Fig. 10.4). When creating time slices the data of parallel or other area-covering GPR profiles is aggregated in a three-dimensional dataset. First, the individual profiles have to be processed in order to enhance the signal-noise ratio and to normalise the data. For slice-mapping the resulting 3D dataset is cut into sections of wave amplitudes at determined travel times. Usually the amplitudes of the GPR signals are added up in a defined time window. The thickness of the slices is based on the travel time and can be converted in real depth values, if the wave velocity of the energy movement through the material is calculated or estimated. Amplitude slices represent the spatial distribution of all reflected wave amplitudes indicating changes in the specific properties of sediments, soils and buried objects. Amplitude slices are not necessarily constructed horizontally. Orientation and thickness of slices can vary depending on the archaeological question.

In order to convert travel time-based amplitude slices to depth slices, a knowledge of the specific wave velocities in the ground is needed. For velocity estimation several ways are available. In case of open excavations, the depth of reflectors in the radargram can directly be determined and used for the calculation of an average velocity of the overlying materials. Without direct access to the reflecting layers, velocity can be determined by analysing the shape of clearly recognisable reflection hyperbolas. Another more complex method is the application of common midpoint measurements using two separate antennas (Conyers and Lucius 1996; Jol 2008). Most GPR manufacturers offer appropriate software packages for all steps of processing of GPR data, including velocity analysis and the creation of amplitude slices, together with the technical equipment. Additionally, there are comprehensive programmes created for data processing of all kinds of GPR data (*REFLEXW*, *GPR-SLICER*).

When interpreting amplitude slices, the general guideline is that high amplitudes show boundaries where the decisive material parameters for electromagnetic wave propagation change. Sharp contrasts, as observed at the boundary between soil matrix and firm construction material or metal objects produce high amplitudes, thus, architectural elements like walls, foundations and cavities are relatively well visible in GPR data. Boundaries of stratigraphic layers differing in the content of clay, water, organic material or layers of different soil texture in many cases show only weak reflection amplitudes. The absence of any reflection signal patterns in amplitude slices can be of manifold origin. When the top layers are of high electrical conductivity, the propagation of electromagnetic waves into deeper layers of the ground is restricted. In such cases, GPR prospection should be dropped and replaced by other prospection methods (magnetic or electrical). Therefore, a study of the geological conditions and soil properties is highly recommended before starting extensive GPR measurements.

When interpreting GPR data, it is suggested that interpretation maps of individual amplitude slices are separately elaborated and displayed (Fig. 10.5). In case of complex 3D structures, another step of abstraction is possible. By defining threshold values for reflection amplitudes and "picking" reflection in different depths, 3D bodies of archaeological structures can be derived from GPR data (Fig. 10.6). An appropriate tool for the presentation of such interpretation maps is the open source and multiplatform application ParaView. These three-dimensional threshold maps may serve as a base for 3D models of archaeological structures (Meyer et al. 2007).

The interpretation within the investigation of mainly regular architecture in the field of classical and medieval archaeology is relatively well manageable due to the mainly sharp contrasts in the material parameters. When investigating prehistoric sites with apparently irregular patterns of post holes, pit houses, ditches and thin cultural layers, the interpretation of GPR data requires a

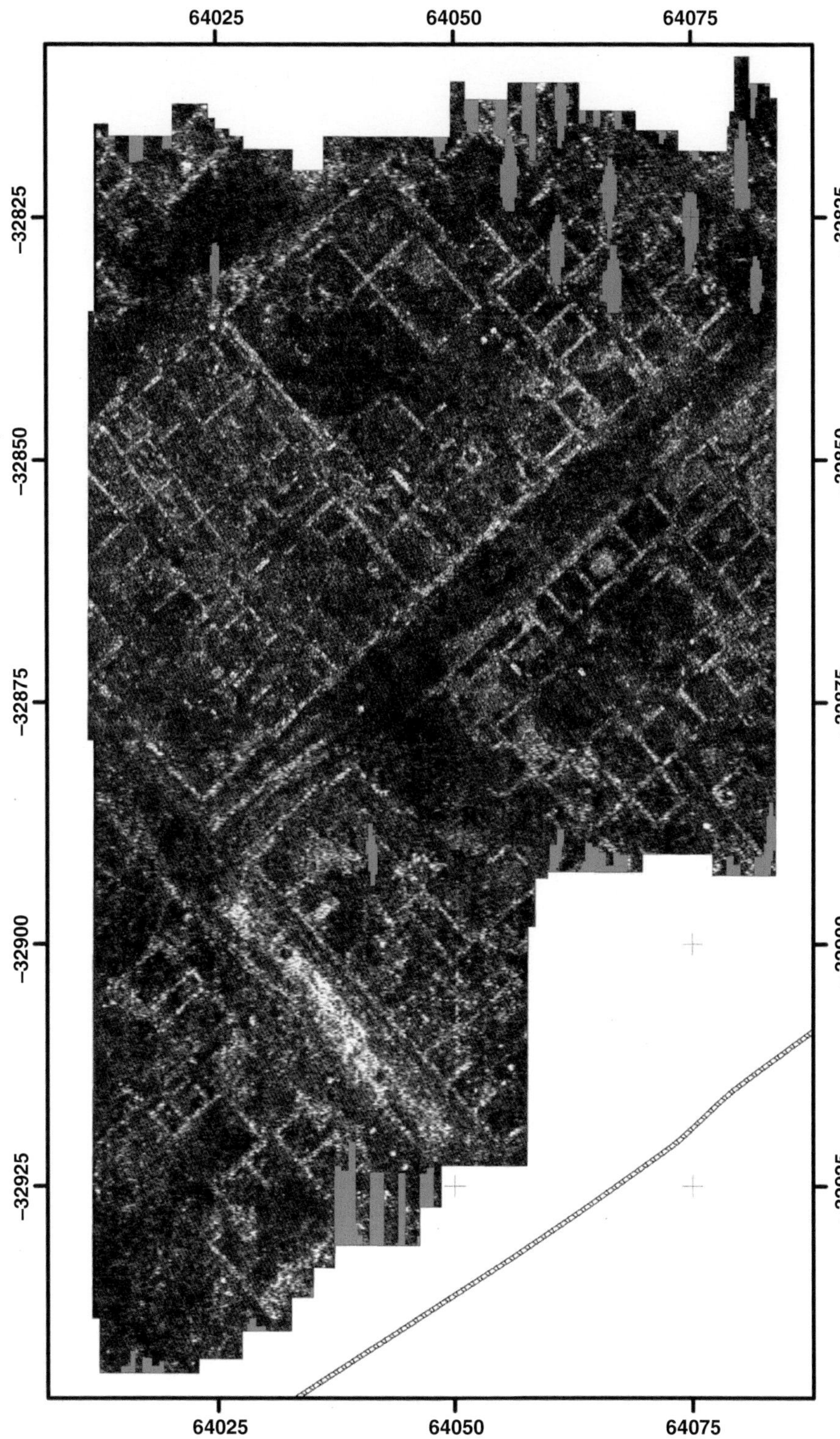

Fig. 10.4 Amplitude slice of GPR data, *Ammaia*, Portugal (Verdonck and Taelman 2012)

Fig. 10.5 Interpretation of one depth slice of GPR data, *Ammaia*, Portugal (Verdonck and Taelman 2012)

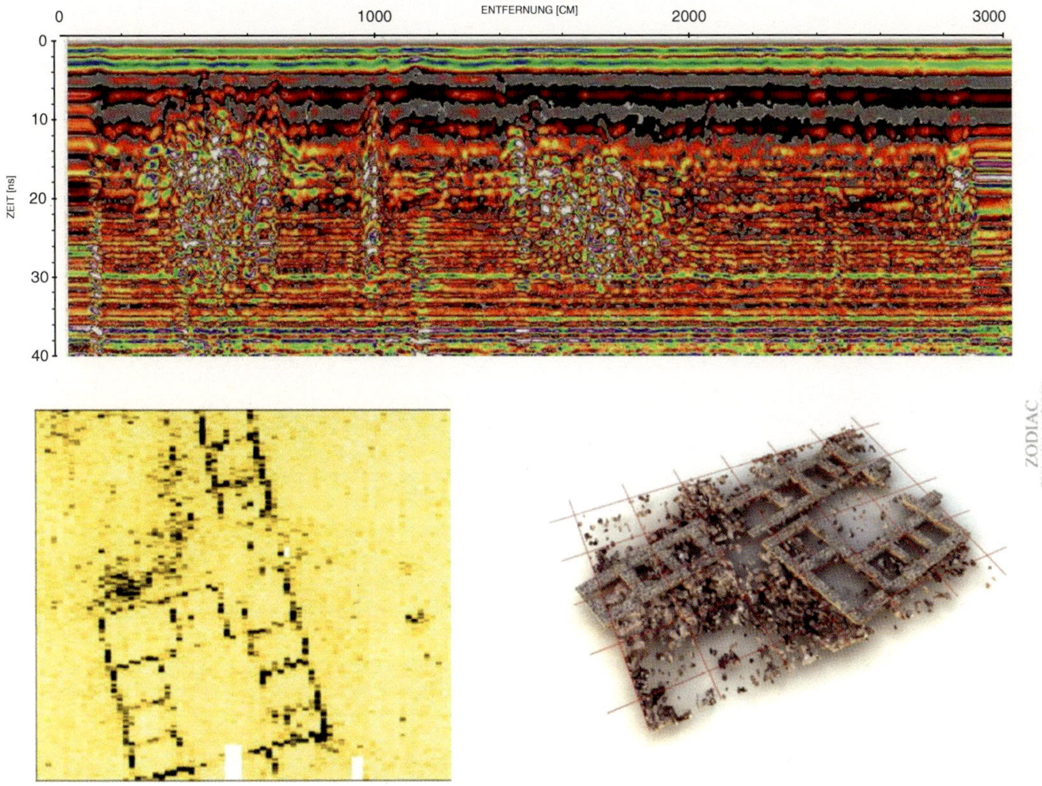

Fig. 10.6 GPR data presentation: single 2D profile, amplitude slice and 3D reconstruction (Meyer et al. 2007)

high level of experience in archaeological prospection and knowledge on the physical principles of electromagnetic wave propagation. The uncertainty of GPR data interpretation, especially in the case of low reflectivity contrasts, can be reduced by the integration of subsurface interpretation from ground truthing (cores, excavations or augers).

10.2.5 Electrical Resistivity Data

Electrical resistivity data can be presented as two-dimensional mapping data, vertical profiles and three-dimensional tomography data. Mapping data are often displayed as images of the measured values of the apparent specific resistivity. In case of vertical profiles and 3D data, it is essential that the apparent resistivity values are converted to real resistivity values using numerical inversion algorithms (Ullrich et al. 2007a, b, 2008; Tsokas et al. 2009). Inversion is an indispensable precondition for meaningful interpretation, since the field data (apparent electrical resistivity) does not provide any direct depth information. Furthermore, the values of the apparent resistivity depend strongly on the array configuration of field measurements. The image-like nature of data display (pseudosections, horizontal slices) can be misleading and thus, causing misinterpretation of electrical resistivity data.

The inherent nature of electrical resistivity data is a lower spatial resolution compared to GPR and magnetic data. The typical sampling intervals in archaeological prospection of 0.5 to 1 m usually result in "smoothed" data images with few details. However, electrical resistivity

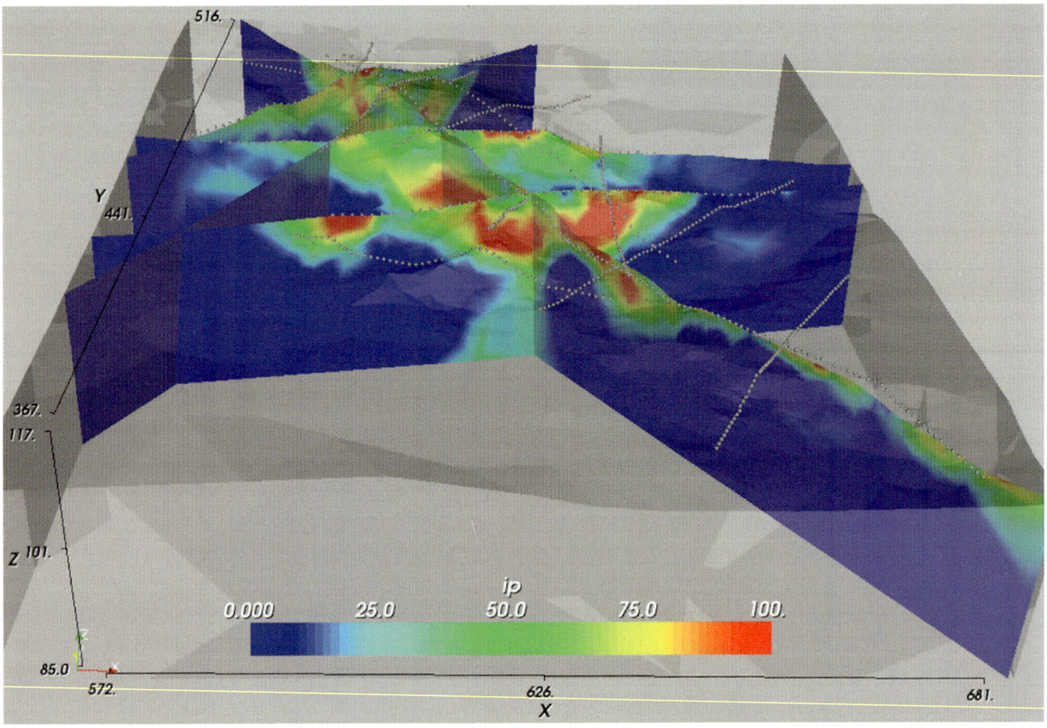

Fig. 10.7 Three-dimensional presentation of ERT data: IP effect measured on a slag mound in Ain Al-Hajar, Morocco (Ullrich et al. 2007b)

measurements have a crucial advantage over GPR and magnetics, because they reach a larger depth range. Today, with the availability of automatically multiplexed multielectrode equipments and mobile arrays, large sets of data can be collected. The variety of configurations allows suiting the measurements to the archaeological questions. Hence, for the interpretation of electrical resistivity data and its presentation, various possibilities turn out in archaeological practice.

An important aspect in the interpretation of electrical resistivity data is the large bandwidth of the physical parameter. Typical electrical resistivity values of subsurface material can range over many decades from 10^1 Ωm in water-saturated and clayey sediments to values of 10^8 Ωm or more in solid rock material like marble or granite (Samouëlian et al. 2005). Thus, a direct identification of materials and structures is difficult and needs additional data from coring or excavations. The large range of electrical resistivity mainly depends on the water content of the subsurface material. Hence, the results of measurements at one and the same site may vary from season to season, sometimes even from one day to another (Fry et al. 2012).

Interpretation of resistivity mapping data with high contrasts is less complex, compared to three-dimensional datasets. Construction materials, in most cases, are recognisable by high resistivity values. Fillings of pits and ditches usually produce low resistivity values, especially when they bear a high content of organic material. Even though electrical methods are less appropriate for the investigation of complex architecture layers, they are a powerful tool in archaeometallurgical research. Remains of metal productions like slag mounds and bases of melting furnaces, can be detected by electrical measurements. Since materials related with metal production often show low resistivity values and only a low contrast to the surrounding soil matrix, a second electrical parameter is used. By measuring the effect of induced polarisation, the complex electrical resistivity can be determined, composed by the scalar resistivity σ and the phase shift φ (Fig. 10.7).

Especially material bearing remains of ore and metal like slags can be identified by an increased phase shift (Ullrich et al. 2007b). But also wood structures in a saturated environment show significant anomalies of the complex electrical resistivity (Schleifer et al. 2002; Schleifer 2004; Weller et al. 2006).

10.2.6 Typical Interpretation Pitfalls

The inherent ambiguity of geophysical data and the diversity of physical effects in the ground always tempt the interpreter to misinterpreting geophysical anomalies or ignoring important information. Of course, without additional information, a magnetic anomaly of a Neolithic pit house can be identical to an anomaly of a modern rubbish pit. But, there are a number of typical failures which can be avoided when following the general guidelines to data interpretation.

The most typical failure in the interpretation of magnetic data is to get bogged down in details. Attributing importance to even the slightest magnetic anomaly leads to a certain disarray of the interpretation. Therefore, it is recommended not to go from anomaly to anomaly, in both interpretation drawing and survey report, but rather to focus on the most important anomalies and structures first and then going more into details without getting lost in the typical numerousness of anomalies, especially in large-scale surveys.

Another pitfall is the misinterpretation of anomalies caused by natural effects. With a certain frequency, anomalies caused by lightning-induced remanent magnetisation (LIRM) appear in magnetic data (Jones and Maki 2005; Maki 2005). These anomalies show strong amplitudes, in most cases of dipole type and vary considerably in shape and complexity. Especially in complex situations they can easily mislead the interpretation of archaeological structures.

In GPR data, there various effects can falsify the existence of archaeological structures. Multiple reflections coming from near-surface reflectors can simulate objects in larger depth. Therefore, a careful mapping of all surface changes has to be carried out when doing a GPR survey (Conyers 2012). Furthermore, too ambitious data processing may result in weakening or complete disappearance of archaeologically relevant reflection signals. For that reason, it is recommended to compare the results of processing continuously to the raw data.

In electrical resistivity survey, it has to be considered that for inversion of electrical data, the principle of equivalence is valid. It is based on the general inversion problem of the theory of potentials and expresses that a particular parameter distribution, obtained from measured potential data at the surface can be caused by infinite real subsurface parameter distributions (Dabas et al. 1994; Barker 1997). In order to reduce the ambiguity of electrical resistivity data, additional subsurface information from coring or excavation is vitally needed. Cutting edge inversion techniques, like boundless electrical resistivity tomography (BERT) take advantage of the application of arbitrary geometries to decrease the ambiguity of electrical data and observe the principle of Occam's razor, in order to find the most conclusive resistivity distribution in the ground (Günther et al. 2006; Rücker et al. 2006).

A general rule for the interpretation of any geophysical data is following: there is no negative evidence, meaning, the lack of anomalies in a dataset supports no conclusions in view of the absence of archaeological structures in the ground. Statements on the probability of buried archaeological in a certain area necessarily require the consideration of all obtainable surface and subsurface data.

10.3 Survey Reports

10.3.1 Structure of Survey Reports

The survey report is the end product of any geophysical investigation. It is a roundup of all relevant technical information regarding the applied survey methods. It includes an intelligible description of the data analysis and interpretation. The terminology should be consisted and understandable for both specialists and nonspecialists. A reasoned survey report can work as a stand-alone

Table 10.2 Essentials of reports

Category	Content
Title page	Title of report, authors, contracting entity or research partner, report number, date
Summary	Short abstract
Introduction	Location, site description, survey history and objectives, description of general site conditions, fieldwork date, overview of survey areas
Methods	Justification of method choice, physical principles of methods, description of equipment, survey parameters (e.g. sample intervals, configuration), data processing steps
Results	Description of results, peculiarities, interpretation
Conclusions	Assessment of accomplishment of survey objectives, summary of archaeological results, implications and recommendations
References	List of sources (papers, maps, photographs) referred to
Appendices	Plots and maps of survey area location (small scale 1:2,000 min), plots of raw and processed data (middle scale: 1:1,000 min), interpretation maps (1:1,000 min)

paper, so constituted that it can be transferred into a publication with little effort. The following Table 10.2 gives a generalised list of the required contents of a adequate survey report.

10.3.2 Description of Methodology

The method statement should include a brief but conclusive motivation for the choice. The connection between the registered physical parameter, the expected underground structures and the estimated parameter contrast has to be outlined. Table 10.3 subsumes the three most common survey methods, their decisive physical parameter and the possible archaeological targets of each method.

Additionally, the specification of the used equipment should be listed (type, manufacturer). Specific data explaining sensibility and precision of the used devices have to be added. Beside this, it is important to describe how the data was gathered, which sample intervals were used and

Table 10.3 Common geophysical survey methods and typical targets

Method	Physical parameter	Survey objectives
Magnetic prospection	Magnetic susceptibility X	Fillings of pits and ditches, construction material (if there is a significant contrast in magnetic susceptibility between material and soil matrix), slag, fireplaces and kilns
Ground-penetrating radar (GPR)	Permittivity ε (and magnetic susceptibility X)	Construction material (stone walls), fillings (debris, sand), cavities
Electrical resistivity	Electrical resistivity ρ	Ditches, walls, foundations, cavities
	Phase angle φ	Slag deposits, wooden constructions

what kind of software was used for primary data processing. A list of the applied processing steps from raw data to the eventually presented data is also part of the methodological statement.

10.3.3 Presentation of Survey Results

Usually geophysical survey data are presented in maps of the individual datasets accompanied by an instructive description in the textual part of the report. For the presentation of the datasets, some basic principles are to be taken in account: the most instructive way to present mapping data (magnetic data, GPR time slices or electrical resistivity mapping) is a greyscale or a bicoloured image. In some cases, structural information can be accentuated by using contour lines. Vertical profiles from both electrical resistivity methods and GPR obtain best comprehensibility by using multicolour scales. In case of presentation of 3D bodies of data, a combination of colour scales with threshold values is recommended in order to hide data from archaeologically not affected zones. When vertical profiles or 3D bodies are separately presented, it is important that the position of profiles and 3D areas is highlighted in the overview maps of the surveyed area. Self-evident is the addition of a colour or greyscale bar with threshold values of the presented physical parameter, a north arrow and a scale bar.

For the interpretation plots it is recommended to use identical scales and allocations as used in the data presentation. Similar schemes facilitate orientation in the maps. In the interpretation maps the survey area should be adumbrated since vivid colours should be reserved for interpreted structures. The division between objective reasoning and more subjective assumptions has to be made distinct. Ambiguity is in the nature of geophysical prospection, whereas interpretation of geophysical data is always a combination of clearly determined facts and more subjective assumptions. For this reason the degree of uncertainty of the interpretation must clearly be recognisable by the reader.

10.3.4 Data Archiving

A careful archiving of survey data is not only an indispensable part of any archaeological prospection activity but also safeguarding a future resource (Schmidt 2002). Interpretation of geophysical field data can be fairly subjective. Users wishing to reuse these data may want to go back to raw results, rather than use someone else subjective interpretations. Therefore, reasoned strategies for archiving of both raw and processed data including the survey reports are needed. Data storage strategies have to be elaborated already in the beginning of a survey project with due regard to storage security, long-term accessibility and legal aspects. Attempts to create both national and international databases have been made in the UK during the last decade, among them the OASIS project of English Heritage (http://oasis.ac.uk).

An essential part of the survey data is the corresponding metadata. If the raw data have to be archived and reused, they have limited value without the field notebooks used to record location, measuring configurations and field conditions. This metadata often comes in paper format and requires some time and effort to digitise and rationalise for general reuse. Inadequate documentation during data creation is the single biggest barrier to the future reuse of data. Indispensable elements of metadata include:

- Equipment used (model and manufacturer)
- Equipment settings
- Assessment of accuracy
- Methodology
- Used software
- Processing carried out
- Data formats/file formats

Survey reports should not only contain the project's outcome, they also have to include the relevant data documentation and a description of metadata, preferably in tabular form. Additionally, the file formats of raw data have to be evaluated in view of their metadata content. For further information, see the Archaeology Data Service of York University website (http://archaeologydataservice.ac.uk/), where useful guidelines for data storage, archiving and data refreshment are given (Austin and Mitchem 2007).

References

Aspinall A, Gaffney CF, Schmidt A (2008) Magnetometry for archaeologists. AltaMira Press, Lanham

Austin T, Mitchem J (2007) Preservation and management strategies for exceptionally large data formats: 'big data'. archaeology data service and English heritage., http://archaeologydataservice.ac.uk/attach/bigData/bigdata_final_report_1.3.pdf. Accessed January 2013

Barker RD (1997) Electrical imaging and its application in engineering investigations. In: McCann DM, Fenning PJ et al (eds) Modern geophysics in engineering geology, Engineering geology special publications no 12. Geological Society, London, pp 37–43

Clark AJ (1996) Seeing beneath the soil. Prospecting methods in archaeology. Routledge, London

Conyers LB (2012) Interpreting ground-penetrating radar for archaeology. Left Coast Press, Walnut Creek

Conyers LB, Lucius JE (1996) Velocity analysis in archaeological ground-penetrating radar studies. Archaeol Prospect 3:25–38

Dabas M, Tabbagh A, Tabbagh J (1994) 3-D inversion in subsurface electrical surveying – I. Theory. Geophys J Int 119:975–990

Faßbinder JWE (1994) Die magnetischen Eigenschaften und die Genese ferrimagnetischer Minerale in Böden im Hinblick auf die magnetische Prospektion archäologischer Bodendenkmäler. VML Verlag Marie Leidorf, Rahden

Fry R, Stott D, Boddice D, Beck A (2012) What a difference a year makes: preliminary DART datasets

from Cherry Copse, Cirencester. AARG News 45: 42–48
Gaffney C, Gater J (2003) Revealing the buried past: geophysics for archaeologists. Tempus Publishing Ltd., Stroud
Gaffney CF, Gater JA, Linford P, Gaffney VL, White RH (2000) Large-scale systematic fluxgate gradiometry at the Roman city of Wroxeter. Archaeol Prospect 7:81–99
García Sanjuán L, Wheatley DW (eds) (2002) Mapping the future of the past: managing the spatial dimension of the European archeological resource. Universidad de Sevilla, Historia y Geografia, Seville, no 78
Günther T, Rücker C, Spitzer K (2006) 3-d modeling and inversion of dc resistivity data incorporating topography – Part II: Inversion. Geophys J Int 166(2): 506–517
Jol HM (2008) Ground penetrating radar: theory and applications. Elsevier, Amsterdam
Jones DM (ed) (2008) Geophysical survey in archaeological field evaluation. English Heritage Publication. http://www.english-heritage.org.uk/publications/geophysical-survey-in-archaeological-field-evaluation/. Accessed January 2013
Jones G, Maki DL (2005) Lightning-induced magnetic anomalies on archaeological sites. Archaeol Prospect 12:191–197
Kvamme KL (2003) Geophysical surveys as landscape archaeology. Am Antiq 68(3):435–457
Leckebusch J (2001) Die Anwendung des Bodenradars (GPR) in der archäologischen Prospektion 3D-Visualisierung und Interpretation. VML Verlag Marie Leidorf, Rahden
Linford N, Linford P, Martin L, Payne A (2010) Stepped frequency ground-penetrating radar survey with a multi-element array antenna: results from field application on archaeological sites. Archaeol Prospect 17:187–198
Maki DL (2005) Lightning strikes and prehistoric ovens: determining the source of magnetic anomalies using techniques of environmental magnetism. Geoarchaeology 20:449–459
Marzoli D, Pardo FL, Padilla JS, Wagner CG, Mielke DP, Martín CL, Cabrero LR, Thiemeyer H, Ortiz MT (2010) Los inicios del urbanismo en las sociedades autóctonas localizadas en el entorno del estrecho de Gibraltar: investigaciones en los Castillejos de Alcorrín y su territorio (Manilva, Málaga). Menga 1:153–183
Meyer C, Plesničar R, Pilz D (2012) Extramural magnetic survey. In: Corsi C, Vermeulen F (eds) *Ammaia* I: the survey. A Romano-Lusitanian townscape revealed. Academia Press, Ghent, pp 105–114
Meyer C, Ullrich B, Barlieb C (2007) Archaeological questions and geophysical solutions: ground-penetrating radar and induced polarization investigations in Munigua, Spain. Archaeol Prospect 14:202–212
Neubauer W (2001) Magnetische Prospektion in der Archäologie. Austrian Academy of Sciences Press, Wien
Neubauer W, Eder-Hinterleitner A, Seren S, Melichar P (2002) Georadar in the Roman civil town Carnuntum, Austria: an approach for archaeological interpretation of GPR Data. Archaeol Prospect 9(3):135–156
Robinson C, Campbell D, Hodgkinson A (2011) Archaeological maps from qGIS and Inkscape: a brief guide, 3rd edn. Documentation. Oxford Archaeology North. http://library.thehumanjourney.net/366. Accessed January 2013
Rücker C, Günther T, Spitzer K (2006) 3-d modeling and inversion of dc resistivity data incorporating topography – Part I: Modeling. Geophys J Int 166(2):495–505
Samouëlian A, Cousin I, Tabbagh A, Bruand A, Richard G (2005) Electrical resistivity survey in soil science: a review. Soil Tillag Res 83(2):173–193
Schleifer N (2004) Anwendung der Spektralen Induzierten Polarisation (SIP) in der archäologischen Prospektion. VML Verlag Marie Leidorf, Rahden
Schleifer N, Weller A, Schneider S, Junge A (2002) Investigation of a Bronze Age plankway by spectral induced polarization. Archaeol Prospect 9:243–253
Schmidt A (2002) Geophysical data in archaeology: a guide to good practice. Oxbow Books, Oxford
Tsokas GN, Tsourlos PI, Papadopoulos NG (2009) Electrical resistivity tomography: a flexible technique in solving problems of archaeological research. In: Campana S, Piro S (eds) Seeing the unseen: geophysics and landscape archaeology. CRC Press/Taylor & Francis Group, London, pp 83–104
Ullrich B, Meyer C, Weller A (2007a) Geoelektrik und Georadar in der archäologischen Forschung: geophysikalische 3D-Untersuchungen in Munigua (Spanien). In: Wagner GA (ed) Einführung in die Archäometrie. Springer, Berlin, pp 76–93
Ullrich B, Günther T, Rücker C (2008) Electrical resistivity tomography methods for archaeological prospection. In: Posluschny A, Lambers K, Herzog I (eds) Layers of perception. Proceedings of the 35th international conference on computer applications and quantitative methods in archaeology (CAA, Berlin, April 2–6, 2007), 25 (Abstract). Kolloquien zur Vor- und Frühgeschichte, vol. 10. Dr. Rudolf Habelt GmbH, Bonn
Ullrich B, Weller A, Günther T, Rücker C (2007b) Geophysical prospecting of ancient slag deposits in Munigua (Spain) and Ain-al Hajer (Morocco) using complex electrical resistivity tomography. In: Archaeometallurgy in Europe 2007. Selected papers of the 2nd international conference (Aquleia, 17–21 June 2007). Associazione Italiana di Metallurgia, Milano
Verdonck L, Taelman D (2012) Ground penetrating radar survey at *Ammaia*. In: Corsi C, Vermeulen F (eds) *Ammaia* I: the survey. A Romano-Lusitanian townscape revealed. Academia Press, Ghent, pp 69–81
Weller A, Bauerochse A, Nordsiek S (2006) Spectral induced polarisation: a geophysical method for archaeological prospection in peatlands. J Wetl Archaeol 6:105–125

Part III

Topographic and Geoarchaeological Surveys

Intra-site Artefact Surveys

11

John Bintliff

Contents

J. Bintliff
Department of Classical and Mediterranean Archaeology, Leiden University, Leiden, The Netherlands
e-mail: j.l.bintliff@arch.leidenuniv.nl

11.1 The Development of Survey Methodology

Since Western European travellers in the nineteenth century began regularly to tour the ancient towns of the Mediterranean, surface finds have been collected and standing architecture noted and drawn. During the first three quarters of the twentieth century what is usually referred to as *extensive field survey* gradually systematised approaches to such sites: good site plans were drawn, finds were illustrated and artefacts of all periods could be sought from prehistory to medieval. Typical for the quality of such work are the numerous, almost one person, extensive surveys of Richard Hope Simpson in Southern Greece (e.g. Waterhouse and Hope Simpson 1960) and the pioneer extensive team survey of South Etruria led by John Ward Perkins from the 1950s into the 1970s (summarised in Potter 1979). South Etruria was one of the first regional, multidisciplinary projects whose main focus was on the study of surface traces of past settlements of all periods. In the United States regional survey was already more advanced, since the landmark survey of the settlement pattern history of the Viru Valley in Peru had been carried out during the 1940s by Gordon Willey (1953). American archaeologists in the Near East were also influential in popularising the regional project model for extensive survey (such as Braidwood from the 1930s and later McAdams in the 1960s to 1970s).

By the 1960s–1970s, regional projects were the leading sector for surface survey in the

C. Corsi et al. (eds.), *Good Practice in Archaeological Diagnostics*, Natural Science in Archaeology, DOI 10.1007/978-3-319-01784-6_11, © Springer International Publishing Switzerland 2013

Mediterranean, but in response to the scientific ethos of the contemporary rise of New Archaeology, extensive approaches became heavily criticised, leading to a general shift to *intensive survey*. Whereas the former had covered very large areas and innumerable sites, there was no attempt to study each field or treat each site as a major study. If we now turn specifically to large complex sites in extensive surveys, normally surface collections were usually small and resulted from a nonsystematic walking around the site bagging some diagnostic pieces. Concentrations of density and of finds of particular periods were often noticed and could be mapped onto the site plan, and inferences were drawn on this basis concerning the general scale of occupation for major periods.

In intensive survey American advances were once more central (cf. Flannery 1976). Apart from the general concept of trying to cover large areas of landscape as fully as possible, or at least deploying an explicit sampling scheme, in both of which field-by-field cover was attempted wherever feasible, larger surface collections were called-for in which all kinds of ceramics were sought, not merely the most diagnostic. However, if we turn to Mediterranean *urban* survey, it appears that regional teams found such sites problematic to study in comparison to the smaller, more manageable rural site. In some cases, funding limitations and the need to complete a regional survey to a certain timescale meant that the big sites could not be given appropriate attention if the vastly more numerous smaller sites were to be located and analysed. For example, the Argolid Project in Greece (Jameson et al. 1994) possessed two ancient cities in its region, but relied on localised excavation and standing remains for reconstructing the occupation history of *Halieis* and a limited nonsystematic collection of 96 sherds and some standing architecture from the 23 ha city of *Hermione*. The Kea Project also in Greece (Cherry et al. 1991) had a single urban site, *Koressia*, but the surface collection from this 18 ha site consisted of just over 100 ceramic pieces.

Koressia raises a second issue apart from sample size. One might now suggest that American New Archaeology overstressed theoretical concepts over empirical research, and in surface survey, I believe that the introduction of *sampling theory* has had a debilitating effect on Mediterranean survey practice. Perhaps due to the usual restrictions of time and money (the Kea Project was designed to conduct a single field season and a second study/revisit season), both rural survey and site survey has since the 1970s more often than not chosen to sample landscapes and site surfaces using one or other form of transect or spot grid/circle, rather than attempt complete field-by-field cover at both scales. Thus, the *Koressia* urban survey used small circles scattered over the city as its sherd collection base. The Melos Survey, Greece (Renfrew and Wagstaff 1982), surveyed one-fifth of that island using long separate transects, producing settlement patterns which now seem inadequate to extrapolate from. With just one survey season, the city of ancient Melos was only examined from its architecture and older localised excavation.

When Anthony Snodgrass and I began our Boeotia Project (Greece) in 1978, we also followed American practice by sampling rural sites using small circles, and later a set of thin transects, producing relatively small sherd collections. At the same time however we collected sherds from the whole site area, aware that the rationale for sampling had never been adequately tested. Comparison of the finds statistics (Bintliff and Snodgrass 1985) was highly revealing. Setting up circle or transect samples was more time-consuming than a simple overall grid, but more worrying was the fact that total collection in the restricted samples was mostly undiagnostic, while the intended back-up collection from the entire site (of the same numerical size), because it was focussed more on diagnostic pieces, revealed not only as to be expected a far greater percentage of datable pieces but a wider range of periods. In fact we had already changed our site study approach to a grid, total or near total, within which the whole surface was examined. It remains our belief that subsampling not only lacks experimental support, but now that we have become aware of the highly variable patterning of finds of different periods on complex sites, is likely to miss the rarer phases. What is very surprising is that all the major Mediterranean intensive urban surveys have been conducted after the publication of the most impressive and well-thought-out urban survey hitherto – that of Teotihuacan in Mexico (Millon 1964), which showed the clear advantages of a total grid survey. But very few Mediterranean urban projects adopted that approach, preferring partial sampling methods with their inherent risks of poor representativity.

There might be exceptions to this judgement however. The ongoing survey of the city of Knossos on Crete (cf. Whitelaw et al. 2006–7), has gridded this several kilometre square prehistoric and historic town in 20×20 m square units, but collects all material from a 10 sq. m area within each unit (thus 2.5 %). While it is now likely that the sample from each square may well omit certain rarer phases which were actually present in the other 97.5 % of the square, overall the picture for different *sectors* of the town may be broadly accurate owing to the immense number of grid units (by 2007 around 10,000 squares had been sampled). Moreover, the database of some half a million sherds from the urban survey, even allowing, given total collection, for a high percentage of undiagnostic pieces, offers confidence in at least the overall picture being offered for the development of the site over time. Finally, we must add that Knossos is unique in the high number of large and small excavations that have been undertaken in the last century of research at the site, offering detailed subsurface clues to aid in the interpretation of the surface finds. Another obvious exception could be large urban sites where occupation was essentially confined to one or two, or just one, period, so that local variability would not be a hindrance to spot – rather than area –sampling. The Italian site of *Portus* could represent such an unusual situation (Keay et al. 2005).

My own experience on the Boeotia Project, where we have undertaken survey at six urban sites sequentially between 1982 and 2010, has allowed us to adjust our methods and evaluate their relative strengths and weaknesses over a time period rarely possible with regional projects operating at a timescale of a few years. At the first urban site, the small town of *Askra* (Greece), we took the existing irregular fields as sample units in 1982, and collected from all their surfaces less than 500 sherds over 11 ha. Reflecting our growing concerns about sampling, we revisited the town in 1985 and set additional, medium-size squares into many fields, from which an overall diagnostic collection was made, boosting the site total to more than 2,000 sherds. Nonetheless, it was clear by the late 1980s (Bintliff and Snodgrass 1988a) that although this size of collection gave a good picture for the main periods of occupation, the data recovered for periods where either the activity level was low, or the recognisability of sherds were poorly developed, was inadequate to avoid considerable uncertainty. This realisation had come too late for the second urban survey of the larger city of *Haliartos* in 1985, 40 ha in size. Here, although we also adopted a two-scale sampling strategy of large 50×50 m units with nested 30×10 m units within them, in order to see if collecting the same amount of pottery over a larger area produced similar results to a subsample, we did not realise that the total collection of 650 sherds could be insufficient for long-term activity reconstruction at the site. Indeed although the site maps for the main periods produced clear results (Bintliff and Snodgrass 1988a), those for minor eras or those with less clear diagnostic material are not reliable.

In all our later urban surveys, at *Hyettos*, *Tanagra* and *Koroneia*, we stayed with total urban grid survey, but gradually shifted from the two-scale collection to single grid units, whose entire surface was collected from. With the aid of palmtop computers, GPS-linked setting such small units up is very simple and they are easily plotted and measured digitally. As noted for the Knossos Survey, and at other recent surveys (e.g. Mallia on Crete: Müller-Celka 2007), a rising concern about the representativity of pottery samples has led us and other recent projects into a striking rise in collection sizes (Mallia has 40,000 sherds) and an increase in the number of grid units (in our case and Knossos settling at 20×20 m squares), which in combination allow us to have far more reliable information and with better spatial resolution.

11.2 Collection Size

Let us now consider this problem. How much should we bring home from a large, complex urban site? The fact is that there has been a central failing in Mediterranean survey to compare different methods on the same data, or use artificial simulations for the same goal. The decision on how much to collect per grid unit, and for an entire large site, has not rested on any statistical calculation of the parameters involved. The recent increase in collection sizes certainly reflects unease on this issue, as well as the fact that over the last generation pottery specialists have developed a much wider range of questions they wish to ask of ceramic assemblages, usually questions with a quantitative aspect.

For two of our Boeotian cities, I have been conducting some experimental calculations to shed some light on the effects of different sample sizes (cf. Bintliff 2012). At *Thespiae*, a large city where some 180 ha were surveyed in 1985–1986, we deployed the two-scale total gridding of the town, with smaller units of 30 × 10 m nested inside more than half of the larger 60 × 50 m units. By now a more respectable total collection of some 14,000 sherds was brought back, of which only 30 % was undiagnostic. Nonetheless, occupation ran from the Neolithic to the early nineteenth century AD, and 27 distinct ceramic phases could be identified. Simply put: was the collection adequate to allow reliable maps of activity across the site to be drawn for each of these phases? This might seem an unanswerable question but there are indirect means of dealing with this problem. For many sites there exist standing architecture, visible grave monuments or localised excavation records, and indeed some projects have been using a combination of urban surface survey over large areas combined with limited test excavations (e.g. *Celti-Peñaflor*: Keay et al. 2000; *Leptiminus*: Stone et al. 2011). Thus, on a city-wide scale, but not on the detailed level, thin or ambiguous data for a particular phase might be assisted by evidence from these alternative sources for burials, public or private buildings, industrial activity and so on.

A second means of approaching the many less well-represented eras on complex sites is through internal analysis of the data themselves. One reason we adopted a two-scale sampling approach at several cities was due to the uncertainty as to how much the area one covers affects the material found for any locality within the town. At *Thespiae*, since we collected roughly equal amounts of sherds from the large and small units it was possible to compare their data directly. Interestingly the density counts for pottery differed little, but the period maps produced from the large and small samples respectively across the town showed significant variation. Some periods looked the same, for others it looked as if different cities were being presented. On the advice of Martijn van Leusen, I tested whether divergent or convergent maps for each period were primarily a result of sample size, and this proved to be the right explanation. Per phase, as the total number of sherds collected rose, so the maps from small and large samples grew more alike. Although we do not possess what statisticians desire, the original 'populations' of pottery to compare our samples with, we can at least hypothesise that the point where numbers bring agreement to the period maps, regardless of sampling method, might be a working minimum, to guide us in how much material we might need to collect in future from complex urban sites with an equivalent range of phases. That borderline figure for *Thespiae* was 600 sherds, meaning that a large site might need that number at least *for each occupation phase* to allow reasonable confidence in spatial representation of the original distribution. For that city this would indicate a minimum total collection of more than 16,000 sherds (our actual sample was that large), but even that is unrealistic as it assumes that all 27 periods are present at the right level. Clearly all complex sites have a small number of phases which dominate site use or at least the surface assemblages, and they are well known to all but submerge periods of slighter evidence, especially the oldest and deepest levels of the site. The implications of these tentative estimates require serious future study.

A second numerical result from *Thespiae* concerns exactly the weakest point of complex site survey – early occupation, particularly prehistoric or protohistoric phases. Apart from a Neolithic tell, always rich in surface finds, the remainder of this large city provided a puzzling Bronze Age surface collection, whose ceramics appeared in small numbers, irregularly, over the entire site. This phenomenon is now well known from many city surveys, and usually no satisfactory interpretation of site use has been possible from such patchy and fragmentary distributions. Encouraged by the debate raised by our 'hidden landscape' theory (Bintliff et al. 1999), we suspected that this thin spread implied a far greater original amount of Bronze Age activity, of which just a tiny proportion survived, was available on the surface or appeared in the samples collected. For example, if just 10 % of the total collection represented 6,000 years of prehistoric site use, the fact that our sample came from an area with probably one million total surface sherds (from density counts) might not unreasonably allow extrapolation by a factor of 63 for all available prehistoric sherds on the site. We can go further, because it has been shown (Bintliff 2012) that the prehistoric sherds at *Thespiae* are heavily filtered by the sample size collected from each grid unit. Any unit with less than 30 sherds collected has little chance of including prehistoric finds.

In addition we must consider how indeed artefacts from earlier occupation levels find their way to the surface of complex sites, usually dominated by pottery and massive building foundations of Greco-Roman age. It is well known that eroded areas are most likely to reveal early occupation levels, as well as recent road cuts, but it is usually suggested that the chief form of bringing older material through later levels onto the contemporary surface is through premodern disturbances of underlying deposits: the preparation of mud brick for houses using earth excavated from the construction site (cf. Slane 2003 for Corinth), or the digging of foundations – Roman cities in particular seem to be typified by almost unparalleled practices of levelling for foundations and the insertion of major building elements deep into the ground. Prehistoric sherds and lithics have been recorded in the construction of historic buildings, and the author has seen such material in the wall of a nineteenth-century mud brick house in Orchomenos, Greece. For our purposes, it seems clear that these processes cannot be assumed to operate universally over an entire city site, so that not only would we not normally expect huge amounts of older finds on top of historic sites, but they may not accurately reflect the prehistoric surfaces from which they have originated. Once more, the implications of these insights remain to be further explored.

Apart from mapping finds of each phase across a complex site, current survey practice is interested in isolating functional areas. This begins with associations of pottery types and lithic finds, but regularly includes architecture and evidence from sporadic excavation 'windows'. Archaeologists have ideas on functional assemblages which can be looked for, and indeed it is common for ceramics to be classified not just by date and type but by function, for example, storage, food preparation, tableware and transport (cf. Whitelaw 1998). We hope for clear signals from surface data which would allow us to define cemetery areas, industrial or commercial zones and properties, public buildings and domestic residences. Here again a statistical question arises for which very little data have been analysed: what are the likely parameters limiting the recognition of such assemblages localised to different parts of a city? As part of an EU Project (CEEDS) a large part of the ceramic database (ca. 12,000 sherds) from our city survey at *Tanagra*, Greece, was subjected to 'data mining' by Eckehard Olbrich and colleagues at the Max Planck Institute for Mathematics in Leipzig (personal communication, September 2012). The aim was to use neutral computerised statistical analysis to identify clusters of pot types which were highly correlated with each other in terms of presence, in the 121 survey units where finds had been collected. The team focussed only on types which were reasonably widespread across the grid network and on the Greco-Roman eras. Of 33 recognised pot types only 14 appeared in more than 20 grid units, so the analysis continued with the correlation of these across the city. As noted earlier, a permanent weakness in archaeological statistics is our ignorance of the original 'population' of data from which our samples are being drawn, so the

Leipzig team first simulated several assemblages and tested the statistical probabilities of identifying them in large sample populations. They found that a sample of 10,000 items was inadequate for showing reliable correlations of ten types, whereas 15,000 gave more realistic results. With the actual *Tanagra* sample of 12,000 sherds, three possible assemblages emerged with stronger clustering. A first tentative interpretation of these automatically generated groups is that they conceivably represent a domestic, a rubbish and a production assemblage (V. Stissi, personal communication 2013), but this is the first stage of a longer analysis. In any case this experiment illustrates the problematic complexities of finding functional zoning in surface assemblages.

11.3 What Is Being Collected?

We should now introduce another feature of complex site sampling: total or diagnostic collection. As we have seen, the often immense scale of urban surface collections prohibits *total site collection*, so all projects have employed sampling of surface finds. Aware of the risk of missing rare pieces, or those less visibly interesting, or even types not so far known in the published literature, many surveyors have opted for total collection of small windows (circles, squares) spread across the site and set into a grid of fields or artificial study units. An alternative approach has been to collect from the entire area of the site, in which case either a strictly diagnostic or a broad 'potential diagnostic' methodology has been used (feature pottery, decorated wares, a range of fabric types and a selection of the variety of body sherds). Both models have their problems. The 'window' method seems to miss periods or types present in a grid unit but not in its subsample, but is more likely to detect finds less clearly diagnostic through total recovery; however, the majority of total finds are often undiagnostic, and hence, useful pieces per grid may be few. The total area selective collection in contrast seems more effective at recovering the full range of periods present in a grid unit, but may underrepresent less obviously diagnostic sherds or those of less well-known periods. On balance I think the latter is more reliable, and here one can sensitise fieldwalkers to look for types less attractive at first sight, e.g. cookware and prehistoric plain wares.

Experimental collections made by Mark van der Enden at ancient *Koroneia* (in Bintliff et al. 2009–2010) have been informative. Five of the standard grid units of 20×20 m were first collected from, using the broad diagnostic approach, then a 4×4 m square was placed within them and all material collected. The densities in this part of the city are so huge that we did not expect a serious bias in the secondary sampling over the same ground. Only a few categories were studied from several thousand pieces in the two collections. Having already sensitised students to collection bias, the overall grid sample was generally better performing than the subsample in fabric types (e.g. coarse wares, fine wares). Chronological types were not analysed, but we have earlier argued that they are also better represented by the overall sampling method. Most interesting was the evidence from body sherds as opposed to feature pottery. The total subsample collection naturally contained a dominance of body sherds, while the diagnostic overall sample had far lower quantities, but it appears that no significant extra information was obtained. This, it must be added, is connected to the high-surface pottery densities in this part of the town; surveyors normally collect more pottery of lower diagnostic potential when densities become low. The likely exceptions to the conclusion that survey can largely focus on feature sherds were recognised as prehistoric finds, rare enough to demand total recovery on most historic urban sites, and cookware-coarse wares – which often due to their physical form disintegrate into small amorphous sherds (although as seen these can be positively discriminated for in the overall sampling method).

11.4 Ancillary Aids to Surface Ceramic Survey

Given these many difficulties which beset the surface survey of complex sites, it has become clear that ancillary approaches are almost obligatory in order to achieve a clearer picture of the long-term development of ancient towns by a process of 'cumulative credibility'. The oldest approach predated ceramic survey, and as noted at the start of this paper was practised in the nineteenth century, the study of standing architecture or exposures of the site levels due to recent erosion or human interventions. In some periods surface ceramics *can* be indicative of functional areas alone, for example, some wares are primarily associated with funerary or sanctuary contexts, or one can find kiln furniture and slag as industrial waste in or around manufacturing locations (cf. Stone et al. 2011). Yet these are not foolproof clues: in Greek culture indeed certain fine ware concentrations and types are rarely found on domestic contexts, but Roman funerary and religious landscapes are largely composed of surface finds which are broadly similar to domestic assemblages (cf. Bookidis 2003). The presence of legal and illegal excavations marking cemeteries, or funerary monuments, may be essential to allow their detailed mapping, and the same goes for religious structures, after Greek culture is replaced by Roman. Late antique graves become progressively less well furnished with artefacts, and to cause further confusion, Christian burials are allowed into the city core in and around churches.

Revolutionary in its assistance for large complex site survey is the battery of techniques termed remote sensing. In favourable conditions, aerial photography and LiDAR scans can produce detailed building plans from the former (cf. Vermeulen 2003) and alignments in the topography not clearly visible on the surface from the latter technique. Geophysical techniques have changed dramatically in the last generation, becoming more varied and most importantly faster. It is now possible to generate an almost complete urban plan from a combination of resistivity, magnetometry and georadar within just a few field seasons. At ancient *Tanagra* city in Greece, the team of Branko Mušič and Bozidar Slapšak not only revealed the detailed town plan within the standing city walls but also discovered that the city insula grid plan extended northwards outside the circuit wall (Bintliff 2006). Here the problem identified just now, where Roman cemeteries leave little distinct ceramic evidence, had left us uncertain whether the high ceramic densities in the immediate northern extramural area were town halo (burials, industry, rubbish dumping, market-garden manure) or domestic areas outside the fortification. The discovery of an earlier city wall beyond this northern insula extension confirmed that the city contracted from its Greek grid plan maximum into a smaller fortified area and that the standing enceinte was in fact a Late Roman rebuild with mostly recycled Classical Greek masonry.

Architectural survey is rarely able to offer a complete urban plan, and over time Mediterranean cities have undergone almost continual recycling of building material. It is a commonplace that Late Antique towns used spolia from abandoned earlier structures, especially pagan temples and gravestones, for construction purposes, so that the current location of individual items often does not mark its original placing in the cityscape. It is often the case that medieval and postmedieval activities on ancient towns continued this practice of recycling material from a wide area of the previous city, moving them into more confined locations. Again it is 'cumulative credibility' that allows us to link ceramic evidence, architectural finds, occasional excavation windows and remote sensing, in order to present plausible town plans at different phases of the site's history.

Confusing our attempts is the unsurprising fact that long-lived complex sites see changes in land use as the town rises and falls, or its infrastructure is altered. Areas once cemeteries can be absorbed into expanding domestic zones, typical for boom periods such as Classical Greek times and the Early to Mid-Roman Empire. In Late Antiquity many towns contract, causing abandonment of peripheral domestic areas and also peripheral public buildings, and these sectors may be used for industrial or burial use, or rubbish dumping. More problematic is the internal reorganisation of town centres, especially in Late Antiquity, when fora, theatres, baths, stoas and mansions may be converted into new more mundane uses such as workshops, press rooms, small apartments and also for rubbish dumping. When we consider the city's 'cultural biography' over time, not only of the complex site as a whole, but of each insula or building plot, we might expect that the surface evidence compresses these diverse land uses into a single assemblage of finds which should be suitably inhomogeneous. It is easy to agree with this, and there are many examples of such local changes from excavation, but approaches to teasing the single surface layer out into its contrasted period strands are barely developed.

11.5 Case Studies

We have seen that the survey of complex sites in the Mediterranean has advanced greatly since the nineteenth century, and the resolution of their internal development is currently improving dramatically. However, a large number of problems remain, as has been indicated so far in this contribution. Let us finally look, through case studies, at some of these areas of uncertainty, where we need focussed research to resolve weak points in our approaches. In focussing on particular projects, my comments are in no way meant to be criticisms; rather, to emphasize common problems complex site surveyors have to overcome to increase the detail and reliability of our reconstructions.

Thus, appropriately I will start with my Boeotia Project in Greece. We are publishing the city surveys in reverse order to their field study, aware that our methods and data improved continually between 1982 and 2010. However, as mentioned previously, it is already obvious that the sample collection from *Askra*, surveyed in 1982 and 1985 (ca. 2,000 sherds), despite its small area of 11 ha, is far too small to allow clear spatial representation of the many rarer periods of this small town, occupied from the Bronze Age through to the fourteenth century AD. As for *Haliartos*, surveyed in 1985, with ca. 650 sherds from a 40 ha city, we have severe problems in evaluating the settlement's development except for the times of the town's maximum activity levels between Archaic and Hellenistic times. Even when we move forwards in time to *Thespiae*, surveyed 1985–1986 (ca. 15,000 sherds from 180 ha); *Hyettos*, surveyed in 1990–1991 (13,500 sherds from 26 ha); and *Tanagra*, surveyed 2000–2004 (ca. 20,000 sherds from 50 ha), the calculations presented earlier in this paper indicate central problems with either the spatial representation of minor period uses, or with detecting functional zoning (all three sites have prehistoric beginnings and continue in use into medieval times or later). *Tanagra* at least produced excellent geophysical results, and *Haliartos* is giving good geophysical and aerial photo data for its Classical town plan. *Thespiae* was a geophysical failure for geological and site spoliation reasons, and we intend a large-scale geophysical survey at *Hyettos* for 2014–2015. Revisiting to carry out a more exhaustive architectural survey at Hyettos is in process and was already done in recent years at *Thespiae*, once we realised how vital such extra information could be to assist the ceramic record. Our latest urban survey at *Koroneia* (2006–2010) took full account of our numerical sample concerns and has around half a million sherds collected using the broad 'potential diagnostic' approach. *Koroneia* is a rugged terrace hill and only limited areas are suitable for geophysics, but they already produced good results in 2012. Just to mention one obvious conundrum from the sample size statistics: the widespread or very local but usually low-level presence of Bronze Age and Early Iron Age material at all these city sites is currently extremely difficult to interpret given the factors discussed earlier in this paper, in terms of scale and type of activity which this might represent. On the other hand, the main lines of these towns' evolution are probably reliably reconstructed even from relatively small collections, at least for their major historic phases.

The survey and test-excavation project at *Celti-Peñaflor* in Spain, covering ca. 24 ha (Keay et al. 2000) illustrates likewise both the strengths and weaknesses of our current approaches to complex sites. The programme comprised large-scale surface survey of most of the site, smaller areas of geophysical survey, yet smaller areas of test excavation and the utilisation of earlier excavation and monument records. The geophysical work revealed elements of a street grid from the main Roman phase of the site and buildings which proved difficult to positively characterise. The surface ceramic survey took a window approach of 3×3 m squares set into a grid of 30×30 m, within which 20 min were allotted for as total a collection as possible (thus 1 % of the gridded surface was collected from). The periods represented were fortunately few: Late Iron Age-Republican Roman, Early Imperial Roman and Late Roman. Near-total collection meant a considerable portion proved undatable, and the small samples added to the interpretative challenge. Thus, the authors comment: 'Relatively little could be deduced from the analysis of the distribution of surface materials by their broad chronology' (Keay et al. 2000, p. 29). However, with sharp insight they suggested that concentrations of fine ware could be rubbish from adjacent residential areas and those of amphorae and coarse wares marked a port area. Special problems beset the pre-Roman occupation, whose finds were widespread but hard to interpret. The accumulation of occupation seems to have created a tell, making the accessibility of older levels for exposure at the surface even more problematic. Exactly what kind of indigenous settlement preceded large-scale Roman replanning remained elusive. The test excavations opened small areas but only for the early Roman centuries could plans be drawn up of structures. Even here the nature of the buildings was disputed – public or private, one or two houses, etc. In a concluding chapter, the authors admit that for pre-Roman times, it was 'difficult to gain a clear understanding of the development and spatial organization of the settlement' (Keay et al. 2000, p. 197). Subsequently, in the Roman Republican era to the first century AD, again 'the remains are particularly difficult to interpret' (Keay et al. 2000, p. 199). Even for the best represented and doubtless most flourishing period of the town, the later first to third centuries AD, 'the archaeological evidence has some limitations' (Keay et al. 2000, 203). Neither a clear town plan nor reliable interpretations of the few excavated buildings were possible. Much the most successful part of the project appears to have been a non-contextual study of the ceramics from the site, which could be fitted easily into wider studies of production, distribution and consumption of Iberian wares and their economic correlates in terms of imports and exports. This is a meticulously executed and published report inspiring admiration, but in the end, the nature of the site and the methods applied conspired to prevent the team from achieving their initial goals, 'to chart the development of a major urban centre' (book cover) except in the most general terms.

Next comes *Veii*, a major Etruscan and minor Roman town in South Etruria. As part of Ward Perkins' South Etruria Survey, this giant site (almost 200 ha) underwent extensive ceramic surface survey in the 1950s (Ward Perkins 1961; Potter 1979). Five main sherd clusters around the fringes of the city plateau, linked to extramural cemeteries, were identified as discrete sub-settlements during Iron Age (Villanovan) times, which should have cohered into a larger but still loose-knit community linked by streets in Etruscan times. Much of the plateau interior was used for intensive farming. By Roman times a smaller but nucleated town occupied the plateau centre. However, a more systematic survey by Guaitoli (1982) mapped some four times as many Villanovan sherd foci distributed across the entire plateau (see the comparison in Spivey and Stoddart 1990, p. 47, fig. 18), suggesting that the settlement was already a continuous unified nucleation in the Iron Age. Most recently the original Ward Perkins data have been re-examined and the finds restudied by Patterson et al. (2004), assisted by recent localised Italian excavations. In fact the early survey was remarkably detailed for its time, with almost the entire hill being covered with 84 recorded ceramic findspots (Patterson et al., p. 13, fig. 2). The database was large, although the recent reworking significantly does not offer relevant statistics for the total collection, and the figures published suggest less than 10,000 sherds (5,400 Etruscan and Roman sherds are cited). If correct this raises the limitation noted earlier to the scope for interpretation of such data (which we have commented on already), while the authors also identify major restrictions caused by the original survey's focus on fine wares and a few other diagnostic types. The Villanovan phase remains a very partial cover, still with the densest clusters around the plateau edge, although 69 foci are now mapped (but some are cemeteries). No attempt is made to address the controversial issue of the form of this settlement. For the Etruscan phase 73 foci were mapped but the plateau remains patchily covered and the denser concentrations are discrete from each other. Indeed the interpretation resembles that of Potter (1979) with the suggestion of separate zones of occupation and of specialised functions (domestic, ritual and manufacturing). Ambiguity appears with the description of 'a diffused and homogeneous distribution over the entire plain, but also some substantial empty areas' (Patterson et al., pp. 15–16). For Roman times essentially the original Ward Perkins-Potter view of nucleation and contraction is supported.

Rather as with the *Celti-Peñaflor* case study, the main novelty of the reworking was improved analysis of the ceramic finds out of context, which may reflect the fact that Keay and Patterson are primarily pottery specialists. While the original interpretation of the city's history appears to this reader little altered by the reworking of the data, central issues regarding the 'reading' of urban surface finds are still not being addressed. Which brings me to the final and perhaps greatest current problem with ceramics from large complex sites: surface finds as *behavioural residues*. Returning to the pioneer Mexican 20 sq. km urban survey of Teotihuacan (Millon 1973; Millon et al. 1973), a careful correlation was made between types of surface finds, surface architecture or building material, and test excavations, to set up interpretative categories for discrete ceramic foci or buildings. Despite being heavily influenced by this model, our own early surveys in Boeotia took density and chronology as primary guides to defining the successive built-up areas of a town. Creating borders around the densest concentrations of sherds, combined with clues such as contemporary city walls, appeared a reasonable way to mark the expansion and contraction of a community, while patchy versus concentrated ceramic spreads could indicate multifocal as opposed to nucleated settlement plans (Bintliff and Snodgrass 1988a). The *Veii* survey remains in this tradition.

As we have already seen, there is now an accumulation of insights which undermine the model we used. The excavation of parts of ancient towns repeatedly reveals that the 'cultural biography' of individual building plots tends to produce quite dramatic shifts in land use. Gardens or open space become houses; houses become workshops or are abandoned, places to dump rubbish; and these changes can run in any order; burials can be intra- and extramural and cemeteries are not respected by secular building and vice versa. A dramatic example of these processes based on systematic excavation test windows has been provided by Fentress (1994) for Roman Cosa in Italy. Ceramic signatures for such different spatial uses may be clear in some periods but not at all in others (as we noted for burials and ritual areas). Even city walls may indicate the urban border only for certain phases.

We must add to this our wider understanding of surface scatters in long-occupied Mediterranean landscapes. Inspired by work in the Near East by Tony Wilkinson on manuring carpets radiating out of small and large settlements (cf. Wilkinson 1989), our Boeotia Survey recognised two related phenomena in ancient Greek landscapes: *haloes* and *offsite sherd carpets* (Bintliff and Snodgrass 1988b; Bintliff 2006). Rural farmsteads and villages, when total density studies were made across their surfaces and into surrounding countrysides, consisted of a core of occupation and a wider 'halo' of densities intermediate between that core and the offsite landscape beyond. These haloes appear to represent one or more of the following factors: rubbish dumping, market gardens, craft areas, burials and dispersal by weather and cultivation. The second phenomenon was confined to urban-scale settlements and consisted of very extensive (usually 1–2 km radius) sherd carpets radiating from the urban core well into the countryside, systematically declining with distance (Bintliff 2006; Bintliff et al. 2007). These are the product mainly of large-scale manuring of extramural market gardens and then further out of open fields, secondarily in the immediate surroundings of the built-up town they reflect extramural industry, rubbish dumping, cemeteries and scattered extramural residences. These urban-focussed carpets are effectively the haloes of large sites. I would underline Wilkinson's comment that urban-based manuring appears to be rare in time and may not have been practised at all towns; indeed our Boeotian evidence limits it so far to the Classical Early Hellenistic period. As a result, we have recorded a major difficulty in analysing urban fringes in the six town sites we studied: surface densities remain very high for hundreds of metres beyond the assumed urban core, declining gradually thereafter for 1–2 km.

Thus, when we reconsider the foci of sherds over the 200 ha plateau of *Veii*, we would like to know which are houses, or manured market gardens, or rubbish dumps, and we may expect these functions often to vary for many locations over time, especially as the broader picture (where smallish sherd collections well distributed are still valuable) indicates a clear cycle of urban expansion then contraction over millennia. While industrial activities have been localised, detailed plotting of their debris on Roman sites shows that it was dumped over far larger areas than the production zone proper, both as rubbish and also as road and building fill (cf. Stone et al. 2011; for *Sagalassos* in Turkey: J. Poblome, personal communication 2012). Here is the nub of the Ward Perkins-Potter versus Guaitoli debate: even after the re-evaluation (Patterson et al. 2004), we are no closer to the Villanovan or Etruscan town plan.

The 1980s survey of the Etruscan town of *Doganella* (Walker 1985; Perkins and Walker 1990) also concentrated on clusters of ceramics, although now close-order intensive field walking gave high resolution to such foci. The distribution of ceramic types and indications of storage, industry and generic settlement debris show a complex pattern without clear segregation of such functional clues, although rubbish disposal and problems of stratigraphic representativity are mentioned, as with restudied *Veii*. Deeper issues of activity resolution remain for both major urban sites. The approach taken by the Tiber Valley Project (which included the re-evaluation of Veii discussed above) seems to take urban survey into a new direction, with great gains but clear losses of information potential. For example, the urban site of *Falerii Novi* (Keay et al. 2004) and other Roman towns newly studied by this team have taken maximum advantage of the great advances in geophysics and related technological tools to reconstruct town plans in an impressive and convincing fashion. The role of surface ceramic survey however represents something of a reversion to the methodology of the 1950s. Thus, at *Falerii Novi*, one person surveyed the 30 ha town, noting clusters of artefacts and locations of some closely dated ceramic types. This was followed up by a series of 3 m radius circles placed over town centre buildings whose function was already clear from the geophysical survey, and within which all material was collected. It must in fairness be added that the town was a Roman foundation, so the usual problems of multiphase representation were minimised, while the same circumstance means that the town plan of the Roman town was largely expected to be *the* town plan rather than one of a sequence. At this town's predecessor *Falerii Veteres* (Carlucci et al. 2007) surface conditions and a preference for geophysics led to just 1,100 pieces collected from the 13 ha Etruscan city, so that the authors admitted that the ceramic evidence did not 'merit any really meaningful analysis or interpretation'(Carlucci et al. 2007, p. 89). Given that the site had also a Bronze Age and medieval phase, even the numerous geophysical anomalies could not be clearly dated, and there was a 'lack of clear association' between them and 'datable ceramics from the surface survey' (Carlucci et al. 2007, p. 98). The partial surface artefact survey at *Portus* (Keay et al. 2005) was also dominated by geophysics and contextless analysis of the ceramics from the survey. The approach was again designed to be nonquantitative and impressionistic, with high and medium density concentrations identified by eye-balling, although large diagnostic collections were collected. Little space is given to the ceramic survey maps, which show some artefact types clustered, others almost ubiquitous, some associatable with geophysical anomalies – others not. The team thoughtfully suggest that the finds might emanate from a diverse range of deposits, including in situ levels, through rubbish dumps to burials and just displaced materials. Nonetheless, such speedy and restricted ceramic surveys would be wholly inappropriate to more typical Mediterranean urban sites with lengthy and complex occupation sequences, and significantly the earliest Hellenistic centuries at *Portus* produced finds without any clear structural context. At *Ephesos* in Turkey, (Groh 2012) a similar trend is observable: the poorly known outer sectors of the city have since 2000 been extensively researched through architectural recording (232 ha) and geophysics (53 ha), but surface ceramic survey was experimentally used for just over 2 ha. Remarkably the ceramic collection reached 160,000 pieces but just 3 % proved diagnostic (ca. 5,000).

My next case study, the 125 ha site of *Leptiminus* in Tunisia (Stone et al. 2011) once more highlights the great strengths of current urban survey in the Mediterranean but also areas we practitioners are still struggling with. In many respects, this is a model project with an excellent combination of surface artefact survey, architectural mapping, small-scale selective excavation, large-scale geophysical testing, geomorphological research and importantly surveying that continued through the urban periphery into the countryside (where both offsite scatters and rural sites were analysed). In almost all respects the narrative reconstruction of this Punic and Roman town's 'cultural biography' is convincing and well argued. However, one area where one is not persuaded is once more the reading of surface sherd concentrations in human behavioural terms. The team intelligently emphasise how land uses throughout the town altered significantly over the main 1,000 year urban evolution, such as cemetery areas overlying other uses or being themselves replaced by everyday activities, and the evidence from amphorae that far wider areas received industrial debris than were covered by kilns and workshops. But the underlying interpretative issues we alluded to earlier, regarding older periods and functional zoning, continue to be problematic.

Thus, the Punic and Early Roman phases are honestly admitted to be so buried beneath the structures and then the massive ceramic spreads of the peak Middle Roman Imperial era that their town plans are very speculative. The brave attempt at a Punic town plan is difficult to confirm from the available data, and the subsequent Early Roman plan is equally beset with contradictions between the empirical surface finds and the need to find a reasonable reconstruction. For the central urban florescence era of the Middle Roman Imperial, second-fourth centuries AD, the team offer a basic model of an *urban core* of maximum 45 ha, an *extended suburbium* and then the *rural areas* (Stone et al. 2011, p. 123, fig. 5.3). The *core* contains the main housing areas and public buildings and the port, and at times other activities normally focussed on the *suburbs* (burials, industry and waste disposal). Mapping standing structures and utilising excavation evidence, taken together with localised geophysics, seems to support the core boundaries. The *rural areas* are marked by lower sherd densities but these form a continuous spread around a number of small sites, and are seen as evidence for manuring activities (Stone et al. 2011, p. 187).

But what are we to make of the distribution of amphorae (op.cit. Fig. 6.16) which cover all the core and large parts of the suburbs? Their ubiquity in almost every rural field can at least be explained through the urban manuring model for those agricultural areas. A similar problem we have encountered across the Greek city of Tanagra, where the vast majority of all dated surface finds belong to Late Roman amphorae, locally made and normally assumed to be for transporting olive oil for export (Poblome et al. in Bintliff et al. 2008). At *Leptiminus* cookware (Fig. 5.9a) is also a solid carpet over the core and much of the suburban zone. Other ceramic forms are largely confined to the core, but some are also clearly markers of cemetery areas. It remains unclear whether ceramic spreads are normally marking active domestic, industrial or public areas, or whether urban gardens and areas of intramural rubbish disposal are responsible for significant parts of their distribution. This problem, common to all current urban surveys, is especially prominent in the discussion of the 'suburb'. This far outsizes the core but its characterisation is ambivalent (particularly discussed in op.cit. Chap. 5). It has high-surface densities (we are not told if they are lower than the core), and the basic land use is seen as industrial production and burial interspersed and supplanting each other from phase to phase. There are a handful of suburban villas, a bath complex and an amphitheatre and a number of sectors of ceramic and metallurgical production. However, most of the suburb lacks clear evidence for function, although a ring of cemeteries and kilns seem to demarcate it from the urban core, and it is primarily defined by high sherd densities.

At times the team speculate that there are poor people dwelling amongst the workshops they may have worked in, but in part soil changes suggest otherwise, and geophysics also restricts solid housing – but maybe mud brick served the poor. In general Greco-Roman ceramic workshops on an industrial level do not seem to have included domestic accommodation (V. Stissi, personal communication 2011; Martens 2005), so we need to consider alternative readings of the vast area of discarded pottery in the suburbs, outside of the areas clearly identifiable as burial, production and suburban villa plots. What of extramural market gardens, rubbish dumps or manured field crops? Interestingly, the population reconstructions for the city rely entirely on the area of the urban core. Patterson's (2000) insightful analysis of the textual and archaeological evidence for activities outside the (changing) city wall line of Rome is an excellent start for sensitising us to the overlapping spaces of burial, industry, suburban housing, rubbish dumping, market gardens, food markets and other distribution facilities; as the built-up core of Rome fluctuated back and forth in the long term (interestingly a matter for official administrative changes), such diverse extramural land uses could move in and out of the urban defences as well as replace each other from period to period.

Our final case study is the ca. 30 ha urban site of *Sagalassos* in Turkey (Martens 2005; Martens et al. 2012). This is a model application of complementary techniques: large-scale excavation and geophysics, geomorphological analysis of superficial deposits and artefact survey of two-thirds of the entire residential and public zone (ca. 24 ha). The high quality of the artefact survey work is especially shown in the critical discussion of the results. Firstly Martens cautions us that the absence of post-occupation cultivation has given exceptionally low pottery densities. Moreover, despite virtual total collection, less than 2 % of the collection proved diagnostic. For example, finds datable to one of the three main urban phases of Early, Middle or Late Roman came to a mere 1,152 sherds (J. Poblome, personal communication 2012). Secondly, the absence of cultivation caused the surface of the site to be dominated by the abandonment debris of Late Antiquity; test excavations below this layer showed richer earlier Roman levels underrepresented at the surface. On the other hand, those subsurface levels in turn could be missing the final occupation layers through processes of erosion or human levelling, and surface finds revealed significant pre-Roman activity as well as post-Roman not previously indicated by the lengthy excavation and architectural programmes. Martens insightfully demonstrates that all the complementary methods deployed at this city have their strengths and weaknesses and none can adequately characterise the changing urban plan of a town on their own.

Conclusions

If I were to attempt a list of recommendations to advance the ongoing and future survey methodology and interpretation of large, complex sites in the Mediterranean, for my own team and other urban surveyors, based on the preceding review, it perhaps might run as follows:

1. Only integrated evidence from large-scale remote sensing (dependent on their local suitability: air photos, LiDAR, geophysics, geochemistry), surface architectural recording, surface artefact collection and either test excavation or widespread exposures in the past or present of the subsurface levels is likely to offer a reasonable hope of reconstructing the main lines of urban development from the earliest to latest occupation at such sites.

2. It appears probable that an overreliance on just one of these approaches is inadequate. For example, remote sensing provides little basis for dating features revealed, surface architecture can be ambivalent in function and sometimes displaced from its original context, while surface pottery requires very large collections and theoretical filtering to adjust period and type representations from the usually biased proportions found at the surface.
3. The constraints on funding and the need to offer quick and impressive results tend to encourage short-cuts in urban analysis, which can mean small and selective sherd collections, focussing on the most flourishing era of a site where the plans and finds are easiest to map, and relying primarily on rapid evaluations achieved through air photos or geophysics (which likewise tend to illustrate just limited phases of the site's complex evolution). We might pause to consider that the many thousands of people who occupied such towns over a millennium or more perhaps deserve a deeper and longer period of study to bring them adequately back into history.

Acknowledgements This paper rests on research chiefly supported by the Belgian Fund for Scientific Research and the EU Project CEEDS.

References

Bintliff JL (2006) The Leiden University Ancient Cities of Boeotia Project: 2005 season at Tanagra. Pharos 13:29–38

Bintliff JL (2012) Contemporary issues in surveying complex urban sites in the Mediterranean region: the example of the city of Thespiai (Boeotia, Central Greece). In: Vermeulen F, Burgers G-J, Keay S, Corsi C (eds) Urban landscape survey in Italy and the Mediterranean. Oxbow, Oxford, pp 44–52

Bintliff JL, Snodgrass AM (1985) The Cambridge/Bradford Boeotian Expedition: the first four years. J Field Archaeol 12:123–161

Bintliff JL, Snodgrass AM (1988a) Mediterranean survey and the city. Antiquity 62:57–71

Bintliff JL, Snodgrass AM (1988b) Off-site pottery distributions: a regional and interregional perspective. Curr Anthropol 29:506–513

Bintliff JL, Howard P, Snodgrass A (1999) The hidden landscape of prehistoric Greece. J Mediterr Archaeol 12(2):139–168

Bintliff JL, Howard P, Snodgrass A (eds) (2007) Testing the Hinterland: the work of the Boeotia Survey (1989–1991) in the southern approaches to the city of Thespiai. University of Cambridge, MacDonald Institute Monographs, Cambridge

Bintliff, JL, Farinetti E et al (2008) The Tanagra project: investigations at an ancient city and its countryside (2000–2002). Bull Corresp Hell 128/129(2004–5): 541–606

Bintliff J, Slapšak B et al (2009–2010) The Leiden-Ljubljana Ancient Cities of Boeotia Project 2009 seasons. Pharos 17(2):1–63

Bookidis N (2003) The sanctuaries of Corinth. In: Williams CK, Bookidis N (eds) Corinth, the centenary: 1896–1996. American School of Classical Studies at Athens, Princeton, pp 249–259

Carlucci C, de Lucia Brolli MA et al (2007) An archaeological survey of the Faliscan settlement at Vignale Falerii Veteres (Province of Viterbo). Proc BSR 62:39–111

Cherry JF, Davis JL et al (eds) (1991) Landscape archaeology as long-term history. University of California Los Angeles, Institute of Archaeology, Los Angeles

Fentress E (1994) Cosa in the empire: the unmaking of a Roman town. J Rom Archaeol 7:209–222

Flannery KV (ed) (1976) The early Mesoamerican village. Academic Press, New York

Groh S (2012) Strategies and results of the urban survey in the Upper City of Ephesus. In: Vermeulen F, Burgers G-J, Keay S, Corsi C (eds) Urban landscape survey in Italy and the Mediterranean. Oxbow Books, Oxford, pp 62–71

Guaitoli M (1982) Notizie preliminari su recenti ricognizioni svolte in seminari dell'Istituto. Quaderni dell'Istituto di Topografia Antica dell'Università di Roma 9:79–87

Jameson MH, Runnels CN et al (eds) (1994) A Greek countryside. The southern Argolid from prehistory to the present day. Stanford University Press, Stanford

Keay S, Creighton J et al (eds) (2000) Celti-Peñaflor. The archaeology of a Hispano-Roman town in Baetica. Oxbow Books, Oxford

Keay S, Millett M et al (2004) New approaches to Roman urbanism in the Tiber Valley. In: Patterson H (ed) Bridging the Tiber. Approaches to regional archaeology in the Tiber Valley, Archaeological monograph 13. The British School at Rome, London, pp 223–236

Keay S, Millett M et al (2005) Portus. An archaeological survey of the Port of Imperial Rome, Archaeological monographs 15. The British School at Rome, London

Martens F (2005) The archaeological urban survey of Sagalassos (South-West Turkey): the possibilities and limitations of surveying a 'non-typical' classical site. Oxf J Archaeol 24:229–254

Martens F, Mušič B et al (2012) The integrated urban survey at Sagalassos. In: Vermeulen F, Burgers G-J, Keay S, Corsi C (eds) Urban landscape survey in Italy and the Mediterranean. Oxbow Books, Oxford, pp 85–93

Millon R (1964) The Teotihuacán mapping project. Am Antiq 29:345–352
Millon R (1973) The Teotihuacán map, vol 1.1. University of Texas Press, Austin
Millon R, Drewitt B et al (1973) The Teotihuacán map, vol 1.2, maps. University of Texas Press, Austin
Müller-Celka S (2007) Terroir et territoire à Malia à l'époque protopalatiale. Bull Corresp Hell 131:855–860
Patterson JR (2000) On the margins of the city of Rome. In: Hope VM, Marshall E (eds) Death and disease in the ancient city. Routledge, London/New York, pp 85–103
Patterson H, di Gennaro F et al (2004) The re-evaluation of the South Etruria survey: the first results from Veii. In: Patterson H (ed) Bridging the Tiber. Approaches to regional archaeology in the Tiber Valley, Archaeological monograph 13. The British School at Rome, London, pp 11–28
Perkins P, Walker L (1990) Field survey of the Etruscan city at Doganella. PBSR 58:1–144
Potter TW (1979) The changing landscape of South Etruria. Elek, London
Renfrew C, Wagstaff M (eds) (1982) An Island polity. The archaeology of exploitation in Melos. Cambridge University Press, Cambridge
Slane KW (2003) Corinth's Roman pottery. Quantification and meaning. In: Williams CK, Bookidis N (eds) Corinth, the centenary: 1896–1996. American School of Classical Studies at Athens, Princeton, pp 323–335
Spivey NJ, Stoddart S (1990) Etruscan Italy. Batsford, London
Stone DL, Mattingly DJ et al (2011) Leptiminus (Lamta) report no. 3: the field survey. J Rom Archaeol, Supplementary Series Number 87 Portsmouth Rhode Island
Vermeulen F (2003) The Potenza Valley Survey: preliminary report on field campaign 2002. BABesch 78:71–106
Walker L (1985) Survey of a settlement: a strategy for the Etruscan site at Doganella in the Albegna Valley. In: Haselgrove C, Millett M, Smith I (eds) Archaeology from the Ploughsoil: studies in the collection and interpretation of field survey data. Sheffield excavation reports. J. R. Collis Publications, Sheffield, pp 87–94
Ward Perkins JB (1961) Veii. The historical topography of the ancient city. Proceedings BSR 29:1–123
Waterhouse HE, Hope Simpson R (1960) Prehistoric Laconia: part I. Annu BSA 55:67–107
Whitelaw T (1998) Colonisation and competition in the polis of Koressos. In: Mendoni L, Ainian AM (eds) Keos-Kythnos: history and archaeology. Proceedings of an international symposium (Kea-Kythnos, 22–25 June 1994). National Hellenic Research Foundation, Athens, pp 227–257
Whitelaw T, Bredaki M et al (2006–7) The Knossos urban landscape project. Archaeol Int 10:28–31
Wilkinson TJ (1989) Extensive sherd scatters and land-use intensity: some recent results. J Field Archaeol 16:31–46
Willey GR (1953) Prehistoric settlement patterns in the Virù Valley, Peru. Bureau of American Ethnology, Bulletin 155. Smithsonian Institution, Washington, DC

Site Discovery and Evaluation Through Minimal Interventions: Core Sampling, Test Pits and Trial Trenches

12

Philip Verhagen

Contents

P. Verhagen
VU University Amsterdam, Amsterdam, The Netherlands
e-mail: j.w.h.p.verhagen@vu.nl

12.1 Introduction

The Radio-Past Project has focused on the application of non-invasive techniques for measuring and mapping characteristics of the soil and the archaeological remains therein. This includes techniques like geophysics, aerial photography, remote sensing and LiDAR. High-end non-invasive technology is becoming cheaper, easier to apply and generally more successful in detecting and evaluating archaeological remains. But these techniques still have a major drawback: they do not give us any *direct* information about what is in the ground. What does a geophysical anomaly tell us about soil conditions at a certain depth? And how should we interpret the features recognized on an aerial photograph, a satellite image or a LiDAR-based elevation model? Inevitably, at some point we need ancillary data to support the interpretations made on the basis of non-invasive survey data.

In practice, the majority of archaeological surveys nowadays are still done using techniques that rely on direct observation of the soil and the archaeological remains found in it. This is either in the form of a field survey, which does not enter the soil, or in the form of 'minimal interventions': invasive techniques like core sampling, test pitting and trial trenching that allow us to observe a limited portion of the subsoil. In this chapter, I will give an introduction to the existing invasive minimal intervention techniques, together with a description of their potential and limitations, and

C. Corsi et al. (eds.), *Good Practice in Archaeological Diagnostics*, Natural Science in Archaeology,
DOI 10.1007/978-3-319-01784-6_12,

I will discuss best practices for using them in conjunction with non-invasive techniques.

12.2 Minimal Interventions

The term 'minimal intervention' refers to any kind of technique that takes one or more samples of the subsoil to map the occurrence, extent and nature of archaeological remains. Minimal interventions are either designed to target future survey and excavation efforts on the spots that are most likely to contain archaeological remains of interest (purposive survey: Banning 2002); or they are used for statistical purposes, to obtain a reliable estimate of, e.g. the density of features or ceramics in a site or the amount of archaeological sites in an area (probabilistic survey). In practice, these two categories are not mutually exclusive. A purposive survey will produce results that may also be used for statistical estimates, and probabilistic sampling may be used for mapping of archaeological sites as well. However, since these forms of survey have different goals, their requirements for *optimal* sampling strategies are different. In this chapter I will focus on the purposive survey, since it relates more closely to the ways that non-invasive methods are used for detection and evaluation of archaeological sites. For an introduction to probabilistic survey theory and methods, the reader is referred to Mueller (1975), Banning (2002, 113–132) and Orton (2000, 67–111).

Three major techniques of minimal intervention can be distinguished: core sampling, test pit sampling and trial trenching. They all have their specific domain of application, depending on the level of detail of information desired and the prevailing soil conditions.

Core sampling (also known as coring or augering) takes very small soil volumes. It has a long tradition in the geosciences as a cheap, reliable and versatile way to take soil samples, to describe their lithological and pedological characteristics and to take additional samples if necessary, for example, for palynological or mineralogical analysis. The information obtained is traditionally used to infer the horizontal and vertical extent of geological or pedological units, but can also be used to map the extent and nature of archaeological sites and features (Fig. 12.1).

Fig. 12.1 Core sampling (Photo: courtesy RAAP Archeologisch Adviesbureau)

Standard manual coring equipment uses a 7 or 12 cm diameter auger (for use in sandy soils) or a 3 or 5 cm diameter gouge (for use in clayey soils or peat). Certain types of mechanical coring machines can take larger samples of up to 20 cm in diameter, but they are more expensive to operate. Core sampling only takes a small volume of soil and does not allow for in situ observation. The soil has to be brought to the surface and inspected there. As a result, archaeological features (specific soil colorations) are usually hard to recognize on the basis of core samples. Furthermore, artefacts may be missed altogether, and sieving[1] is necessary to increase the probability of observing them. However, as it is the only method capable of easily penetrating the soil at depths of more than about 1.5 m, it is an indispensable tool for archaeological surveying in areas where a strong accumulation of peat or sediments is found and where archaeological sites may be buried at considerable depths. Gouges can reach depths of up to 10 m, although at these depths the use of mechanical equipment is more convenient.

Test pit sampling[2] involves digging of small test pits by hand (usually 1 × 1 m or 50 × 50 cm), accompanied by systematic sieving of the

[1]Known as screening in the USA.

[2]Known as shovel testing in the USA.

artefacts. Small excavators may also be used to dig the pits. Test pitting allows for in situ observation of the soil in the test pit itself, so stratigraphic sequences and archaeological features may be recognized more easily. For this reason, test pits are frequently applied in soil science as well. The soil volume taken out of the test pit is substantially larger than with core sampling, so the risk of missing artefacts is lower. It can also be a useful technique to take larger soil samples, for example, for pollen analysis or OSL dating. It is very popular as an archaeological survey method in the United States and Canada, and elsewhere it is regularly applied in situations where digging larger trenches is impractical, for example, in dense forest. However, test pits cannot be applied at depths larger than 1.5 m, and they usually do not provide sufficient information for correctly interpreting archaeological structures.

Trial (machine) trenching[3] is currently the most widely used invasive prospection method in European archaeological heritage management. Excavators are used to dig trenches that are usually 2–4 m wide and have lengths that can vary between 10 and 50 m. It can sample large areas in a short period of time, and the area coverage is usually between 2.5 and 10 %. The soil removed from the trenches is usually not sieved because of the large volume taken out, and the observation of artefacts is only done in the trench itself. In situ observation of the soil over a larger distance allows for easier interpretation of the stratigraphic sequence and of the features and structures found. It is also regularly applied to study geological cross sections. It is possible to dig trenches deeper than approx. 1.5 m, but this takes more time and requires extra security measures (Fig. 12.2).

In practice, all three intervention techniques are primarily used to systematically sample larger study areas (usually of the order of a few hectares), with sample locations (boreholes, test pits or trenches) positioned in a regularly spaced pattern. However, targeted application is possible as well; in those cases the samples are located in places where we want to have more specific information.

Fig. 12.2 Trial trench (Photo: courtesy RAAP Archeologisch Adviesbureau)

12.3 Minimal Interventions and Site Discovery

The literature on the best ways to apply invasive methods for discovery and evaluation of archaeological remains is relatively limited and mainly focuses on statistical aspects of site discovery (purposive survey). This was first discussed in American publications of the early 1980s, where test pitting became an important technique for site discovery in archaeological heritage management (Krakker et al. 1983; Nance 1983; McManamon 1984; Lightfoot 1986; Nance and Ball 1986; Kintigh 1988). A good overview of the relevant issues is found in Zeidler (1995). Banning (2002) would be the best starting point for an introduction to the subject, and Orton (2000) provides a lot of background on the statistical aspects. For core sampling, the theoretical

[3] Known as backhoe trenching in the USA.

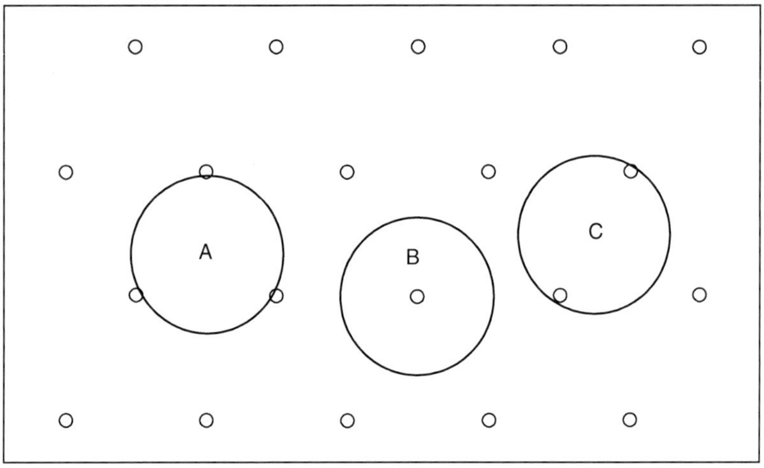

Fig. 12.3 The number of core samples inside a circular target can vary according to its position (Source: Tol et al. 2004, reproduced with permission)

framework largely follows that for test pitting and is discussed in Tol et al. (2004; in Dutch) and Verhagen (2005). For trial trenching, the number of publications is limited as well, with the major theoretical issues being discussed by Champion et al. (1995), Hey and Lacey (2001) and Verhagen and Borsboom (2009) (Fig. 12.3).

Krakker et al. (1983) introduced the concepts of *intersection probability* and *detection probability* for modelling the probability of site discovery given a certain survey method. They stated that the probability that an archaeological site will be discovered is the product of the probability that it is *intersected* by boreholes, test pits or trenches and the probability that the artefacts and/or features in the site will actually be *detected.* When the detection probability is equal to 1, then intersecting the site is sufficient for discovery. However, in most cases this probability is (much) lower. The reciprocal of discovery probability, or the probability of non-discovery, is known as the *consumer's risk* (Gilbert 1987).

12.3.1 Intersection Probability

The intersection probability of a site of a certain size is directly related to the spacing between trenches, test pits or boreholes. Intersection probabilities are most easily calculated for continuous trenches (equivalent to walking lines in a field survey). In that case, the following equation applies for all elliptical targets (Davis 1986; Sundstrom 1993):

$$P(\text{int}) = \frac{p}{\pi D}$$

where
p = the perimeter of the ellipse and
D = the distance between trenches.

This can be simplified to

$$P(\text{int}) = \frac{2\sqrt{a^2 + b^2}}{D}$$

where
a = the major semi-axis of the ellipse
b = the minor semi-axis of the ellipse.

In the case of a circle this equates to

$$P(\text{int}) = \frac{2r}{D}$$

where
r = the radius of the circle.

For linear targets the following equation applies:

$$P(\text{int}) = \frac{2l}{\pi D}$$

where
l = the length of the line.

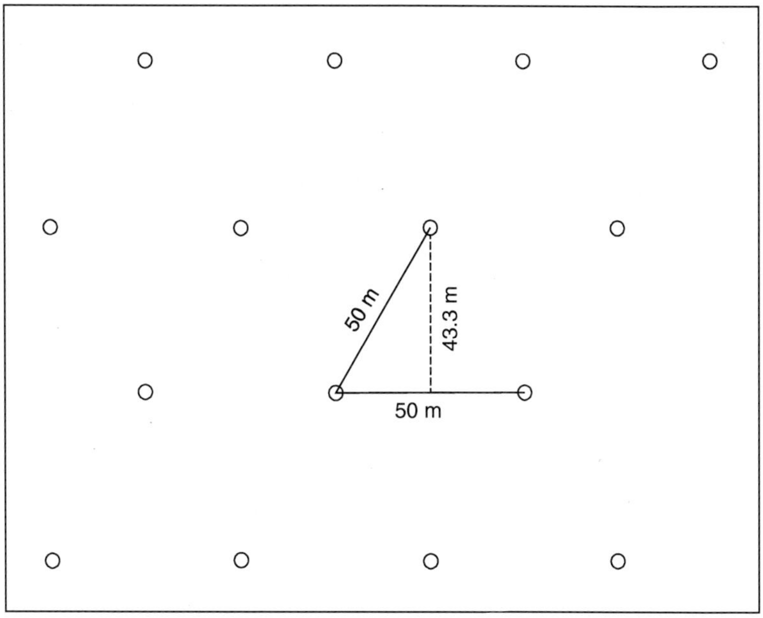

Fig. 12.4 An equilateral triangular sampling grid, with distances of 50 m between the sampling points (Source: Tol et al. 2004, reproduced with permission)

The probability that a site will be intersected by the points of a regular point sampling grid of boreholes or test pits is given by (Drew 1979):

$$P(\text{int}) = \frac{A}{i \times s}$$

where
A = the area of the site
i = the distance between sampling points
s = the distance between sampling rows.

The optimal point sampling layout was proved to be an *equilateral triangular grid* by Krakker et al. (1983; see also Kintigh 1988; Fig. 12.4). This provides the highest intersection probabilities for both circular and elongated targets. However, when the orientation of elongated targets is more or less known, then a rhombic grid is more efficient, with the long axis of the rhombus parallel to the long axis of the ellipse (Banning 2002, pp. 97–100).

These equations only give true intersection probabilities for site sizes *smaller* than the spacing between sampling points or lines. In fact, the number calculated by the equations is the mean of the probability distribution of hits inside a site of a certain size, given a certain spacing between sampling points or lines. The equations are therefore equally applicable for sites larger than the line or point spacing (see also Sundstrom 1993). Only in the case of circular targets and continuous trenches an 'intersection probability' of 1 implies that the site will never be missed. When using an equilateral triangular point sampling layout, the 'probability' at which a circular site will always be intersected is 1.21. For an elliptical site with a ratio of major to minor axis of 2:1, this figure equals 1.67 (Verhagen 2005).

For trial trenches, no similar equations are available because of the complexity of possible trench configurations. A systematic trial trenching pattern is described by four basic parameters (Verhagen and Borsboom 2009):

- Trench width (W)
- Trench length (L)
- Trench spacing
- Trench configuration

Trench spacing and length are the primary factors determining the intersection probability of archaeological sites. Trench spacing can be specified in two directions: parallel to the trenches (the interval, I) and perpendicular to the trenches (the distance, D). It is common practice to keep the two equal, but this is not necessary. Most

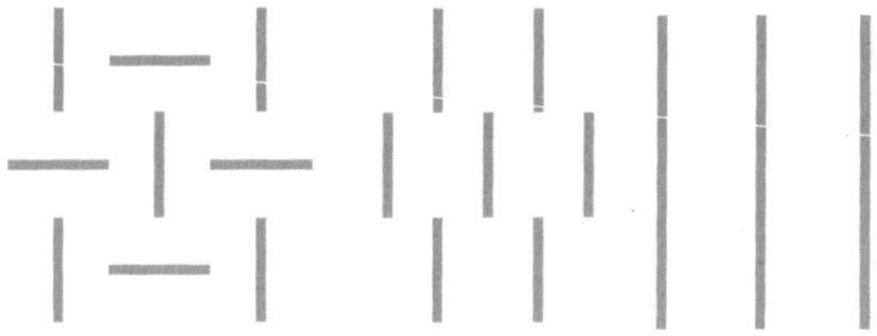

Fig. 12.5 The three most commonly used systematic trial trenching patterns. From *left* to *right*: the standard grid, the parallel staggered grid and the continuous parallel trenching pattern (Source: Verhagen and Borsboom 2009)

archaeologists tend to decide on trench length first, rather than on trench spacing. The two factors are, however, closely related and influence each other.

Trench configuration also has an effect on the intersection probability. The three most commonly applied configurations are (Fig. 12.5):

- A pattern of continuous parallel trenches
- A pattern of staggered parallel trenches
- A pattern of staggered trenches that are rotated alternately by 90° (the English 'standard grid')

The staggered parallel system is the most widely used in Europe, and the standard grid is only applied in England. For the standard grid and staggered parallel trenches, the most commonly used systems have equal interval, distance and trench length. Increasing or decreasing the trench length then also implies increasing or decreasing the trench spacing. Interval, distance and trench length can, however, be manipulated independently, leading to a wide array of potential configurations apart from the 'normal' systems (Fig. 12.6).

The only way to calculate the intersection probabilities of trench patterns is by means of simulation. Verhagen and Borsboom (2009) present the results of such simulations and conclude that:

1. A continuous trenching pattern is the *least efficient* for site intersection when using the same area coverage compared to other configurations. Furthermore, it has a substantially higher risk of missing linear targets that are running more or less parallel to the trenches (Fig. 12.7).
2. The *most efficient* configuration when using the same area coverage is the 'standard grid'. This was already recognized by Hey and Lacey (2001) and follows from the fact that this pattern is much better at intersecting linear targets because of the alternating orientation of the trenches. For circular targets, the difference with the staggered parallel grid is very small and depends on the size of the target relative to the trench spacing.
3. Instead of choosing between a standard or staggered parallel grid, however, it is much *more effective* to manipulate the trench spacing. Reducing I and D relative to L clearly increases the intersection probabilities while maintaining the same area coverage. The most efficient strategy is one where the interval and distance are twice the trench length ($I=D=2L$) (Fig. 12.8). Care should be taken, however, when linear targets are expected, as these are most effectively intersected by a standard grid where $I=D=L$ (Fig. 12.9).

12.3.2 Detection Probability

Detection probability determines whether an intersected site will actually be detected. In the case of clearly visible and spatially continuous occupation layers, there will be no problem detecting the phenomenon concerned, even when using core samples. Archaeological features, however, are usually relatively small and widely spaced, so they need to be viewed in context from above or in a section in order to be recognized with certainty. Core sampling and test pitting are therefore not very well suited for detecting features, and trenches are preferable. When small unit sampling is chosen as a survey method, the probability of detecting an archaeological site is predominantly dependent on the artefact density. The lower this density is, the more difficult it will be to detect an archaeological site (Fig. 12.10).

The probability of encountering an artefact in a small size sample unit is given by an exponential distribution (Stone 1981; Verhagen and Tol 2004):

$$P(\det) = 1 - e^{-A \times d \times W}$$

where

e = the base of natural logarithms (= 2.711828)

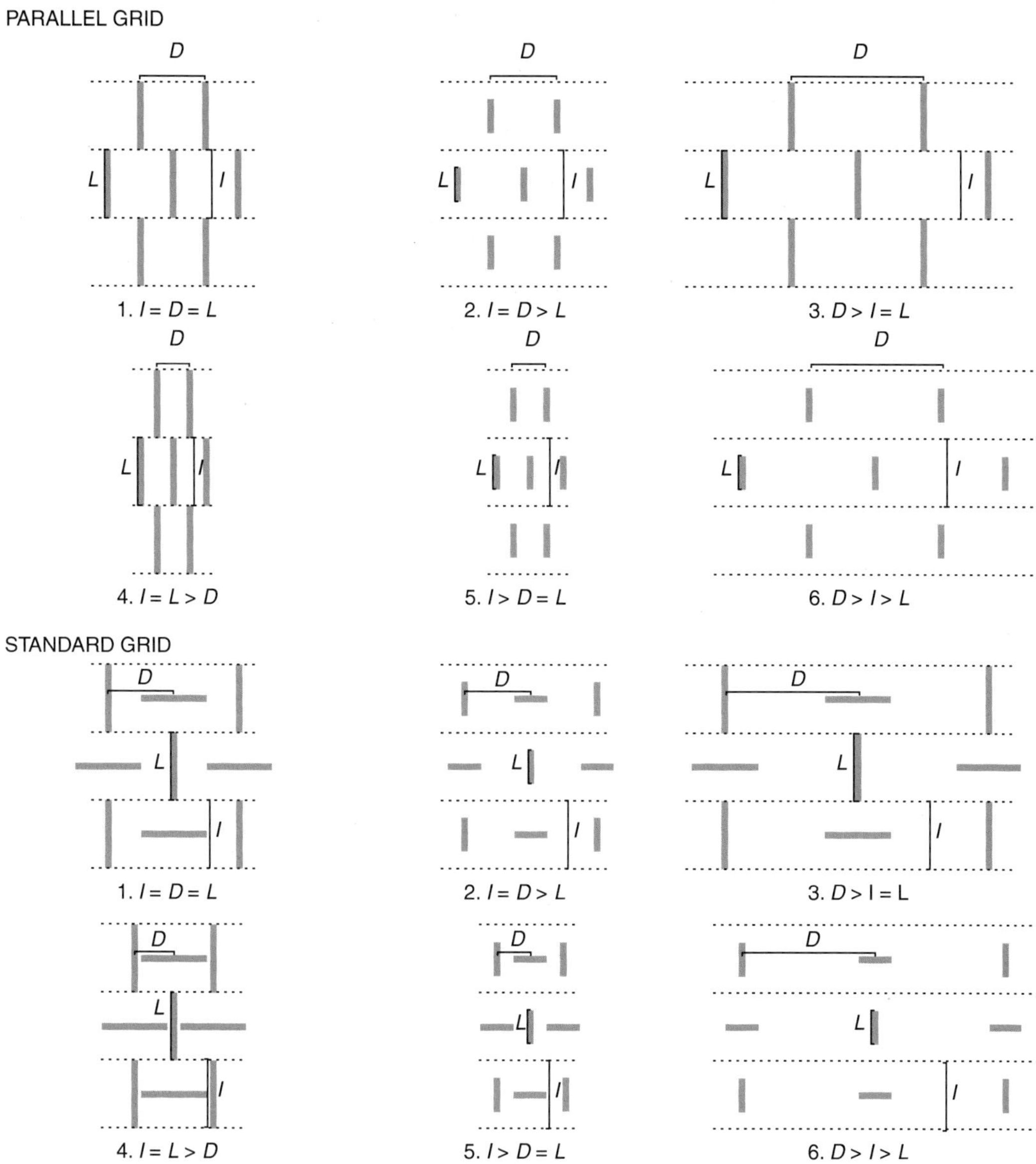

Fig. 12.6 Manipulating trench length (*L*), interval (*I*) and distance (*D*) can lead to a potentially endless array of systematic trenching patterns (Source: Verhagen and Borsboom 2009)

A = the sample unit's area
d = the density of artefacts per area unit
W = the probability that the artefacts will actually be observed.

This distribution allows us to calculate the probability of detecting a certain artefact density, given a particular sample unit size.

Detection probability is directly related to the density of the phenomena to be observed (in most cases artefacts). In general, it can be stated that the observation window of trial trenching is sufficiently large to permit the observation of artefacts without sieving. However, for test pitting and core sampling, the small sample unit size is much more problematic. Typical test pit sizes in the United States are 30×30 cm (0.09 m^2), and for core sampling diameters of 7 cm (0.004 m^2) or 3 cm (0.0007 m^2) are used. Sieving is therefore necessary to increase the detection probability of artefacts for test pits and core samples. A conservative estimate would be that the amount of observable flints increases by at least a factor of

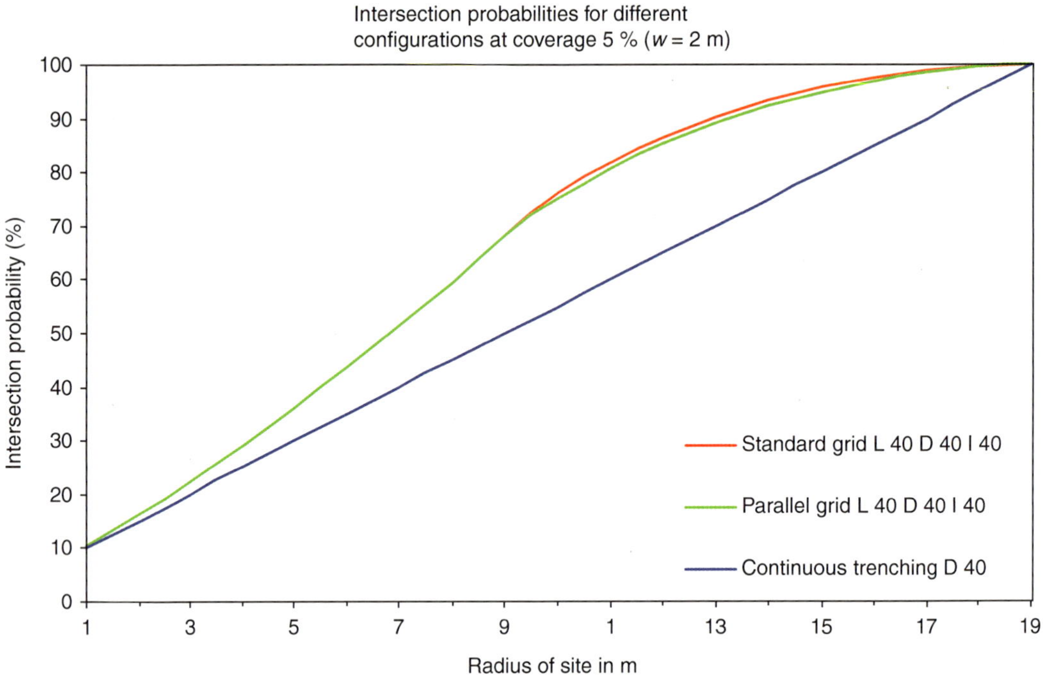

Fig. 12.7 Effect of trench configuration on intersection probability, using the same area coverage of 5 % (Source: Verhagen and Borsboom 2009)

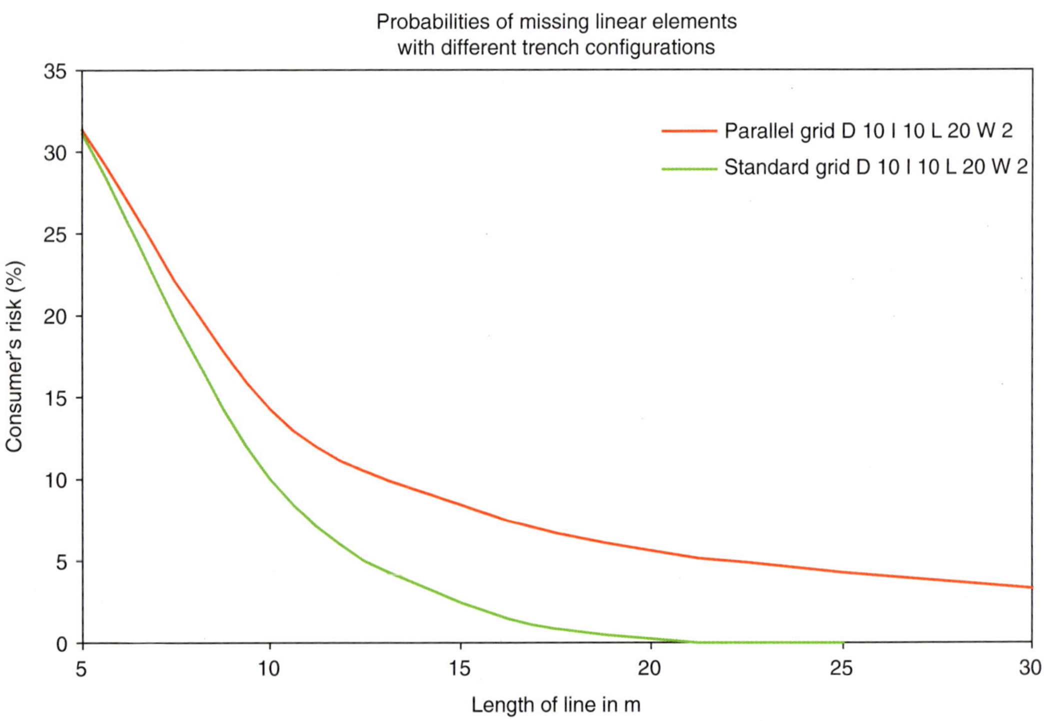

Fig. 12.8 Probability of missing a linear element with different trench configurations, using the same area coverage (Source: Verhagen and Borsboom 2009)

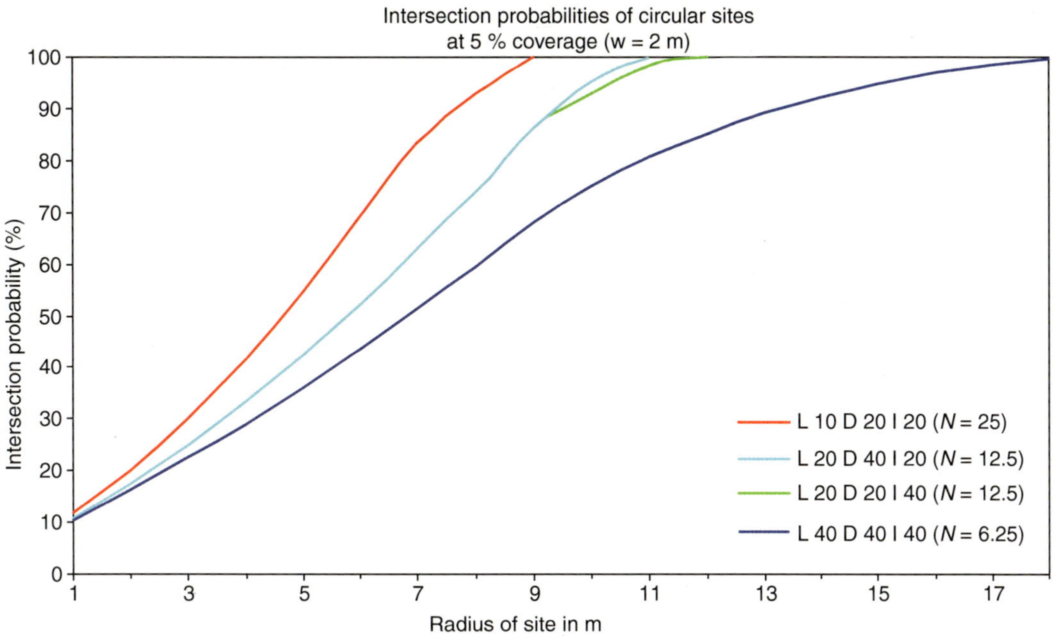

Fig. 12.9 The effect of manipulating trench spacing and trench length on intersection probability for parallel grids while maintaining the same area coverage (Source: Verhagen and Borsboom 2009)

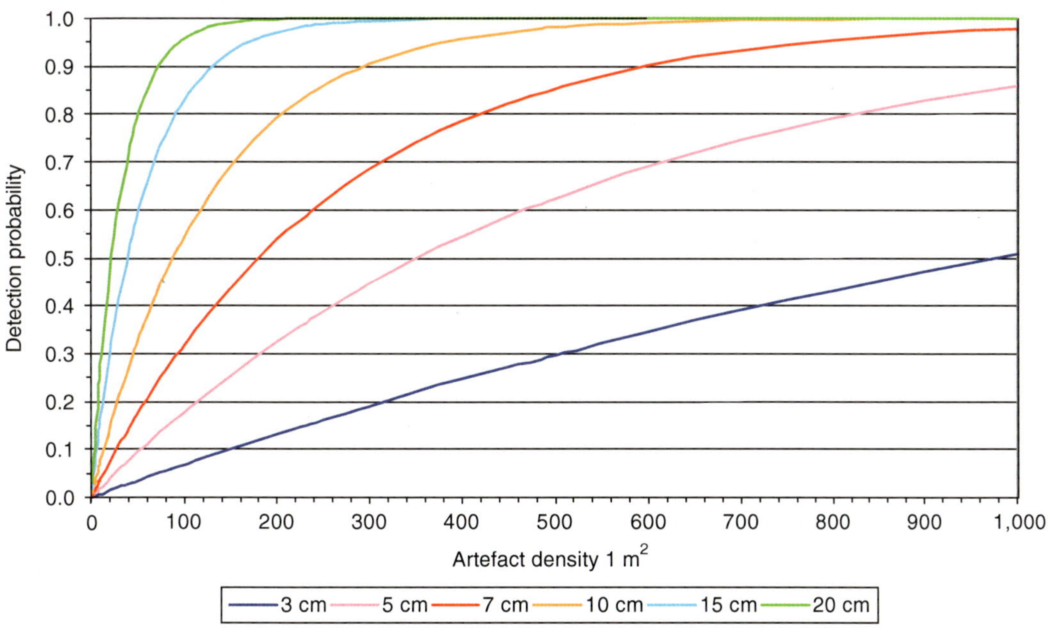

Fig. 12.10 Detection probabilities for different core diameters (Source: Tol et al. 2004, reproduced with permission)

10 when sieving at 1 mm instead of observing with the naked eye (Verhagen et al. 2012).

The theory presented by Krakker et al. (1983) assumes that artefacts will be randomly distributed within a site, but this is not very likely. Artefacts will appear as concentrations with density decreasing from the site centre to zero at the edge or to a background noise of 'off-site' scatter. This decay will occur at varying distances and will follow varying decay curves. Kintigh (1988) attempted to model different spatial artefact distributions, but these distributions will probably not bear a great similarity to reality. Simulations carried out on 12 Stone Age sites in the Netherlands and Flanders by Verhagen et al. (2012) clearly show that the detection probability of flints can be considerably lower than the random distribution suggests because of the effect of artefact clustering.

Finally, it should be pointed out that the assumption put forward by Krakker et al. (1983) that discovery probability is the *product* of intersection probability and detection probability is not correct. Unlike the equations for intersection probability, the equation for detection gives a true probability. In other words, there is always a possibility that taking two or more samples in a site will result in non-detection. The correct equation for calculating *discovery probability* on the basis of samples that might not reveal evidence of artefacts is therefore:

$$P(\text{dis}) = 1 - \binom{n}{x} P(\text{det})^{x} \left(1 - P(\text{det})\right)^{n-x}$$

where
n = the number of samples taken inside the site
x = the number of empty samples, not revealing any evidence; this should be set to 0.

12.3.3 Optimal Strategies for Site Discovery

On the basis of the theory discussed in Sects. 12.3.1 and 12.3.2, Tol et al. (2004) proposed a basic subdivision of archaeological sites into 'prospection groups'. Each group has its own prospection characteristics (in terms of size, shape and artefact density) for which an effective and efficient strategy is recommended that will guarantee a probability of discovery of 0.75 (Tol et al. 2006; see also Verhagen and Borsboom 2009; Table 12.1). This latter figure is not based on a (quantitative) cost-benefit analysis of the investment in research and the corresponding archaeological output but reflects a general agreement among Dutch archaeologists that this is a reasonable and feasible rate of success for a survey. The task of designing a survey strategy is then redefined to specify what prospection groups are expected in a study area and choosing the appropriate accompanying strategy. A comparable system of recommended strategies for trial trenches was designed by Borsboom and Verhagen (2009; Table 12.2). There is, however, one main weakness in this system: the prospection characteristics of specific site types, and in particular artefact density, are not very well known and vary per region. We don't have very reliable data on, e.g. the average regional flint densities of Mesolithic sites (see also Verhagen et al. 2012) so there is still a certain amount of guesswork involved in making the specified predictions. However, in Dutch archaeological heritage management, this approach is preferred to conducting surveys either indiscriminately or on the basis of expert judgment only (see also Sect. 12.5).

Nicholson et al. (2000) discuss the problem of trying to estimate the risk that archaeological remains of a certain size may be missed, given a specific research intensity. The probability of not intersecting remains with a size of 1 % of the study area (e.g. a 100 m^2 site in a 100 × 100 m survey area) is still 61 % when taking 50 samples. However, when we start a survey with a specific hypothesis in mind about the type of sites we are looking for, we can use Bayesian statistics to come up with a more realistic estimate.

For that, we need to be able to specify the smallest area of archaeological remains that we want to detect. The problem of imperfect detection is tackled by dividing this area by the detection probability involved. So, in the case of a 200 m^2 site with an 80 % detection probability of

Table 12.1 Overview of standard strategies for core sampling survey in the Netherlands for different site types

Type/dating	Lithology	Sampling grid	Auger diameter	Sieving	Sampling grid	Auger diameter	Sieving
Stone Age	Flint scatter				Cultural layer		
Very small (< 50 m²)							
Low find density (40–80/m²)	Irrelevant	–	–	–	N/A		
Very low find density (< 40/m²)	Irrelevant	–	–	–			
Small (50–200 m²)							
Low find density (40–80/m²)	Irrelevant	4×5 m	15 cm	3 mm	N/A		
Very low find density (<40/m²)	Irrelevant	4×5 m test pits	–	–			
Medium sized: 200–1,000 m²							
Medium-high find density (>80/m²)	Irrelevant	13×15 m	12 cm	3 mm	20×25 m	3 cm	–
Low find density (40–80/m²)	Irrelevant	8 x 10 m	15 cm	3 mm			
Very low find density (<40/m²)	Irrelevant	4×5 m test pits	–	–			
Large: >1,000 m²							
Medium-high find density (>80/m²)	Irrelevant	20×25 m	12 cm	3 mm	40×50 m	3 cm	–
Low find density (40–80/m²)	Irrelevant	13×15 m	12 cm	3 mm			
Bronze Age–Middle Ages	Ceramic scatter				Cultural layer		
Small, 500–2,000 m²	Sand	30×35 m	15 cm	4 mm	30×35 m	3 cm	–
	Clay/loess	20×25 m	12 cm	4 mm			
	Clay/loess	17×20 m	12 cm	–			
Large, >8,000 m²	Sand	80×90 m	15 cm	4 mm	80×90 m	3 cm	–
	Clay/loess	60×70 m	12 cm	4 mm			
	Clay/loess	40×50 m	12 cm	–			
Unspecified							
	Sand	20×25 m	15 cm	4 mm			
	Clay/loess	13×15 m	12 cm	–			

Sources: Verhagen and Borsboom (2009) and Verhagen et al. (2012)
Flint and ceramic scatters can only be effectively detected when artefact densities are >40/m², for lower densities core sampling is not recommended. Sieving is used to increase the detection probability of artefacts. Cultural layers are distinct lithostratigraphical units that can be recognized directly as archaeological relics and hence have a detection probability of 1

artefacts, we should reduce the 'site area' to $200 \times 0.8 = 160$ m². We also have to specify an assessment of the probability that these remains are present at all. For the purpose of illustration, let's assume that earlier research indicated that these remains were actually found in 10 % of cases. This means that the initial probability of such sites being present is 3.7 %. When taking 50 'empty' samples in the survey area, this risk is reduced to 1.0 % (for the actual mathematics involved, see Nicholson et al. 2000). The risk that we missed two of these is then 0.3 %. Such an approach seems helpful in analysing the risks involved with purposive surveying, but it implies that sufficient data should be collected to estimate our prior assumptions on the presence and size of archaeological remains.

12.4 Minimal Interventions and Site Evaluation

In many cases, minimal interventions will not just be aiming at the discovery of sites. Equally important is the capability of the technique to delimit the areas of archaeological interest and to allow for an easy interpretation of what is found. Evaluation of sites using core sampling or test pitting is more problematic, since both methods rely largely on the detection of artefacts in the soil sample. When we have struck an artefact, this will usually be a reason to look more closely, by taking more samples in the vicinity of the find location and trying to establish a site contour. This approach is also known as adaptive

Table 12.2 Overview of recommended strategies for trial trenching in the Netherlands for different site types

(1) Finds only	Trial trenching in general not suitable. alternatives: test pitting, core sampling, field survey		
(2) Features only	Feature density (average density)		
	<1 % (0.5)[a]	1–10 % (2.0)	>10 % (10)
Small (500–2,000 m²)	W4 L15 D15 I30 (C 13 %)	W4 L10 D20 I20 (C 10 %)	W4 L15 D30 I30 (C 6 %)
Medium sized (2,000–8,000 m²)	W4 L10 D20 I20 (C 10 %)	W4 L15 D30 I30 (C 6 %)	W2 L12.5 D25 I25 (C 4 %)
Large (>8,000 m²)	W4 L15 D30 I30 (C 6 %)	W2 L15 D30 I30 (C 3 %)	W2 L25 D50 I50 (C 2 %)
(3) With features and finds	Feature density (average density)		
	<1 % (0.5)[a]	1–10 % (2.0)	>10 % (10)
Small (500–2,000 m²)	W4 L15 D15 I30 (C 13 %)	W4 L10 D20 I20 (C 10 %)	W2 L10 D20 I20 (C 5 %)
Medium sized 2,000–8,000 m²)	W4 L15 D30 I30 (C 6 %)	W2 L10 D20 I20 (C 5 %)	W2 L15 D30 I30 (C 3 %)
Large (>8,000 m²)	W2 L12.5 D25 I25 (C 4 %)	W2 L25 D50 I50 (C 2 %)	W2 L50 D100 I100 (C 1 %)
(4) Linear sites	Unknown orientation	Orientation more or less known	
	'Standard grid' or 'H'-configuration	Continuous trenching or parallel staggered grid with $L = D = I$	
(5) Point sites	Unknown location	Location more or less known	
	[b]	Cross, X-configuration)	
(6) 'Off-site'	No settlement known or predicted	Settlement known or predicted	
	W2 L25 D100 I 100 (C 0.5 %)	Continuous trenching, D100)	

Source: Verhagen and Borsboom (2009)

Unless specified otherwise, strategies are based on a parallel staggered grid, with parameters *W* (trench width), *L* (trench length), *D* (trench distance) and *I* (trench interval) in metres. Coverage (*C*) is given for reference

[a]Very small sites with very low feature densities can be discovered by means of trial trenching, but interpretation might be (very) difficult

[b]Other methods might be better suited

sampling (see Orton 2000, pp. 34–38). However, since we are dealing with imperfect detectability of the artefacts, the neighbouring samples may be empty, even when they are inside the site's boundaries.

Trial trenching is much more suitable since it uncovers larger areas and allows us to observe archaeological features and structures much more closely. Studies into defining optimal trial trenching strategies for purposes of site evaluation have not come up with very clear guidelines, however. Given the current state of knowledge, it seems impossible to adequately model the requirements for site evaluation in quantitative terms. Even highly experienced archaeologists cannot specify the exact amount or proportion of features that needs to be uncovered for evaluation, as this also depends on the nature of the features themselves and how easily they can be recognized and interpreted (see also Hey 2006). Therefore, only general guidelines can be supplied in this respect, and for the moment the conclusions presented by Hey and Lacey (2001, pp. 34–51) remain the best available estimate. They showed through simulation on 11 sites in the UK that 5 % coverage will in many cases be sufficient for site evaluation, although they warn that in the case of pre-Roman sites greater coverage might be necessary. Coverage of more than 10 %, however, did not lead to a substantial increase in information gained in the cases investigated.

Coverage is, however, not the only factor playing a role in successful evaluation. Champion et al. (1995, pp. 36–37) had already concluded that the best evaluation strategy aims for the maximum number of trenches possible when using a specific area coverage. From a statistical point of view, a larger number of trenches will be more useful for purposes of prediction of, e.g. feature and artefact densities, and interpolation between sampling locations is easier when the trench distances are smaller. In other words, the trenches should be just large enough to allow for the detection and recognition of features. Champion et al. (1995, pp. 53–55) also showed that test pit sizes under 10 m^2 will insufficiently guarantee the detection of features.

The coverage obtained through a systematic trial trenching pattern can be easily calculated using this equation (Verhagen and Borsboom 2009)

$$C = \frac{L * W}{D * I}$$

For continuous trenches this can be simplified to

$$C = \frac{W}{D}$$

In many cases a systematic trial trenching strategy aiming at site discovery will also be sufficient for site evaluation, since this will also achieve the required coverage needed for evaluation. However, we have to take into account that, for example, 5 % coverage of a study area is not necessarily the same as 5 % coverage of a site. Depending on how the trenches are positioned and the shape of the site itself, trenches will overlap only in part or completely with the site itself. Site coverage may therefore show considerable variation within the same trenching strategy because of chance positioning of the trenches (see also Hey 2006, p. 118). This effect is largest when the site size is small relative to the trench spacing. It is therefore advised to always keep trenches 'in reserve' to help interpretation of the features found.

12.5 Minimal Interventions and Non-invasive Techniques

We can safely state that, given the choice, archaeologists prefer to use trial trenching for discovering and evaluating archaeological sites. The area uncovered by trial trenches is relatively large, and they allow for easier detection and interpretation of archaeological remains. However, since archaeological heritage management is also about spending money on surveying both effectively *and* efficiently, large-scale machine trenching may not always be the best way to spend money on archaeological surveying. Opening up large tracts of soil that do not reveal any archaeological finds can be seen as a waste of precious money and time. It is within this debate on the effectiveness of survey techniques and their relative costs that a trade-off has to be made between the different survey options available, including non-invasive techniques like geophysics and remote sensing.

12.5.1 Survey Techniques and Heritage Management Systems

The role that is given to remote sensing, geophysics and invasive survey methods in archaeological heritage management clearly differs between European countries. In most cases this distinction is not based on a good understanding of the effectiveness and efficiency of these techniques for purposes of site discovery and evaluation. Instead, it is largely governed by the way in which national (and in some countries federal) heritage management systems have developed since the implementation of the Valletta Convention. We can distinguish three major trends in this respect, which can be broadly defined as the 'English', 'French' and 'Dutch' approach – although these are fairly crude generalizations.

In the English system of archaeological heritage management, the decision whether to do archaeological investigations at all is mainly determined by the availability of prior archaeological evidence. A survey will not be undertaken if no archaeological evidence has been found in the vicinity of the area under development. A similar system is in place in, for example, Italy, Belgium, Japan and a number of German federal states. In practice, this means that a survey will only be undertaken in a limited number of cases, most frequently in the form of trial trenching, with geophysics being regularly applied as well. Wilcox (2012) notes that in the county of East Anglia, only 2 % of development plans have any form of archaeological intervention because of this. A major disadvantage of this approach is also that it is site based and not landscape based (see, e.g. Powlesland 2011). For this reason, remote sensing in English heritage management is mainly used to supplement the existing sites and monuments record. English Heritage has invested in a National Mapping Programme that mostly relies on aerial photography to detect archaeological sites (Horne 2011). In a similar vein, the German state of Baden-Württemberg aims to map all sites that can be recognized on high-resolution LiDAR images (Böfinger and Hesse 2011).

In France the situation is different: trial trenching will be performed in the majority of cases where developments take place, regardless of whether there is previous evidence of occupation (Demoule 2004). This means that there is little demand for targeted surveys. Trial trenching done the French way is a landscape archaeological research method and aims to obtain a representative sample of the archaeological remains in the area under development. Only in cases where trial trenching is not a feasible alternative, especially under forest, targeted, site-based surveying may take place. Recent French case studies point to an increased interest in LiDAR as a site detection technique in forested areas (Georges-Leroy 2011; Opitz et al. 2012).

In the Netherlands, heritage management is based on a system of increasingly targeted surveys, starting with the development of predictive maps. In many cases, a preliminary survey using core sampling is then undertaken before applying trial trenching. In practice, 20–25 % of developments in the Netherlands will have some form of archaeological survey (Isarin et al. 2009). A similar system is applied in Slovenia (Rutar and Črešnar 2011) and in large parts of the USA. The approach is more landscape based, and especially LiDAR plays an important role in defining zones of archaeological interest (e.g. levees in a floodplain) rather than just as a site detection tool.

The Italian BREBEMI project (Campana 2011) illustrates that applying the principles from one heritage management system to a country with a different system can lead to unexpected (political) consequences. They employed a combination of digital resources and prospection methods, including large-scale geophysics, as tools for targeted surveying of a motorway development project. Campana refers to this as 'total archaeology', as opposed to the prevailing system in Italy that relies on surface stripping ('rescue excavation') over the whole development area. Despite the generally positive results of the project, the regional Italian authorities have refused to apply this method more widely. This shows that established legislative and administrative frameworks may impose approaches to archaeological surveying that are not or no longer the best available.

12.5.2 Combining Invasive and Non-invasive Techniques: On the Way to Best Practice

High-resolution remote sensing, in particular LiDAR, offers unparalleled and detailed coverage of the whole landscape. It is currently mainly used as a tool for direct detection of archaeological sites and has certainly proved its potential in this respect. However, remote sensing is not capable of detecting all archaeological site types, since this very much depends on the presence of (spectral) contrasts in vegetation and relief. Beck (2011) notes that in this respect, there is still a large degree of uncertainty as to what soil conditions actually produce a spectral contrast. Remote sensing also has great potential for predictive modelling, especially when combined with other digital resources like geological and pedological mapping, and should therefore be the first source to be used before setting up any kind of targeted field survey campaign.

In field surveys, geophysics is now becoming a serious contender for traditional minimal intervention techniques aiming at site discovery and evaluation. It can be expected that geophysics will eventually become the first choice for most archaeological surveys. It is the only field technique that can cover the whole study area quickly (provided it is accessible to the equipment) and provides information on the subsoil without entering it. Surprisingly, however, very little attention has been given to the investigation of the effectiveness and efficiency of combining minimal interventions and non-invasive methods. The study by Hey and Lacey (2001, pp. 14–33) is, as far as known, the only one where a side-to-side comparison of magnetometry and minimal interventions was made for the purposes of site evaluation. They concluded that the method could not compete with trial trenching in this respect. However, this was long before ground-penetrating radar had developed into the powerful three-dimensional survey technique it is now. However, geophysical methods will certainly not detect and allow for unambiguous interpretation of all archaeological phenomena of interest, since much depends on the type of remains involved and the contrast they provide with the surrounding soil. Geophysics is much more effective at finding stone-built remains and ditches than at finding features that only show up as soil colorations in a trench or sites that only show up as artefact concentrations. They will be most effectively applied when supplemented by minimal interventions to help interpret the anomalies found (see e.g. Kvamme 2003). Local soil conditions will influence the detection capabilities of geophysics as well. Kattenberg (2008), for example, analysed the applicability of magnetometry in the Netherlands and concluded that it has clear limitations in areas that have been inundated by seawater and in sandy soils. Conyers and Leckebusch (2010) state that the potential and limitations of geophysical methods are now fairly well known. However, this is certainly not the case outside the circle of experts involved. Field archaeologists rely to a high degree on expert knowledge, often provided by private companies, to decide whether or not to apply a geophysical survey and to interpret its results.

If we want to move beyond the application of particular survey techniques because of grown traditions, personal preferences or existing legal frameworks, we therefore have to improve the knowledge base of the advantages and disadvantages of each technique and to better understand how they can supplement each other. On the one hand, this entails a concerted effort to improve the quantitative knowledge base regarding prospection characteristics of archaeological sites, in terms of size, shape, artefact and feature densities, and the specific soil characteristics that produce spectral signals and geophysical anomalies. On the other hand, it requires a better understanding of the results of archaeological surveying itself: under what circumstances has a technique been successful, and when has it failed to produce good results? Improving the knowledge base underpinning the choice for an optimal survey strategy should therefore be a top priority in archaeological heritage management, where the available budgets are always at odds with the protection of our cultural heritage.

References

Banning EB (2002) Archaeological survey. Kluwer Academic/Plenum Publishers, New York

Beck A (2011) Archaeological applications of multi/hyper-spectral data – challenges and potential. In: Cowley DC (ed) Remote sensing for archaeological heritage management. Europae Archaeologia Consilium, Brussels, pp 87–97

Böfinger J, Hesse R (2011) As far as the laser can reach … Laminar analysis of LiDAR detected structures as a powerful instrument for archaeological heritage management in Baden-Württemberg, Germany. In: Cowley DC (ed) Remote sensing for archaeological heritage management. Europae Archaeologia Consilium, Brussels, pp 161–172

Borsboom A, Verhagen P (2009) KNA Leidraad Inventariserend Veldonderzoek. Deel: Proefsleuvenonderzoek (IVO-P). SIKB, Gouda

Campana S (2011) 'Total archaeology' to reduce the need for rescue archaeology: The BREBEMI Project (Italy). In: Cowley DC (ed) Remote sensing for archaeological heritage management. Europae Archaeologia Consilium, Brussels, pp 33–42

Champion T, Shennan SJ, Cuming P (1995) Planning for the past, vol 3, Decision-making and field methods in archaeological evaluations. University of Southampton/English Heritage, Southampton

Conyers LB, Leckebusch J (2010) Geophysical archaeology research agendas for the future: some ground-penetrating radar examples. Archaeol Prospect 17:117–123

Davis JC (1986) Data analysis in geology, 2nd edn. Wiley, New York

Demoule J-P (2004) De l'archéologie de sauvetage à l'archéologie préventive. Aménager le territoire, préserver notre histoire. In: Demoule J-P (ed) La France archéologique. Vingt ans d'aménagements et de découvertes. Éditions Hazan, Paris, pp 7–22

Drew LJ (1979) Pattern drilling exploration: optimum pattern types and hole spacings when searching for elliptical shaped targets. Math Geol 11:223–254

Georges-Leroy M (2011) Airborne laser scanning for the management of archaeological sites in Lorraine (France). In: Cowley DC (ed) Remote sensing for archaeological heritage management. Europae Archaeologia Consilium, Brussels, pp 229–234

Gilbert RO (1987) Statistical methods for environmental pollution monitoring. Van Nostrand Reinhold Company, New York

Hey G (2006) Scale and archaeological evaluations: what are we looking for? In: Lock G, Molyneaux BL (eds) Confronting scale in archaeology. Issues of theory and practice. Springer, New York, pp 113–127

Hey G, Lacey M (2001) Evaluation of archaeological decision-making processes and sampling strategies. Kent County Council/Oxford Archaeological Unit, Oxford

Horne P (2011) The English Heritage National Mapping Programme. In: Cowley DC (ed) Remote sensing for archaeological heritage management. Europae Archaeologia Consilium, Brussels, pp 143–152

Isarin R, Verhagen P, Goudswaard B (2009) Archaeology as a risk in spatial planning: manoeuvring between objectivity and subjectivity. In: Kamermans H, van Leusen M, Verhagen P (eds) Archaeological prediction and risk management. Alternatives to current practice. Leiden University Press, Leiden, pp 41–48

Kattenberg AE (2008) The application of magnetic methods for Dutch Archaeological Resource Management. Vrije Universiteit, Amsterdam

Kintigh KW (1988) The effectiveness of subsurface testing: a simulation approach. Am Antiquity 53:686–707

Krakker JJ, Shott MJ, Welch PD (1983) Design and evaluation of shovel-test sampling in regional archaeological survey. J Field Archaeol 10:469–480

Kvamme KL (2003) Geophysical surveys as landscape archaeology. Am Antiquity 68:435–457

Lightfoot KG (1986) Regional surveys in the Eastern United States: the strengths and weaknesses of implementing subsurface testing programs. Am Antiquity 51:484–504

McManamon FP (1984) Discovering sites unseen. In: Schiffer MB (ed) Advances in archaeological method and theory, vol 7. Academic Press, New York, pp 223–292

Mueller JW (ed) (1975) Sampling in archaeology. University of Arizona Press, Tucson

Nance JD (1983) Regional sampling in archaeological survey: the statistical perspective. In: Schiffer MB (ed) Advances in archaeological method and theory, vol 6. Academic Press, New York, pp 289–356

Nance JD, Ball BF (1986) No surprises? The reliability and validity of test pit sampling. Am Antiquity 51: 457–483

Nicholson M, Barry J, Orton C (2000) Did the burglar steal my car keys? Controlling the risk of remains being missed in archaeological surveys. University College London, London. http://eprints.ucl.ac.uk/archive/00002738/01/2738.pdf

Opitz R, Nuninger L, Fruchart C (2012) Thinking topographically about the landscape around Besançon (Doubs, France). In: Kluiving S, Guttmann-Bond E (eds) Landscape archaeology between art and science. From a multi- to an interdisciplinary approach. Amsterdam University Press, Amsterdam, pp 395–412

Orton C (2000) Sampling in archaeology. Cambridge University Press, Cambridge

Powlesland D (2011) Identifying the unimaginable – managing the unmanageable. In: Cowley DC (ed) Remote sensing for archaeological heritage management. Europae Archaeologia Consilium, Brussels, pp 17–32

Rutar G, Črešnar M (2011) Reserved optimism: preventive archaeology and management of cultural heritage in Slovenia. In: Cowley DC (ed) Remote sensing for archaeological heritage management. Europae Archaeologia Consilium, Brussels, pp 259–264

Stone GD (1981) On artifact densities and shovel probes. Curr Anthropol 22:182–183

Sundstrom L (1993) A simple mathematical procedure for estimating the adequacy of site survey strategies. J Field Archaeol 20:91–96

Tol A, Verhagen P, Borsboom A, Verbruggen M (2004) Prospectief boren. Een studie naar de betrouwbaarheid en toepasbaarheid van booronderzoek in de prospectiearcheologie. RAAP Archeologisch Adviesbureau, Amsterdam

Tol A, Verhagen P, Verbruggen M (2006) Leidraad inventariserend veldonderzoek; Deel: karterend booronderzoek. SIKB, Gouda

Verhagen P (2005) Prospection strategies and archaeological predictive modelling. In: van Leusen M, Kamermans H (eds) Predictive modelling for archaeological heritage management: a research agenda. Rijksdienst voor het Oudheidkundig Bodemonderzoek, Amersfoort, pp 109–121

Verhagen P, Borsboom A (2009) The design of effective and efficient trial trenching strategies for discovering archaeological sites. J Archaeol Sci 36:1807–1816

Verhagen P, Tol A (2004) Establishing optimal core sampling strategies: theory, simulation and practical implications. In: Fischer-Ausserer K, Börner W, Goriany M, Karlhuber-Vöckl L (eds) [Enter the past]: the E-way into the four dimensions of cultural heritage: computer applications and quantitative methods in archaeology. Proceedings of the 31st conference, CAA 2003, Vienna, Austria, April 2003. Archaeopress, Oxford, pp 416–419

Verhagen P, Rensink E, Bats M, Crombé P (2012) Establishing discovery probabilities of lithic artefacts in Palaeolithic and Mesolithic sites with core sampling. J Archaeol Sci, http://dx.doi.org/10.1016/j.jas.2012.05.041

Wilcox B (2012) Archaeological predictive modelling used for cultural heritage management. In: Zhou M, Romanowska I, Wu Z, Xu P, Verhagen P (eds) Revive the past. Proceedings of the XXXIX conference on Computer Applications and Quantitative Methods in Archaeology, Beijing, 12–16 Apr 2011. Pallas Publications, Amsterdam, pp 353–358

Zeidler JA (1995) Archaeological inventory survey standards and cost-estimation guidelines for the Department of Defense. Champaign (IL): USACERL. http://oai.dtic.mil/oai/oai?&verb=getRecord&metadataPrefix=html&identifier=ADA311181

Creating and Analysing Digital Terrain Models for Archaeological Research

13

José-Ángel Martínez-del-Pozo,
Victorino Mayoral-Herrera, and Pedro Ortiz-Coder

Contents

13.1 Digital Terrain Models: A Definition

Digital terrain models play a very relevant role in many aspects of archaeological research. The process of data capture, interpolation and application of the results covers a huge realm that is quite difficult to include in the limits of this chapter. In the first section we will explain the main methods of data acquisition, showing the advantages and drawbacks of different techniques currently used. Then we will go into some detail on the crucial question of how we obtain the digital models through several interpolation procedures. Finally, we will give a brief overview of how models are used by archaeologists with different purposes.

Many definitions of digital terrain models (DTMs hereafter) can be found (Felicísimo 1991; Weibel and Heller 1991; Bosque Sendra 2000). Most of them refer to DTM as the digital representation of quantitative and continuous variables that are spatially distributed. Therefore, the diversity of DTM types is as great as the variables whose values change over space. However, the most commonly used variable in DTM creation for archaeological purposes is elevation. These types of model are referred to as digital elevation models (DEMs hereafter).

DTMs were used for the first time in the 1950s (Miller and Laflamme 1958). Since then, creation and analysis methods have been largely improved, becoming a fundamental element when studying any variable or process related with topography (Weibel and Heller 1991).

J.-Á. Martínez-del-Pozo (✉) • V. Mayoral-Herrera
P. Ortiz-Coder
Merida Institute of Archaeology,
Spanish Council of Scientific Research,
Madrid, Spain
e-mail: jangel.martinez@iam.csic.es;
vmayoral@iam.csic.es

C. Corsi et al. (eds.), *Good Practice in Archaeological Diagnostics*, Natural Science in Archaeology,
DOI 10.1007/978-3-319-01784-6_13,

Like other disciplines, archaeology deals with the spatial component of many kinds of data. It implies the need to model different environmental and cultural variables (e.g. the density of finds in a landscape or the distribution of chemical values within a site). One of the most usual needs is to produce DEMs with very diverse scales and purposes. It is essential to keep these objectives clearly in mind, since they determine the use of different techniques for acquiring and analysing data (discussed below). These can range from the detection of buried structures through the identification of topographic anomalies, to the accurate recording of visible remains for conservation and planning purposes. On a broader scale, an accurate terrain model could be useful to search for clues about the behaviour of human groups through the exploration of variables such as resource catchment, visual relationships or movement. Finally, we also need DEMs to understand many factors affecting the representativeness of the archaeological record, since topography is closely linked with erosion and geomorphologic dynamics.

13.2 Data Acquisition

DEMs are created from points or line features with a known position in a reference system and associated with the value of a variable (Miller and Laflamme 1958; Felicísimo 1991). Like any other model, they are a simplified representation of reality. Since it is not possible to gather information on every single location, we must rely on a sampling strategy to select a discrete set of data. Obviously, this implies a generalisation process. Therefore, acquisition of data is the first step in the creation of a DEM. In order to minimise the loss of information, data capture must be methodically done to assure that the variation of the variable to be modelled is correctly recorded (Weibel and Heller 1991).

This process encompasses a series of methods for extracting the value of the target variable from the ground and its codification for later processing. These are called direct when data are recorded in situ. But in the case of large areas demanding massive data capture, it may be more practical to use indirect methods that do not require any physical interaction with the subject of study.

13.2.1 Methods for Data Acquisition

13.2.1.1 Total Station

This is an electronic/optical instrument widely used for topographical studies. Total stations represent the evolution of theodolites, with the integration of a laser to measure distances (distance metre or EDM) and angles. Older devices needed to be operated by two individuals, since the prism reflector was mounted on a rod. This limitation has been overcome in more recent equipment, which is designed to follow the prism and take measurements automatically (Fig. 13.1).

Total stations are a flexible and easy solution for field surveys for archaeological purposes. Once the work is finished, data are organised internally and can be exported to several formats depending on the type of device being used. Normally, the fastest and easiest way is to extract them in a plain text format ("txt" or "csv" extensions). These files store the information in several columns (Fig. 13.2) that represent each measurement with

Fig. 13.1 Working with a total station

Point	Easting	Northing	Z	Code
base	264672,016	4284498,279	448,583	base
1	264742,417	4284498,033	441,178	
2	264742,253	4284488,487	441,162	
3	264742,361	4284498,42	448,218	
4	264742,363	4284498,556	441,26	
5	264742,506	4284498,037	441,251	
6	264742,548	4284498,184	441,331	
7	264742,622	4284498,543	441,241	
8	264742,721	4284498,849	441,25	
9	264742,895	4284498,319	441,187	
10	264743,192	4284498,684	441,168	

Fig. 13.2 A typical plain text file for storing total station measurements

their *X* and *Y* coordinates (also referred to as northings and eastings, respectively). A third column stores the values of the target variable (*Z*), which is the height in the case of DEMs. One or two additional columns can include information about identification and feature coding. Additionally, total station data can be managed and exported to advanced formats like ESRI Shapefiles ("shp"). As the total station can be placed at a point referred to its spatial coordinates, survey results are also georeferenced.

One of the main advantages of working with total stations is that they provide very high accuracy (around 2–3 mm/km) with fairly low costs. However, there are some drawbacks to take into account. First, their efficient use demands specialised training for data capture and management. Regarding their operation, the radius does not exceed 300 or 400 m and is very conditioned to visual connection between station and prism. Moreover, it requires a person moving through the survey area, so a previous knowledge of the topography is required for a correct sampling strategy. Although, as said before, it is a more accurate method, when the coverage of large areas is required, it may be too costly in terms of time and effort. A good solution to this problem is the use of a robotic total station. This device can automatically follow a prism mounted on a wheeling rod, increasing noticeably the density and speed of data capture. A nice example of the archaeological application of this technique is the micro-topographic survey of a great, complex site of the Northern Great Plains developed by the University of Arkansas (Barrat et al. 2000) (Fig. 13.2).

13.2.1.2 GNSS Receivers

This technology has an increasing role in many aspects of archaeological research and has become a common tool for the generation of DEMs. Global Navigation Satellite Systems (GNSS) are large infrastructure (covering the whole planet) facilities comprising a constellation of satellites and a series of terrestrial stations. These can be accessed by receivers in order to continuously calculate their position with an acceptable level of accuracy. The main global positioning systems that are working today are NAVSTAR-GPS (USA) and GLONASS (Russia). The European Galileo system is still not operating (http://www.esa.int/esaNA/galileo.html). All the indications hereafter will refer to the NAVSTAR-GPS system, which is the most commonly used system worldwide at this time.

A GNSS station provides 3D spatial coordinates (geographical or projected) of the receiver with respect to a geodetic reference system or datum. There are several advantages of working with GNSS stations for data capture. They are more flexible, working directly with georeferenced information. Unlike total stations, they do not depend on visibility conditions, and therefore their range is much greater. It can reach up to 5 km between the base station and mobile receiver in Real Kinematic Mode or even be unlimited in post-processing mode (both described below). This ability to be implemented almost anywhere makes it an ideal solution for extensive jobs. Like the total station, it can be very selective in data capture, adapting to the changing complexity of target surfaces. It also allows the definition of thematic attributes using feature coding (Fig. 13.3).

Among the drawbacks, the main ones are the higher cost of equipment and the expertise required (although money can be quickly recouped if the purchase is well justified. Providing external services could be a good strategy for short-term jobs). During the last few

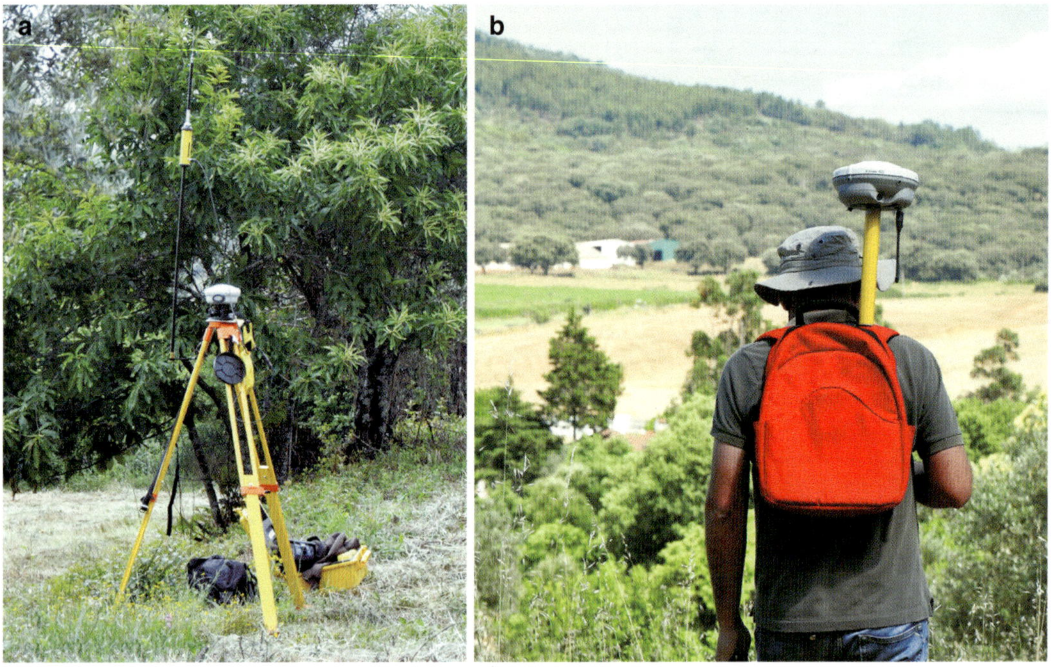

Fig. 13.3 Working with a GNSS in archaeological sites

years, prices have gone down very significantly, and the technology has advanced enormously. This has made it possible for small companies and research groups to have access to this kind of equipment.

Signal reception can be seriously affected by obstacles like vegetation cover and topographical barriers (some devices adapted to receive correction from GPS and GLONASS systems are more efficient in these circumstances). Regarding the time spent on data collection, as we will describe in detail later, it can be much faster than a total station. It may be quite slow for big areas, but there are alternatives that can speed up the process significantly without losing precision, producing fairly dense elevation point clouds.

A critical point in the elaboration of DTMs with GNSS stations is to choose the equipment that ensures the necessary level of accuracy in the measurements. It will always be a bit lower than that obtained with a total station but depending on its operation mode can range from 2–3 cm to 5 mm. Concerning the NAVSTAR-GPS, it is desirable to use receivers that are designed to operate with the two frequencies at which this system works (L1 and L2). These are able to work in Real Time Kinematic mode (RTK) with differential correction or in post-processing mode. The former is the one that provides the best results for massive data capture for MDT creation.

How Does It Work?

We calculate the position of the GNSS receiver measuring the time invested by a radio signal to travel to it from the satellites. Observing this interval for at least four satellites, the system is able to calculate the position of the receiver. Since the radio signal must pass through the atmosphere, it is affected by various factors that introduce errors into the measurements. This makes it critical to use equipment that is able to quantify and eliminate these error sources.

Once we know where we are, the first step in data collection is to have a well-designed sampling strategy, adapted to our technical resources and to local conditions of the area to be surveyed. It is advisable to generate a sampling pattern which ensures a homogeneous coverage of the area. This can be achieved by marking on

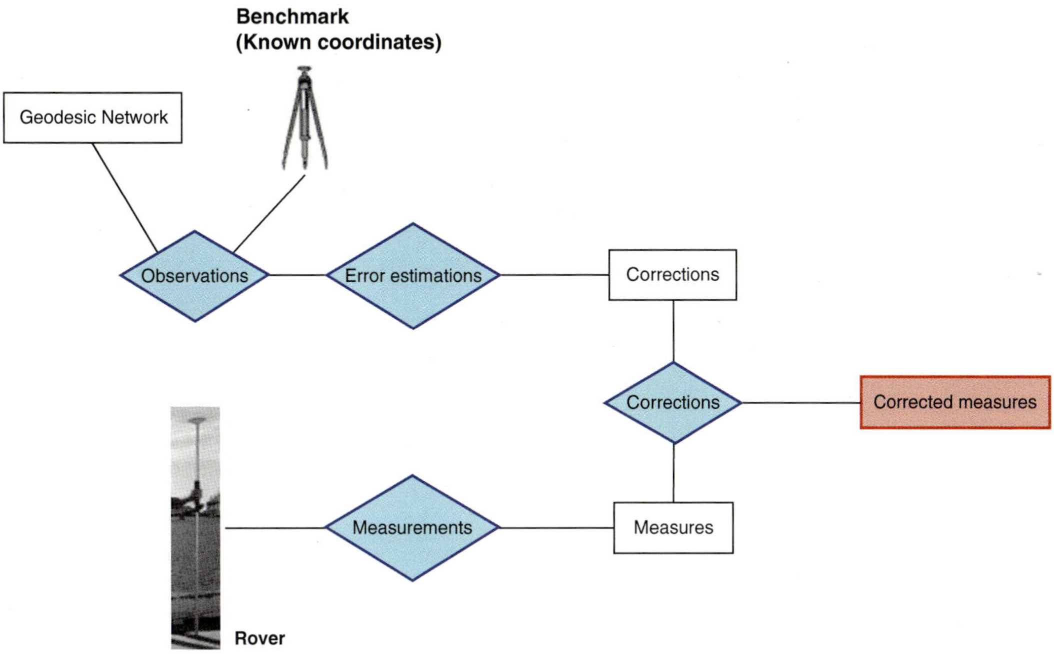

Fig. 13.4 A schematic workflow of RTK survey with a GNSS

the ground a series of transects to guide the operator of the mobile receiver. Also very helpful to pattern data capture is the ability to configure the data logger in order to take measurements at fixed intervals of space or time (Fig. 13.4).

Displacements of the mobile receiver can be performed in different ways. Using a rod, a measurement is taken every time that it is placed at a selected point on the ground. This method ensures that the receiver is placed at a fixed and controlled height above the ground and therefore provides reliable measurements. Its main drawback is that it is a very tiring and time-consuming method. In order to speed up the process, there is the option of mounting the mobile receiver on a backpack. This is a very versatile option for massive data capture, since the operator only needs to cover the terrain following the sampling pattern. A drawback of this method is the need to filter and correct the random introduction of anomalous height values. These errors are due to uncontrolled movement of the receiver as the operator walks through the survey area. There is also the alternative of using a wheeled device. Properly adapted to terrain roughness, these can keep a constant height and alleviate the effects of error produced by the lack of verticality in the receiver. Nevertheless, the use of these gadgets is constrained to flat or topographically low contrast areas.

When choosing one of these methods, it is essential to assess the level of accuracy we need to achieve, as well as the available resources and the size of the area that it is expected to survey.

To obtain the best accuracy in the measurements, the GNSS stations can be operated in two modes:

1. Real Time Kinematic (RTK) mode: the measures are corrected in situ and simultaneously as they are taken. In RTK mode two receivers are used, one of which is positioned over a point of known coordinates, so the receiver is calculating the systematic error in the measurements and sends via radio modem to the other receiver that is capturing the data. This mode of operation allows the immediate availability of the data free of errors. The accuracy obtained in RTK mode is about 2–3 cm of error.

2. Post-process mode: this mode of operation consists in the correction of the measurements with information provided by fixed geodetic stations. The data are collected by a receiver that stores its position along with additional information on the ephemeris of the satellites. The corrections are constantly calculated by the geodetic stations and provided as RINEX files (Receiver Independent Exchange) which can be downloaded and applied to the data with the proper software. Although this mode requires the treatment of the data, so the measurements are not immediately available, the level of accuracy of the measurements once corrected is very high (less than 5 mm).

To sum up, GPS can be a non-exclusive option that can be applied to most surveying tasks where traditional techniques have been applied, from landscape to excavation. Depending on our research goals and needs, most archaeological targets could demand more than one survey method.

13.2.1.3 Digital Photogrammetry

Another method for DEM production is photogrammetric restitution. This has been widely used for mapping purposes using aerial images, and it is a discipline with a long tradition. This method relies on the stereoscopic properties of several frames that share an overlapping area. Image acquisition and processing requires a high qualification and very specialised equipment beyond the capabilities of an average user. Nevertheless, the development of digital photogrammetry and the increasing availability of software packages like ERDAS IMAGINE and ENVI have made it possible for teams without previous specific training to apply this methodology. Nevertheless, these are specialised tools that require some theoretical background and very specific conditions for data capture in order to get valid results. Recent developments like the structure from motion techniques (De Reu et al. 2013; see Chap. 3 by Verhoeven et al. in this volume) bring to almost everyone the ability to extract three-dimensional models from series of pictures or video frames without information about the kind of camera used, its location or orientation.

DEMs derived from photogrammetry can correspond to very different scales, from a pottery vessel or a burial to a landscape. Here what makes the difference is the method used for image capture. At an excavation level, any means of getting a view from above can be suitable, but for more professional solutions we can use devices like a telescopic pole with a remote control camera (Fig. 13.5).

If we want to obtain a higher point of view, we enter the realm of the Low-Altitude Aerial Photograph (LAAP). This is a technique to acquire aerial photographs based on the use of unmanned vehicles with cameras controlled by radio. There is a very wide range of alternatives, like kites, helium blimps, and drones (see Verhoeven 2009) with their advantages and drawbacks. They should be able to provide vertical high-resolution images with known optical parameters (focal length is a critical one). It is also imperative to obtain accurate 3D coordinates (with GNSS positioning or a total station) of an evenly distributed sample of points that are easy to number and recognise in several overlapping pictures. When all these requirements are fulfilled, it is possible to use these images to produce high-quality DTMs through photogrammetric restitution.

13.2.1.4 Secondary Data Sources

Of course topographic equipment or photogrammetry does not exhaust all the possibilities for obtaining three-dimensional models. Digitising contours from analogical maps could be a simple and direct procedure (Bosque Sendra 2000). They are introduced as line features with the help of digitising tablets, adding elevation values for further interpolation and DEM generation. With the increasing availability of digital maps, this system is used less and less, but it could well be worth the effort for recovering historical maps or in areas where we lack other cartographical sources.

13.2.2 Interpolation

So far, we have examined several procedures to obtain datasets consisting of polylines or points containing elevation values. As said before, they

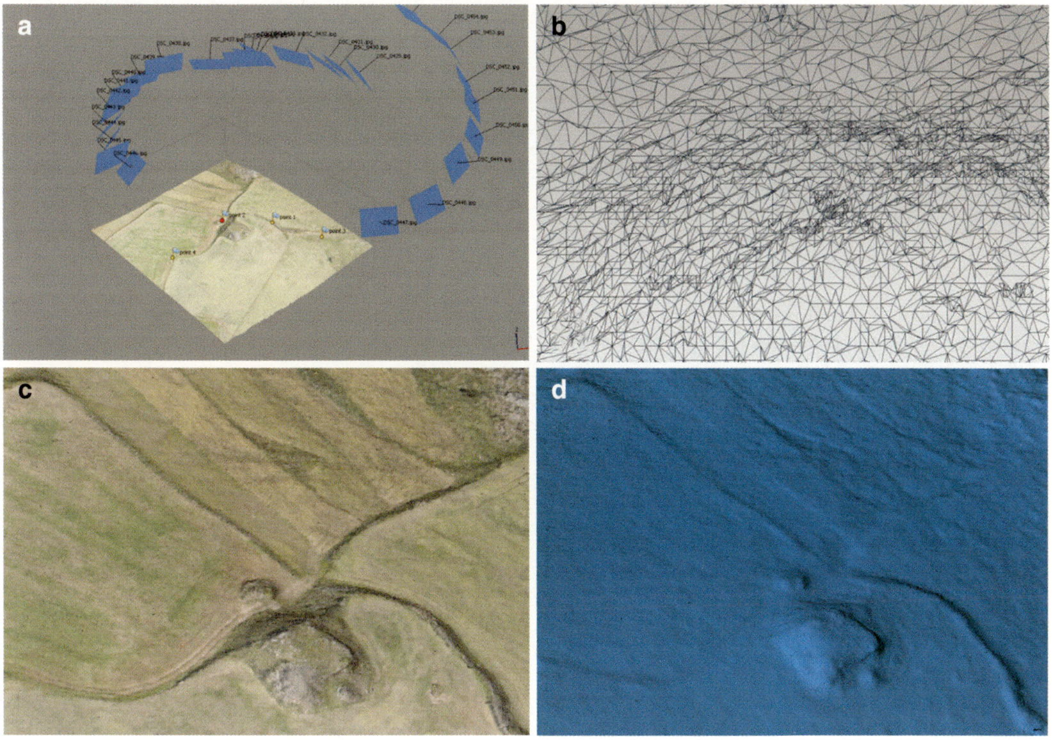

Fig. 13.5 Different stages in the production of DEMs through photogrammetric restitution. (**a**) Orientation of cameras over the photo-texturized DTM. (**b**) Mesh obtained from the point cloud. (**c**) DTM rendered with the orthoimage. (**d**) Non-texturized model with an analytical hill-shading

represent a *discrete* sample of the distribution of a *continuous* variable of the real world. Therefore, we need to go one step further to produce a surface "filling the gaps" of unsampled locations. This must be done using a mathematical procedure called interpolation (Weibel and Heller 1991; Conolly and Lake 2006). The aim of these functions is not merely to generate soft, realistic looking models but to make a reliable prediction of the spatial distribution of the target variable (Fig. 13.6).

Interpolation belongs to the discipline of numerical analysis, and it is widely used in a great variety of studies. It allows the study of the spatial distribution of any quantitative natural phenomena whose values are not randomly distributed. This means that the target variable exhibits positive spatial autocorrelation (Hageman and Bennett 2000; Conolly and Lake 2006). Concerning geographical information, we use the term spatial interpolation to define the procedure of calculating the value of a variable in a specific location (unknown point), according to the values in a set of spatial points (sample known points) around it (Bosque Sendra 2000; Conolly and Lake 2006).

The result of an interpolation process is a digital terrain model (DTM) of the variable we want to model. This can be elevation, but of course we can use many other variables, from soil chemical values to the density of archaeological finds in a discrete area.

There are different data models for the creation of DTMs and specific interpolation procedures, also referred to as interpolators, which can be applied to obtain a raster or vector results (Fig. 13.7).

13.2.2.1 Types of Interpolation Methods

We must carefully choose the type of interpolator to be applied according to the characteristics of the phenomena we are trying to model and the

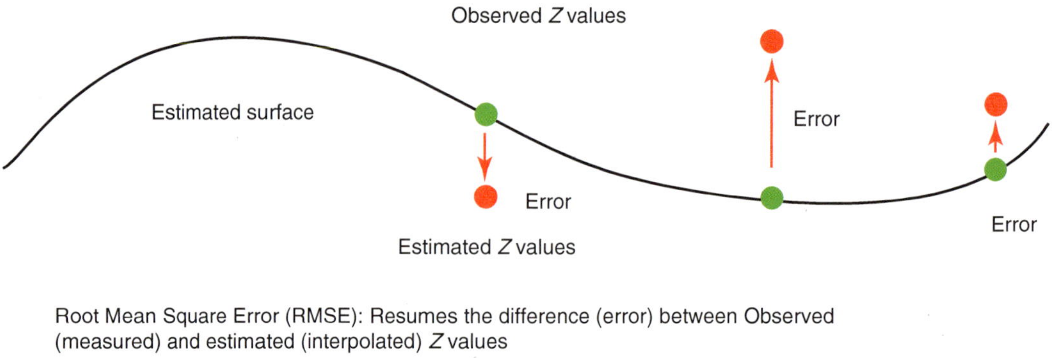

Fig. 13.6 How interpolation works (Graphic by Antonio Uriarte González, CCHS, CSIC)

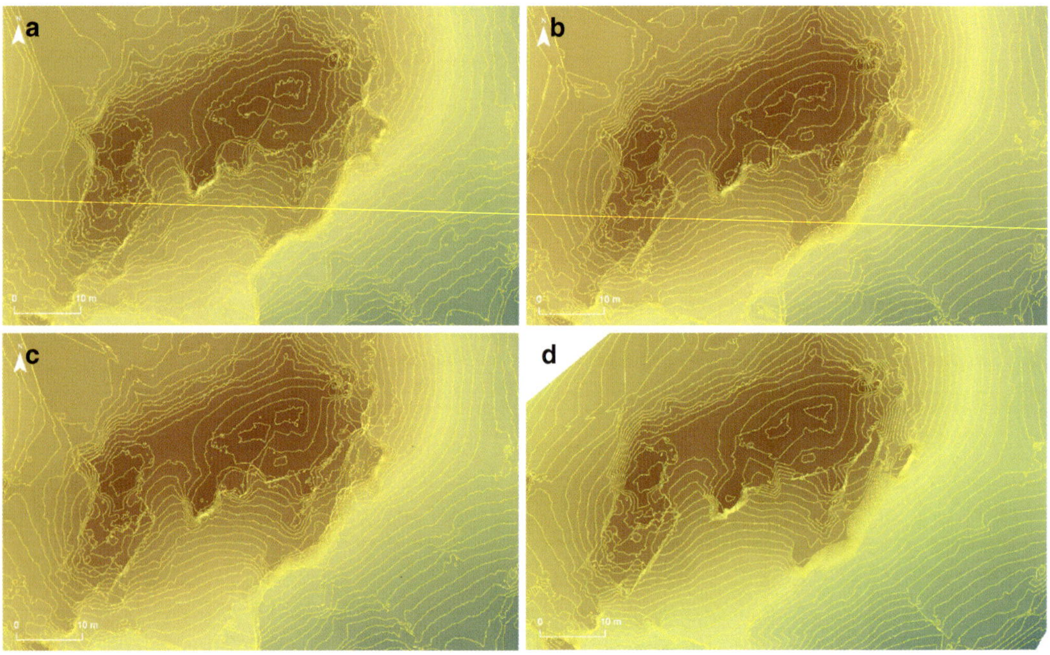

Fig. 13.7 Visual comparison of raster DEMs obtained by IDW (**a**), kriging (**b**), splines (**c**) and TIN (**d**) of an archaeological site (Taken from Martínez del Pozo et al. 2010)

nature of the original data sample. Very often software interfaces encourage a "push button syndrome" that can produce very misleading or simply unreal results.

As a whole, interpolation methods can be classified in several general categories. Attending to the number of sample points involved in the calculation, the methods can be *general* if all sample points are considered or *local* when only a set of known points (neighbourhood) surrounding the unknown location is taken into account. If we consider how the interpolator deals with the values of the sample points, we will distinguish between *exact* and *approximate* methods. In the first ones the target variable values in the sample points are preserved in the final model. In the second ones, the original values are modified to produce a smoothed surface (Lam 1983). Finally, the *direct* or *deterministic* interpolators are those that can be directly applied to the data, while *analytical*, also called *geostatistical*, interpolators work from a previous analysis of the sample data structure and autocorrelation. A geostatistics is defined as the application of stochastic functions to the assessment of natural phenomena where one or more variables vary in space or time (Deutsch 2002), also called regionalized variables (Matheron 1962). A complete list of the criteria used to classify the interpolators can be seen in Hengl (2009).

As a mathematical procedure, there can be infinite algorithms for interpolation. Therefore, only the most commonly used methods will be described here.

Raster Model Based

Inverse Distance Weighted (IDW)

This is one of the oldest interpolation techniques used (Shepard 1968; Hengl 2009). Following Bosque Sendra's (2000) notation, the IDW mathematical formulation is as follows:

$$Z(x_j) = \frac{\sum_{i=1}^{n} \left(Z(x_i) * W_{ij}\right)}{\sum_{j=1}^{n} W_{ij}}$$

where $Z(x_j)$ is the value of the target variable in the unknown location (j) to be interpolated, $Z(x_j)$ are the values of the variable at the known points (i) that surround j and W_{ij} is the distance weight factor that multiplies the value of $Z(x_i)$. The distance between the known points and the unknown point is weighted as

$$W_{ij} = \frac{1}{D_{ij}^{a}}$$

In this case, D_{ij} is the straight distance between i and j, while a is the power that affects the way the distance is weighted. For larger values of a, the W_{ij} for the known points located at great distances is decreasing. Therefore, only the closest known points are relevant in the final calculation. If $a = 1$, then the method is exact and the initial values in the known locations are preserved; otherwise, these values are different once recalculated, so the method is approximate (Bosque Sendra 2000; Conolly and Lake 2006).

Depending on the number of known points involved in the calculation of the unknown locations, IDW can run as a global or as a local method. In the first one all the known points take part in the interpolation. In the second method we use a neighbourhood – based on the distance from the unknown point or a number of points – that constrains the number of points involved in the interpolation. If the data set is large enough, the calculations can be very time-consuming, so it is worthwhile to establish a threshold distance to exclude distant points with a very low influence from the result of the interpolation (Hengl 2009).

The IDW method is suitable in many situations. It is a relatively simple method compared with others, so its implementation and execution are easy without any diminution of the final results. But IDW tends to create artefacts and some blunders in the known points called "bull's-eyes" (Villatoro et al. 2008). These are small depressions that can, however, be reduced by an appropriate set-up of the power with which the distances are weighted.

Kriging

Kriging is a geostatistical method which is based on probabilistic models and pattern recognition techniques (Olea 2009). First described by Matheron (Matheron 1963a, b), it is one of the most frequently used interpolation methods today. A large amount of bibliography can be found that explains its mathematical and statistical foundations (Maynou 1998; Webster and Oliver 2007; Hengl 2009). Unlike deterministic methods (such as IDW), which only take into account the distance between unknown and known points, geostatistical procedures rely on a previous evaluation of the spatial variability of the target variable. Kriging also sets up a model of the parameters which are determined objectively from the probability theory. This is a strength of this method, since we obtain not just an interpolated surface but also a precision and error evaluation of the results (Hengl 2009).

The way in which kriging assigns the weights to the known points depends on the spatial structure of the data and on the degree of spatial autocorrelation of the target variable (Conolly and Lake 2006). Therefore these weights reflect the true spatial autocorrelation structure (Hengl 2009). The analysis of point data is done by calculating and plotting the differences between pairs of neighbouring points, also called semivariance (Matheron 1962; Hengl 2009). The formula for calculating the semivariance is:

$$\gamma(h) = \frac{1}{2N}\left[\left(z(x_i)\right) - z\left(x_{i+h}\right)\right]^2$$

where $\gamma(h)$ is the semivariance between the values of the target variable ($z(x)$) at the point x_i and at other points located at a certain distance (x_{i+h}). N is the number of pairs at a distance h. Therefore, it is expected to find low values of $\gamma(h)$ for those points that are closer, and the values will increase as the h distance goes up.

The spatial distribution of the variable can be explored with the help of a variogram. This is a schematic representation of its spatial variability (Bosque Sendra 2000; Conolly and Lake 2006). The variogram (Fig. 13.8) represents the values of semivariance calculated for each pair of points with the formula (XX) versus the distance (lag) that separates each point of the pairs (Clark 1979).

Figure 13.8 is an ideal representation of the shape expected for a variogram. For pairs of close points, the values of $\gamma(h)$ are low, as each point of the pair has similar values of the target variable. However, as the distance between the points rises, the values of the variable in each location tend to be more different, so the $\gamma(h)$ function increases with the lag that separates the points of the pair. The ratio of rise of $\gamma(h)$ shows an asymptotic shape, due to the fact that when the lag is large enough the differences tend to be the same.

Graphically, the variogram presents three characteristics used for the analysis of the data:

1. Range: this is the distance (lag) at which the variance is stabilised (Bosque Sendra 2000). The spatial autocorrelation of the variable exists only within the range of the variogram. The values of those points separated by larger distances are independent (Bosque Sendra 2000).
2. Sill: this is the value of $\gamma(h)$ at which the range is reached and so provides information about the maximum variability of the sample.
3. Nugget: this is the crossing between the variogram and the Y axis. In an ideal situation, there should not be a nugget, because two points separated by an $h=0$ should not have any variance. The nugget occurs when there are microscale variations of the target variable that cannot be measured by the data acquisition method.

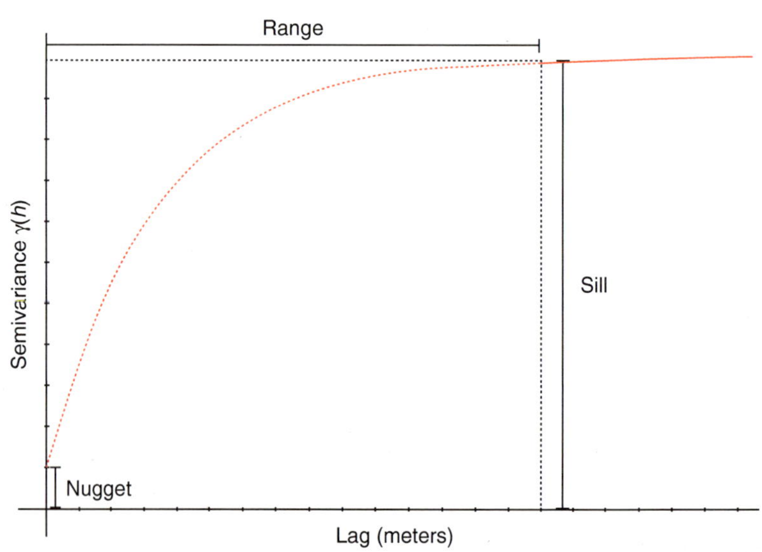

Fig. 13.8 Representation of a theoretical variogram

Basically, when applying kriging to a set of data, several steps have to be followed:

1. Analyse the data structure. Some kriging methods require that the data accomplishes some initial conditions (normality, stationarity, etc.)
2. Calculate the experimental variogram cloud.
3. Fit a theoretical variogram to the experimental variogram.

   There are several types of theoretical variograms that can be used to fit the experimental variogram. The theoretical functions that are used to fit the data are spherical, exponential, etc. The semivariance of two points (*i* and *j*), separated by a distance, can be calculated with this model, and so the covariance *C*(*h*) between those points, for both terms, is related by the expression (Hengl 2009; Isaaks and Mohan Srivastava 1989)

$$C(h) = sill - \gamma(h)$$

4. The result is a covariance matrix used to calculate the weights of the known data that are included in the interpolation of the unknown point.

This procedure must be repeated for each unknown point, so the weights matrix is recalculated specifically for each point that must be interpolated. Finally, the interpolated value is calculated with the general formula of interpolation:

$$Z(x_0) = \sum_{i=1}^{n} W_i * Z(x_i)$$

As supported by solid statistical foundations, kriging is able to provide very good and accurate results when approximating the interpolated surface to the sample of points. But this strength has, in contrast, some drawbacks and challenges. The main one is the great number of conditions that the initial data must satisfy. Kriging is a decision-making procedure where the data structure defines the approach to the data and the final parameters and variogram models that are needed to obtain the maximum accuracy. To get the best results, an assessment of the errors in the model (through the Root Mean Square Error, RMSE) is needed to adjust the parameters of the interpolator until a good level of error is achieved.

Splines

Interpolation through splines is a deterministic method. It generates a piecewise mathematical function that comprises a series of sub-functions adjusted for each unknown point (Conolly and Lake 2006).

The functions that are used are polynomial (Webster and Oliver 2007), and their adjustment to the surface depends on the degree of the function; hence the final results rely on the type of polynomial used. Splines are an exact method, as the mathematical fundamentals used demand that the sub-functions in the known points must be coincident and also that they must have the same tangent in the node they connect.

Two modalities of splines can be found:

(a) Regularised: this generates a smooth surface that changes gradually (ArcGIS on-line help).

(b) Tension: this method controls the stiffness of the surface and generates less smooth surfaces (ArcGIS on-line help), but by controlling the tension the surface can be more approximated to the original points (Conolly and Lake 2006).

Like IDW, the splines method can be global or local according to the number of points involved in the interpolation.

Vector Model Based

Triangulated Irregular Network (TIN)

The TIN structure is a vectorial format widely used in the representation of the height values of a surface from previous point data (Conolly and Lake 2006). Normally, TINs are based on an irregular distribution of points with their value of height. The TIN interpolation creates a continuous succession of nonoverlapping triangles from the point cloud.

The TIN creation process is depicted in Figure 13.9. From the initial point distribution (Fig. 13.9a), the Thiessen polygons are created (Fig. 13.9b). These polygons are created by establishing its limits at equal distance from a

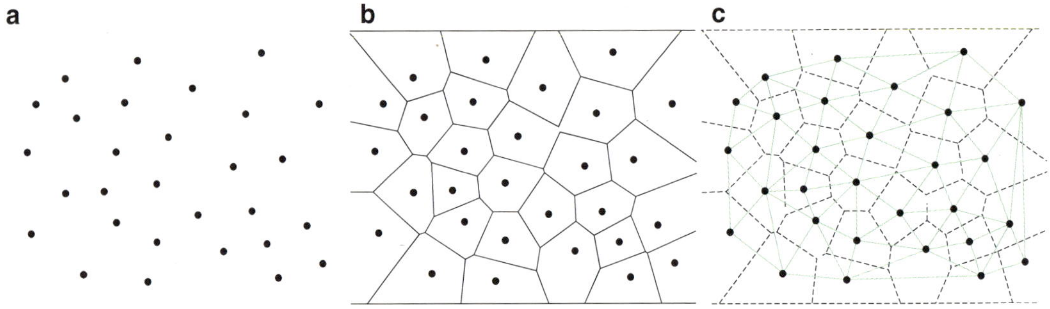

Fig. 13.9 How a TIN model is produced

pair of points, and, finally, the triangulation of Delaunay (Harmon and Anderson 2003) is carried out by connecting points with edges that are perpendicular to the segments of the Thiessen polygons (Fig. 13.9c).

The TIN structure also reflects the topology and the relations between the known points. Inside each triangle, the values of the target value (generally the height) vary linearly. The TIN model offers good results when a good distribution of known data is available, which assures a good adjustment between the model and the real surface. Besides, this procedure does not require too many computational requirements.

The drawback of the TIN model lies in the poor quality of the visual representations, caused by an excessive faceted aspect of the surface. Figure 13.10 shows the TIN model of the *Ammaia* archaeological site. In both views, the orthogonal (Fig. 13.10a) and the perspective (Fig. 13.10b), the faceted aspect of the model is clearly visible.

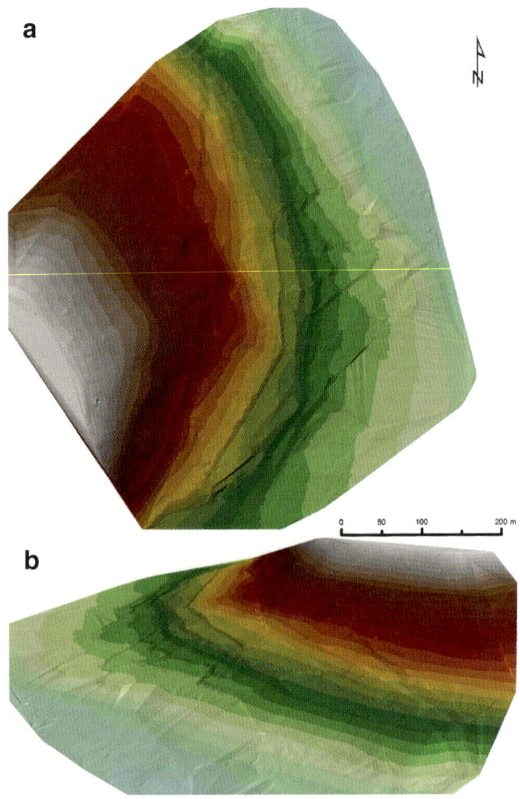

Fig. 13.10 TIN model of the Roman Town of *Ammaia*. The faceted aspect of the model is clearly visible in the orthogonal (**a**) and the perspective (**b**) views. (Marvao, Portugal)

Interpolation from Contour Lines

This could be one of the easier ways to obtain a DTM (Bosque Sendra 2000). The original data is geographic information coded by contour lines with their height values. This procedure is applied to the data derived from secondary sources (see previous section) when contour lines are digitalized. This interpolation method is similar to the interpolation of point maps and offers better results for the good distribution of initial information. However, it is not usual to find information coded in contour lines format because data cannot be sampled in this way.

13.3 How Do We Use Models for Archaeological Research?

Once we have our digital model, the time to use it comes. In this stage we will see that a good understanding of how DTMs have been produced is essential in order to understand their

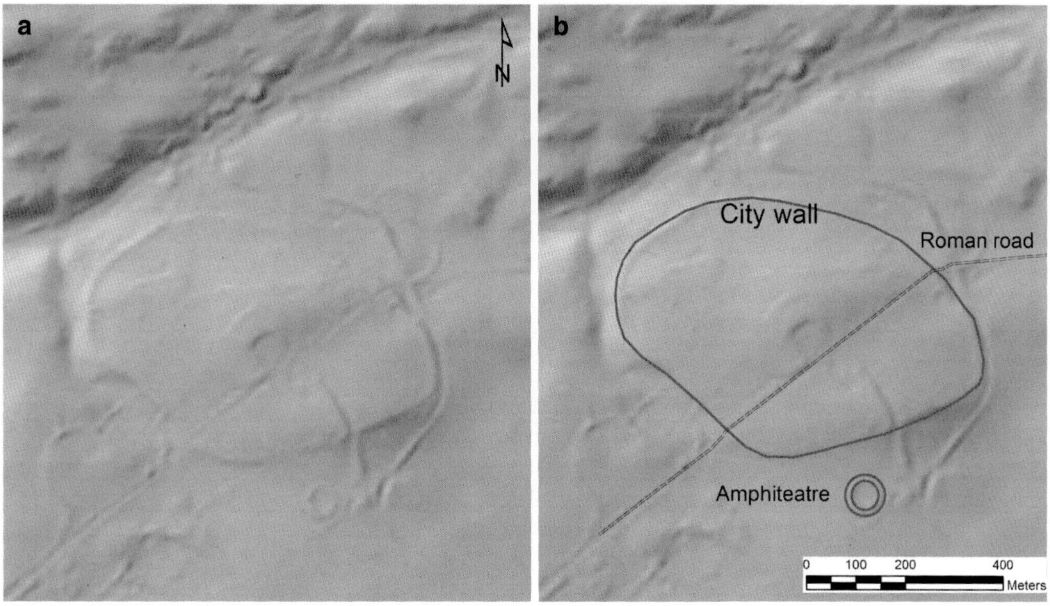

Fig. 13.11 Ventas de Caparra (Cáceres) as seen through the Web map service of the National Geographic Institute of Spain (http://www.ign.es/iberpix2/visor/). This example shows how accessible are via Internet elevation data that could be archaeologically meaningful

possibilities and constraints for the achievement of our goals. As we briefly saw at the beginning of this chapter, DTMs are present in many different archaeological applications of spatial technologies. We can find a good overview of these in general use manuals (Wheatley and Gillings 2002; Conolly and Lake 2006).

Simply for a purely descriptive work, accurate models are a useful tool to elaborate detailed maps and archaeological plans. These can be based on total station, GNSS or photogrammetry data, and are produced by public cartographical services and can be used in most archaeological projects (a good example of a systematic record of this kind of information can be seen in Fig. 13.11). Of course, these high-resolution data are a valuable resource for metric and morphological analysis and provide a very flexible alternative for creating contour maps, vector designs or sections of sites, structures or even small features like burials or other deposits.

This detailed representation of terrain can provide valuable information about buried remains through the identification of small elevation anomalies. Analytical hill-shading allows natural lighting conditions to be recreated. Furthermore, by tuning the parameters of the azimuth or altitude angle of the light source, it is possible to produce "shadow marks" on naked terrain or even "crop marks" on surfaces that can reveal previously unknown features. This is easy to understand through the identification of big earthworks in the landscape, for example, Roman public monumental buildings. The DTM shown in Figure 13.11a is derived from LiDAR data (a sample of 1 point per 5 m) and represents the environment of the Roman town of Capera (Ventas de Cáparra, Guijo de Granadilla, Caceres, Spain). Hill shade (Fig. 13.11b) exhibits a clear elliptical ring that corresponds to the Amphitheatre (as has been demonstrated by excavation) and the outline of the city walls and the Roman road crossing the town.

A wide realm of DTM applications in archaeology corresponds to landscape analysis. Many variables derived from them are linked with the study of locational criteria of past societies. They may be used as rough proxies to characterise decisions taken by human groups (average values of slope give us indications about the potential for agriculture around a site, while relative

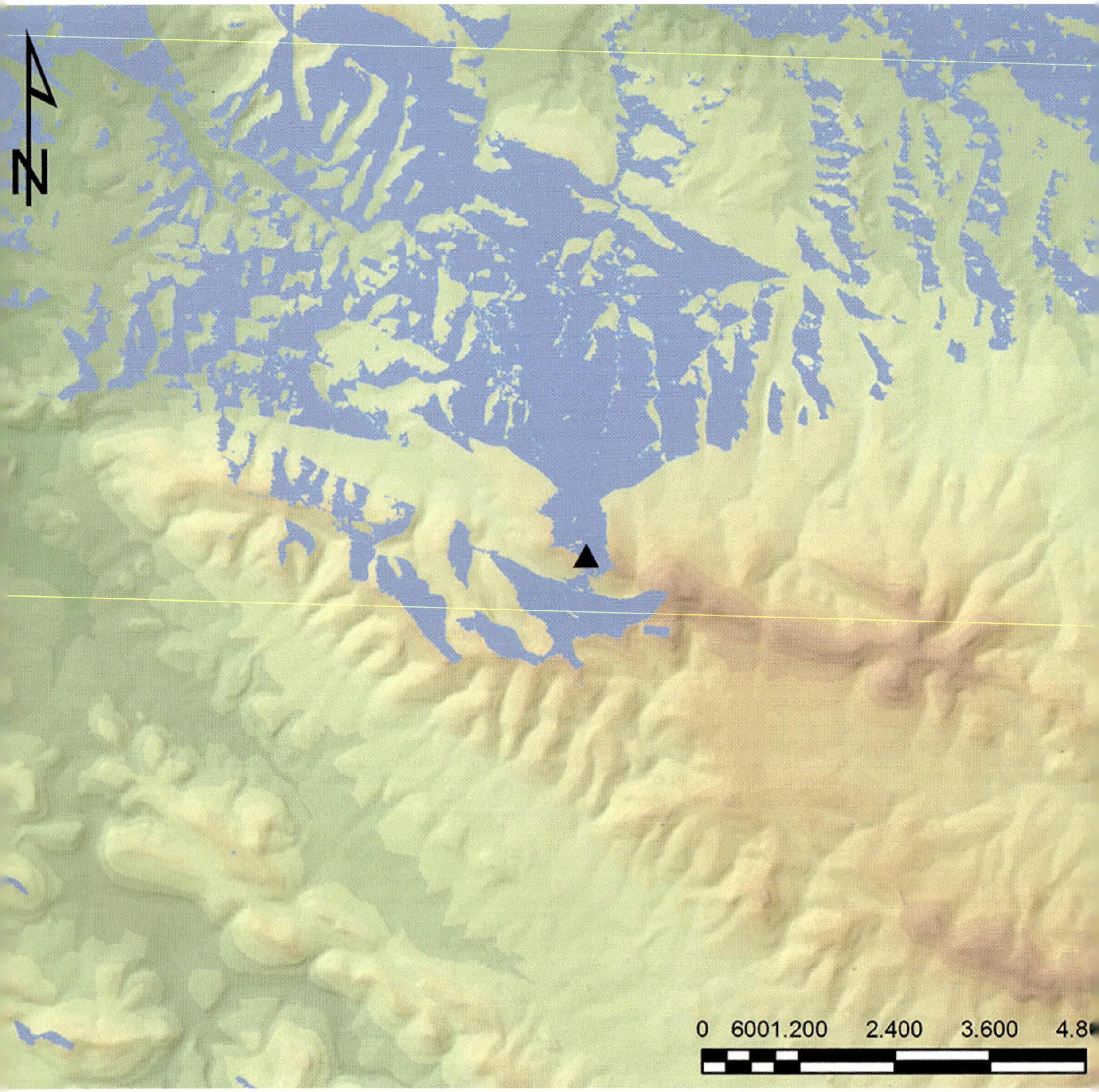

Fig. 13.12 A typical viewshed analysis for archaeological purposes

elevation values draw a picturc of accessibility between settlements and their surrounding territory). Other secondary variables often used by other disciplines like geomorphology can be useful in this sense. Morphometric analysis of terrain features can be a powerful tool to define land forms as discrete entities, and a quantification of their presence in an area of interest could help to understand why people chose certain places in the past (Fig. 13.12).

But isolated variables derived from DTMs may also be combined to produce more sophisticated models in order to assess the suitability of different areas for the development of economic, political or ideological activities. Predictive models usually take into account topography as a key variable.

Also within the study of human behaviour, we find an enormous scientific literature on the analysis of subjective variables like visual relationships or movement across the landscape. Here DTMs play a crucial role, since they determine the final results of the GIS processes involved (view shed analysis, cost surfaces). The main problem with these kinds of application is that they assume the validity of using current topography to explore past processes. In this sense, it is essential to take into account the spatial resolution and the source of data; for example, sometimes the overall land forms of a region can be an useful information while a detailed record of recent alterations introduces a lot of noise for our aims. Nevertheless, like any other model of reality, they should be considered as heuristic, rather than reconstructive exercises, since relief is just one of the many variables related with visibility or human displacement.

DTMs can also be part of research on the effects of processes affecting the representativeness and conservation of the archaeological record. At a landscape scale, elevation data should be included for the elaboration of multi-criteria models to simulate geomorphologic and land-use dynamics. These can become a predictive tool for the past and future evolution of terrain changes. In this sense, the archaeologist may learn a lot from GIS applications developed by hydrology, geomorphology and erosion studies. Another powerful way to explore the temporal dimension can be the cross-evaluation of different models derived from photogrammetric restitutions of historical flights. Cut fill calculations and other map algebra procedures can be used to estimate topographical changes quantitatively. These results may be very useful to assess the reliability of the results of surface surveys and the preservation of archaeological deposits.

The same role as these models can be played by a total station or GNSS survey of photogrammetry at a local scale. Archaeological sites can be analysed as "miniature landscapes". If a detailed record of surface evidence is available, topography can help to explain biasing factors in its distribution. For example, a raster layer expressing slope or potential water flow in a given location can be statistically correlated with the amount or total weight of different artefact categories (Fig. 13.13). This could lead to obtaining explanatory models on the spatial structure of surface assemblages.

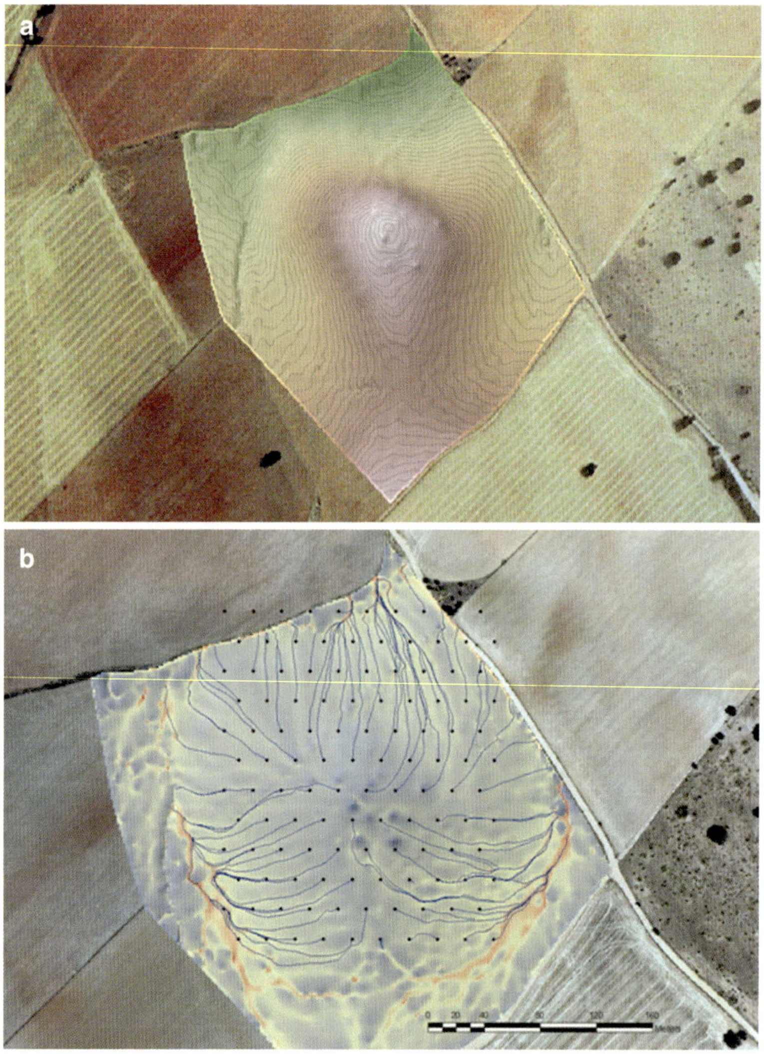

Fig. 13.13 Using a high-resolution topographic model of an archaeological site (**a**) to analyse the behaviour of surface artefacts. In (**b**) an index is derived to estimate potential water flow in the surface. *Black dots* represent survey sampling points, and *lines* are calculated steepest paths from these points

References

Bosque Sendra J (2000) Sistemas de Información Geográfica. Rialp, Madrid

Clark I (1979) Practical geostatistics. Elsevier Applied Science, London

Conolly J, Lake M (2006) Geographical information systems in archaeology. Cambridge University Press, Cambridge

De Reu J, Plets G et al (2013) Towards a three-dimensional cost-effective registration of the archaeological heritage. J Archaeol Sci 40(2):1108–1121

Deutsch CV (2002) Geostatistical reservoir modeling. Oxford University Press, New York

Felicísimo ÁM (1991) Modelos Digitales del Terreno. Introducción y aplicaciones en las ciencias ambientales. Penalfa, Oviedo

Harmon JE, Anderson SJ (2003) The design and implementation of Geographic Information Systems. Wiley, Hoboken

Hengl T (2009) A practical guide to geostatistical mapping, 2nd edn. University of Amsterdam, Amsterdam

Isaaks EH, Mohan Srivastava R (1989) Applied geostatistics. Oxford University Press, New York

Martínez del Pozo JÁ, Uriarte González A et al (2010) Geospatial characterization of archaeological sites in La Serena region (Badajoz) by direct methods (DGPS): capturing and analyzing data. In: Melero FJ, Cano P, Revelles J (eds) Fusion of cultures. Abstracts of the XXXVIII annual conference on computer applications and quantitative methods in archaeology, Granada, pp 537–540

Matheron G (1962) Traité de géostatistique appliquée, Tome I. Editions Technip, Paris

Matheron G (1963a) Principles of geostatistics. Econ Geol 58(8):1246–1266

Matheron G (1963b) Traité de géostatistique appliquée, Tome II, Le krigéage. Editions Technip, Paris

Maynou F (1998) The application of geostatistics in mapping and assessment of demersal resources. *Nephrops norvegicus* (L.) in the northwestern Mediterranean: a case study. Scientia Marina 62(Suppl. 1):117–133

Miller CL, Laflamme RA (1958) The digital terrain model – theory and application. Photogramm Eng 24(3):433–442

Nina Siu-Ngan Lam (1983) Spatial interpolation methods: a review. Am Cartographer 10(2):129–149

Olea RA (2009) A practical primer on geostatistics: U.S. Geological Survey Open- File Report 2009–1103. http://pubs.usgs.gov/of/2009/1103/

Shepard D (1968) A two-dimensional interpolation function for irregularly-spaced data. ACM Press, New York

Verhoeven G (2009) Providing an Archaeological Bird's-eye View: an overall picture of ground-based means to execute low-altitude aerial photography (LAAP) in archaeology. Archaeol Prospect 16:233–249

Villatoro M, Henríquez C, Sancho F (2008) Comparación de los interpoladores IDW y Kriging en la variación espacial de PH, Ca, CICE y P del suelo. Agronomía Costarricense 32(1):95–105

Webster R, Oliver MA (2007) Geostatistics for environmental scientists. 2nd ed. Wiley Online Library. doi:10.1002/9780470517277

Weibel R, Heller M (1991) Digital terrain modelling. In: Maguire DJ, Goodchild MF, Rhind DW (eds) Geographical information systems: principles and applications. Longman, London, pp 269–297

Wheatley D, Gillings M (2002) Spatial technology and archaeology. The archaeological application of GIS. Taylor & Francis, London/New York

14 The Geoarchaeological Approach

Eric Fouache

Contents

E. Fouache
Université Paris Sorbonne Abu Dhabi,
Abu Dhabi, United Arab Emirates

UMR 8385 ENEC,
CNRS Université Paris Sorbonne, Paris, France
e-mail: eric.g.fouache@wanadoo.fr

14.1 Introduction

In the Anglo-Saxon sense of the term, geoarchaeology is often seen as a major discipline of environmental archaeology applied to archaeological issues, together with palaeoecology, archaeobotany and archaeozoology (Gladfleter 1977; French 2003). This "restricted" definition mainly relies on the study of the soils and sediment.

From a broader perspective, archaeology also appears as a general earth sciences discipline which associates the palaeoenvironmental studies on a given archaeological site and in the environs of a site (Bravard and Presteau 1997; Rapp and Hill 2006). A rather similar understanding of the term was proposed by Butzer (1982), who distinguishes geological archaeology or geoarchaeology – which tries to solve archaeological issues thanks to the methods and techniques of earth sciences – from archaeological geology or archaeogeology, which is inspired by geological and geomorphological issues, the results of which may be used in archaeology.

14.2 The Rise of a Concept

It is only in the last third of the twentieth century that geoarchaeology was truly defined both conceptually and institutionally. Karl Butzer (1971) was the first to use the term geoarchaeology, originally spelt geo-archaeology in an essay entitled *Environment and Archaeology: An Ecological*

C. Corsi et al. (eds.), *Good Practice in Archaeological Diagnostics*, Natural Science in Archaeology,
DOI 10.1007/978-3-319-01784-6_14, © Springer International Publishing Switzerland 2013

Approach to Prehistory. The first essay specifically entitled *Geoarchaeology* was the one written by Davidson and Shackley (1976) while the journal *Geoarchaeology* was created in 1986 by Willey's. After the 1970s, geoarchaeology gains more and more ground, and there are more and more works related to the formation processes of the archaeological sites in the 1980s and 1990s as well as until now, particularly by the American school (Butzer 1971, 1982; Schichk 1986; Schiffer 1987; Waters 1992; Rapp and Hill 1998; Stein and Farrand 2001), the British school (Needham and Macklin 1992; Brown 1997; French 2003) or the French school (Courty 1992; Bravard and Prestreau 1997; Sordoillet 1999; Miskovsky 2002; Lenoble 2005; Couchoud 2006).

The Working Group on Geoarchaeology was founded in 1997 on the initiative of Morgan De Dapper as part of the international Association of Geomorphologists. Since then, the working group has regularly organised separate sessions in international and regional conferences of the IAG, the EGU (European Geosciences Union), the INQUA (International Union for Quaternary Research) and the IUGS (International Union of Geological Sciences). It has become a major protagonist of this field of study. From our point of view, geoarchaeology is not a new discipline in itself but rather an interdisciplinary approach that federates and integrates the methods and techniques of natural sciences, geography and geosciences in order to address very precise archaeological issues in close relation with archaeologists.

14.3 General Principles of the Geoarchaeological Approach

Geoarchaeology is too often presented as a catalogue of successive interventions in the fields of geomorphology, sedimentology, pedology, stratigraphy, geochronology, micromorphology, archaeobotany, archaeozoology, palaeoclimatology (Waters 1992; Goldberg et al. 2006) or even as a list of laboratory techniques originally from geology, as shown in the definition of Rapp and Hill (1998): 'the use of laboratory methods originally from geology and prehistory for archaeology'. To this, two visions can be added; they are often exclusive of the archaeological approach while they are in fact complementary, that is, the in-site analysis, which often focuses on the stratigraphic, pedological and micromorphological approaches, and the off-site analysis, which focuses on regional palaeoenvironmental reconstructions.

Similarly, Butzer (1982) distinguishes geological archaeology or geoarchaeology, which tries to address archaeological issues thanks to the methods and techniques of earth sciences, from archaeological geology or archaeogeology, which addresses geological and geomorphological issues that maybe be applied to archaeology.

We wish to promote a third way, that of a synthetic approach which would combine both in-site and off-site analyses, biotic as well as abiotic elements from a transversal, interdisciplinary perspective.

14.4 In-Site and Off-Site Studies

In-site analysis first and foremost focuses on the genesis of the archaeological sites, that is, the formation processes on the scale of the sites themselves and on the factors leading to the fossilisation, preservation or reworking of the archaeological vestiges (Bertran et al. 1995; Bertran and Texier 1997; Lenoble 2005). It is thus complementary with the purportedly purely archaeological, traditional stratigraphic approach. It allows to establish the origin of the archaeological sediment and their evolution by highlighting what is linked with the anthropogenic, cultural and bio-pedological processes as well as the geological depositional (sedimentary) and/or post-depositional (diagenetic) factors, whether it concerns cave sites, rock shelters or open-air sites (Courty et al. 1999; Sordoillet 1999). In-site analysis thus addresses issues related to the formation processes of those sites, in close collaboration with field archaeologists. Such processes can be observed in the stratigraphy, sedimentology and micromorphology as well as in the spatial distribution of the vestiges.

When the geoarchaeological study is not limited to the site, it may concern a larger, consistent area: a plateau, a watershed, the flood plain of a river, etc. This off-site analysis focuses more on the study of the landscapes and their evolution, of palaeoclimates and palaeoenvironments (Waters 1992; Stafford 1994; Waters and Kuehn 1996; Bravard 1997; Helgren 1997). It mainly consists in the reconstruction of the environments within which prehistoric and historic human societies have evolved. This is why the evolution of the environment is considered part and parcel of human history. Such off-site analysis thus offers a major contribution to the study of the biophysical system that places together human societies, physical and biological environments as well as their interactions and evolutions (Bravard and Presteau 1997).

Geoarchaeology, broadly speaking, thus appears as a non-specialised discipline that coordinates and integrates palaeoenvironmental studies on an archaeological site (in-site analysis) and/or in the environs of a site (off-site analysis). From this perspective, we consider geoarchaeology as an interdisciplinary approach in which the methods of geomorphology play an important, but not unique, part. We thus propose a new definition of the geoarchaeological approach by slightly altering that of the International Working Group on Geoarchaeology (Fouache et al. 2010): 'The application, from a diachronic, multi-scalar perspective, of the methods originally from geosciences and geography to the reconstruction of the palaeoenvironments and the biophysical landscape dynamics in relation with human occupancy'.

From our point of view, there cannot be any geoarchaeological approach without archaeologists, but this is not incompatible with the integration of the methods and tools of historical geography and history. Chronologically speaking, geoarchaeology, like archaeology in its broad sense which includes prehistoric archaeology, covers a time span that ranges from the end of the Tertiary to the historical periods even if in this handbook, for practical reason, we will restrict ourselves to a short span that covers the Holocene. From a spatial point of view, the multiscalar approach of geoarchaeology goes from in-site to off-site and from the study of the sediment to the scale of a watershed or even a whole region. Geoarchaeology thus deals with two key concepts, the environment and the landscape, thus underlining the fact that everything that is contact with human societies directly interacts with them.

14.5 Prior to Any Geoarchaeological Study

Any geoarchaeological research is necessarily interested in the nature of the archaeological site under study, in the physical characteristics of the surrounding environment, and raises the question of the scope of the change recorded in the landscape before, during and since the given period of occupation considered under the double action of the environmental dynamics and human actions, which add up through time. Depending on the nature of the archaeological programme, the questions will concern a more or less vast region, also depending on whether it is an excavation or a regional survey, whether we take into account the natural resources available for archaeological purposes or not. In any case, the landscape will be read as a palimpsest, of which the genesis must be retraced by going back in time thanks to a multi-scalar, systemic, regressive and diachronic approach. The traces of this evolution may be either erased, visible on the surface or concealed by superficial formations; they may be palaeo-channels, palaeo-shores, formers fields, palaeosoils, irrigation or drainage channels, former roads and former limits of agricultural parcels, of housing or of any other archaeological vestiges.

For the sake of rationality, of coherence in the embedding of the spatial scales and of cost control, it is often preferable to organise the study from a regional scale to a local scale but also to assess clearly the surface dynamics before embarking on a detailed study of superficial formations. At each stage, the contribution of the tools and concepts originally from geography, geomorphology, geology and of geosciences in general is crucial.

There cannot be any good geoarchaeological study that does not raise one or several relevant questions that connect the environmental dynamics to the archaeological and historical data. The questions raised may originate in:

- The reading of the regional archaeological map, when available
- The mutual reflections carried out between archaeologists/historians, geographers, geomorphologists, geoscientists and palaeoenvironmentalists
- The reflection derived from the results of surface explorations
- The archaeological excavations themselves

Geomorphological dynamics – colluviation, alluviation, tectonic deformation, relative variations of the water levels (sea, lake, marsh) – may account for the fact that the archaeological map, forever evolving because perpetually amended and completed as the discoveries go, does not always reflect the full occupancy of a given land at a given time. Identifying these dynamics is an indispensable prerequisite if we want to put into perspective the chronological interpretations derived from the sole archaeological excavations.

This is why it is preferable to start any geoarchaeological study by a regional geomorphological analysis in relation with the geodynamic context whenever the archaeological programme concerns a specific excavation or exploration on a larger area. Drawing a geomorphological map meets several transversal, complementary aims:

- To identify the zones of sedimentary/detrital accumulation and the erosion zones (sedimentary budget).
- To outline the zones where the geomorphological dynamics are particularly active and the location of palaeoforms.
- To assess the depth of the superficial formations, both in situ and reworked.
- To identify the contexts that help preserve the sedimentary and biotic archive (for instance peat) and/or the archaeological archive (taphonomy).
- To enable discussion of the spatial distribution of the data on the archaeological map. The geomorphological map will allow to build up a strategy to guide exploration, diagnosis and potential sampling (e.g. dating) and to characterise the geomorphological context of a given archaeological site.

Besides, the geomorphological map itself relies on knowledge of topography, mapping (Pavlopoulos et al. 2010) and, generally speaking, of management of spatialised databases (Chapman 2006; Ghilardi 2006). This is why, depending on the available data, any geomorphological map should be integrated to a GIS (Geographic Information System) which might itself include a DEM (Digital Elevation Model) carried out thanks to SRTM (Shuttle Radar Topography Mission) data or differential GPS measurements, to mosaics of aerial photographs and satellite image coloured compositions or to photogrammetric mapping. Any good GIS ought to be designed from the start on a scale that allows to include environmental and archaeological data, which will enable to create specific and dynamic thematic maps.

The drawing of a regional geomorphological map thus constitutes the first stage of a rationally conducted geoarchaeological approach. It may of course be completed simultaneously if necessary with a mapping of the vegetal formations, the soils or the mineral resources. The geomorphological map is like the backbone of stage one of the geoarchaeological approach, which consists in four stages.

14.6 The Stages of the Geoarchaeological Approach

Stage one consists in a mission of exploration leading to the drawing of a regional geomorphological map. This is an essential stage because it allows to assess the scope of the present and inherited geomorphological dynamics. Geomorphological exploration may also, naturally, be oriented thematically depending on the questions that are raised, but it must retain a regional character and lead to define the dimension of the area to be taken into account for the study, in agreement with the archaeologists. With this map in hand, depending on the information obtained, it then becomes possible to propose possible scenarios for geomorphological, taphonomic and palaeoenvironmental evolutions that must then be confirmed by data.

Stage two consists in working out the interdisciplinary scientific approach that will allow to obtain the palaeoenvironmental data necessary to confirm the hypotheses, as quickly as possible and to the lesser cost, whether those hypotheses relate to the study of geomorphological dynamics, to the identification of the physical environments or to issues pertaining to the in-site archaeological or sedimentary structures. This often requires a study of the natural or anthropogenic sedimentary formations, within an archaeological context or not. The techniques used are numerous and range from indirect approach, remote sensing, geophysical exploration (non-destructive techniques) to direct observation (boring, drilling, surface or micromorphological sampling), all carried out at carefully chosen places according to sampling strategies that depend on the expected markers, whether they be pedo-sedimentary, physico-chemical or biological.

Stage three is the stage of laboratory work, of the work by specialists in palaeoenvironments, archaeometry and geochronology who had ideally been associated with the sampling work. Such studies are always long and costly, and it is advisable to resume field work only after the results of such analyses have been included into the research so as to feed stage four.

Stage four will enable to provide a spatial and chronological reconstruction of the results of the programme in terms of palaeogeographic and palaeoenvironmental data and to infer a maximum of information useful to address the archaeological issues. The data included into a GIS will enable to build spatial models of the different palaeogeographical reconstitutions obtained or of the successive states of a given site within its environment. Several years may elapse between stage one and stage four.

14.7 The Contribution of Geomorphology to the Geoarchaeological Approach

The geomorphologist's role consists in studying and reconstructing the off-site environment by including topography, the forms of the landscape, its dynamics of evolution, the superficial formations, the location of mineral and hydrogeological resources and the evolution of the soils, while assessing the place of man in this environment. Then the geomorphologist examines the possible geomorphological dynamics resulting from the interaction of natural dynamics and dynamics originating in the action of man, particularly since the beginning of the Neolithic (Martini and Chesworth 2010). He also takes into account taphonomy and the processes of formation and destruction of the archaeological sites, including the evolution of the different levels of occupation and their artefacts (the material man-made vestiges) and ecofacts (the non-fabricated remnants of man's activities such as food residue or charcoal), as well as the in-site anthropogenic sedimentation linked to human activities (Goldberg and Macphail 2006).

Geomorphology specifically describes the reliefs and forms and interprets their evolution, in relation with the lithological and geodynamic context (structural geomorphology), under the effect of both pre-Quaternary, Quaternary or Holocene tectonics and of erosion agents (dynamic and climatic geomorphology). It establishes the availability of mineral and hydric resources over a given period and allows to assess the impact of human activity such as clearing, agriculture or pastoral farming on the physical environment. For that purpose, the geomorphologist examines the topographic maps, the geological maps, aerial photographs and satellite images. This surface approach is combined with geophysical exploration, boring and the study of stratigraphic sections on the field (Schoeneberger 1998). The geomorphological approach takes into account very diverse scales, from the regional to the local, from the evolution of the landscape as a whole to microscopic analyses of sedimentary units.

The methods of sedimentology, such as micromorphology, or the geochemical analyses allow to characterise the dynamics of deposition and evolution of the sediments. The depositional processes can thus be clarified according to the granulometric and micro-granulometric sorting of the sediments. Thus the application of multivariate statistic analysis, factor analysis of the components and of Passega's CM image technique (Passega 1957, 1964) to alluvial deposits allows to precisely reconstitute the hydrodynamic palaeoconditions in which sediments were deposited and to determine where the alluvium was deposited in the main riverbed or in the flood plain for instance (Brown 1997; Holliday 1997; Gladfelter 2001).

The sediments may be displaced for a number of reasons and through either simple actions or actions resulting from a combination of different processes that may alternate one with the other in time and space, namely, the action of liquid freshwater (e.g. alluviation, rill wash) or of solid freshwater (e.g. glacier), of gravity (e.g. mass movements, colluviation, considering that colluvium can also be deposited thanks to sheet wash), marine and coastal action (e.g. beach deposits, slikke/mud flat deposits, or else the action of the wind (eolian deposits).

In a Karst environment, the flow of water may lead to the formation of travertine either in the open air or in the form of speleothems in underground cavities. In the case of excavations in caves, those speleothems provide valuable chronological and palaeoenvironmental markers, as in the case of the excavations of the White Paintings Rock Shelter in the Kalahari Desert in Botswana (Robbins et al. 2000). In a coastal environment, carbonated cementing called beachrock may lead to the formation of slabs that attest to the existence of fossil sea levels that are sometimes associated with archaeological vestiges (Desruelles et al. 2008). In the case of travertine, they can cement sediment but also fossilise archaeological sites. Thus the monuments of the ancient Greek city of *Poseidonia*, better known under its Roman name of *Paestum* on the coast of Campania to the south of the mouth of the river Sele, have had to be excavated from four generations of thick travertine deposited by the Capodifiume river (Amato et al. 2009).

Pedology, which studies the formation and the evolution of the soils under the influence of bioclimatic conditions, from the bedrock under the action of the vegetal cover and/or of man, holds a specific interest in the geoarchaeological approach (Holliday 2004). Indeed, the archaeological sites are often set on ancient, often fossil, soils which it is essential to characterise. The assessment of the agricultural potential and the study of the transformation of the agricultural soils are an important component of geoarchaeology (Parnell et al. 2002).

The presence of sediment on the archaeological sites may also result from human action. If we examine the sediments or the soils of the natural or archaeological stratigraphies on the field, we may highlight lateral and vertical variations in the facies. However, in order to interpret the interpenetration of anthropogenic sediments and sediments resulting from the dismantling of built structures and to reconstruct the taphonomy of the site, it is necessary to combine a macrostratigraphic study with a micromorphological study (Goldberg 1980; Courty 1992).

Micromorphological analysis consists in studying the arrangement and the nature of the sediments microscopically. For that purpose a sedimentary block is taken as is, directly from a stratigraphic section. The block, which is often nonhomogenous except when there was a natural induration, which is frequent in a cave, is soaked with transparent resin in the laboratory then it is cut in thin plates. It is then possible to study the composition and the inner structure of the sediment. The study can be completed by a petrogeochemical study in order to determine the composition of the rocks, the minerals or the artefacts (e.g. ceramics, building material such as cob).

The geomorphological and sedimentary approach, thanks to field and laboratory work, is thus necessary to understand and interpret an archaeological site and its physical context, the origin and the dynamics of deposition of the sediments that compose it and the post-depositional processes. Only in these conditions can a meaningful sampling of material destined to radiometric datation and to archaeozoological, archaeobotanical or geochemical studies be done.

Conclusion

Geomorphologists, as well as geoscientists in general, play a particularly important role, although they are not the only ones, working hand in hand with archaeologists, first, as well as with different specialists of palaeoenvironments and archaeometry, in order to coordinate the stages of the geoarchaeological approach. When confronted to particularly dynamic geomorphological contexts, they may be a central mainspring in the environmental study. This is particularly true in fluvial geoarchaeology (Brown 1997).

Obviously this protocol and agenda need not be imposed on each and every archaeological programme. A geoarchaeological study can actually start directly with sampling destined to palaeoenvironmental analyses. But this can only be done if and only if it is absolutely certain that the data previously obtained indeed cover the whole of the protocol we have just outlined, otherwise the risk of committing major errors of interpretation or of excavation strategy is real, with a further risk of ending up in a methodological dead end.

Geoarchaeology as a professional practice is becoming more and more structured, with numerous opportunities offered to students in positions of researchers, academics, but also more and more in archaeological services, public or private companies or local authorities specialising in rescue archaeology or the development of the archaeological and environment heritage.

References

Amato V, di Paola G, Rosskopf CM, Avagliano G, Cipriani M, Cinque A, Pontrandolfo A, Santoriello A (2009) Geomorphology and geoarchaeology of the Paestum area: modifications of the physical environment in historical times. Méditerranée 112:129–135

Bertran P, Texier J-P (1997) Géoarchéologie des versants: les dépôts de pente. In: Bravard J-P, Presteau M (eds) Dynamique du paysage: entretiens de géoarchologie. D. A. R. A, Châtillon-sur-Chalaronne, p 59

Bertran P, Francou B, Texier J-P (1995) Stratified slope deposits: the stone- banked sheets and lobes model. In: Slaymaker O (ed) Steepland geomorphology. Wiley, London/New York, pp 147–169

Bravard J-P, Presteau M (1997) Dynamiques du paysage: entretiens de géoarchéologie (Table ronde tenue à Lyon les 1è et 18 novembre 1997). Documents d'Archéologie en Rhône-Alpes n°15. Service Régional de l'Archéologie-Direction Régionale des Affaires Culturelles Rhône-Alpes, Lyon

Brown AG (1997) Alluvial geoarchaeology: floodplain archaeology and environmental change, Cambridge manuals in archaeology. Cambridge University Press, Cambridge

Butzer KW (1971) Environment and archaeology: an ecological approach to prehistory. Aldine, Chicago

Butzer KW (1982) Archaeology as human ecology: method and theory for a conceptual approach. Cambridge University Press, Cambridge

Chapman H (2006) Landscape archaeology and GIS. The History Press Ltd., Stroud

Couchoud I (2006) Étude pétrographique et isotopique de spéléothèmes du sud-ouest de la France formés en contexte archéologique: Contribution à la connaissance des paléoclimats régionaux du stade isotopique 5. Thèse de doctorat. Université Bordeaux 1, Bordeaux

Courty M-A (1992) Soil micromorphology in archaeology. Proc Br Acad 77:39–59

Davidson DA, Shackley M (1976) Geoarchaeology: earth science and the past. Westview Press, Boulder

Desruelles S, Fouache E, Ciner A, Koşun E, Pavlopoulos K, Coquinot Y, Potdevin J-L (2008) Beachrocks and sea level changes since Mid-Holocene: comparison between the insular group of Mykonos-Delos-Rhenia (Cyclades, Greece) and the southern coast of Turkey. Global Planet Change. doi:10.1016/j.glopacha.2008.07.09

Goldberg P, Macphail RI (2006) Practical and theoretical geoarchaeology. Blackwell Science, Oxford

Fouache E, Pavlopoulos K, Stiros S (eds) (2010) Landscape evolution and geoarchaeology. Quat Int 216(1–2):1–162

French C (2003) Geoarchaeology in action: studies in soil micromorphology and landscape evolution. Routledge, London

Ghilardi Mathieu. 2006. *Apport et intérêt de la modélisation numérique de terrain en géomorphologie. Étude du site de Méthoni (Piérie – Grèce).* Mémoire du Laboratoire de géomorphologie et d'environnement littoral de l'Ecole Pratique des Hautes Etudes, 45. Dinard.

Gladfelter BF (1977) Geoarchaeology: the geomorphologist and archaeology. Am Antiq 42(4):519–538

Gladfelter BG (2001) Archaeological sediments in humid alluvial environments. In: Stein JK, Farrand WR (eds) Sediments in archaeological context. The University of Utah Press, Salt Lake City, pp 93–125

Helgren DM (1997) Locations and landscapes of Paleolithic sites in the Semliki Rift, Zaire. Geophys J Roy Astron Soc 12(4):337–361

Holliday VT (1997) Paleoindian geoarchaeology of the southern high plains, Texas archaeology and ethnohistory series. University of Texas Press, Austin

Holliday VT (2004) Soils in archaeological research. Oxford University Press, Oxford

Lenoble, Arnaud. 2005. Ruissellement et formation des sites préhistoriques. Référentiel actualiste et exemples d'application au fossile. (British Archaeological Reports International Series 1363) Oxford: Archaeopress.

Martini IP, Chesworth W (eds) (2010) Landscapes and societies. Selected cases. Springer, New York/Heidelberg

Miskovsky, Jean-Claude, ed. 2002. Géologie de la préhistoire: méthodes, techniques, applications. Association pour l'étude de l'environnement géologique de la Préhistoire, Paris, Géopré, Presses universitaires de Perpignan.

Needham S, Macklin MG (1992) Alluvial archaeology in Britain, Oxbow monograph 27. Oxbow, Oxford

Parnell JJ, Terry RE, Nelson Z (2002) Soil chemical analysis applied as an interpretative tool for ancient human activities in Piedras Negras, Guatemala. J Archaeol Sci 29(4):379–404

Passega R (1957) Texture as a characteristic of clastic deposition. Bull Am Soc Petrol Geol 41:1952–1984

Passega R (1964) Grain size representation by CM pattern as a geological tool. J Sediment Petrol 34:830–847

Pavlopoulos K, Elvepidou N, Vassilopoulos A (2010) Mapping geomorphological environments. Springer, New York/Heidelberg

Rapp GR, Hill CL (1998) Geoarchaeology: the earth-science approach to archaeological interpretation. Yale University Press, London

Robbins LH, Murphy ML et al (2000) Archaeology, palaeoenvironment, and chronology of the Tsodilo Hills White Painting Rock Shelter, Northwest Kalahari Desert, Botswana. J Archaeol Sci 27(11): 1085–1113

Schichk K (1986) Geoarchaeological analysis of an acheulean site at Kalambo Falls, Zambia. Geophys J Roy Astron Soc 7(1):1–26

Schiffer MB (1987) Formation processes of the archaeological record. University of New Mexico Press, Albuquerque

Stafford CR (1994) Geoarchaeological perspectives on paleolandscapes and regional subsurface archaeology. J Archaeol Method Theory 2(1):69–104

Stein JK, Farrand WR (2001) Sediments in archaeological context. The University of Utah Press, Salt Lake City

Schoeneberger PJ (ed) (1998) Field book for describing and sampling soils. Natural Resources Conservation Service (USDA) and National Soil Survey Center, Lincoln

Sordoillet, Dominique. 1999 . Géoarchéologie de sites préhistoriques holocènes. Thèse de doctorat, Dijon: Université de Bourgogne, Centre des sciences de la terre,

Waters MR (1992) Principles of geoarchaeology: a North American perspective. University of Arizona Press, Tucson

Part IV

Visualisation and Site Management

Implementing Best Practice in Cultural Heritage Visualisation: The London Charter

15

Hugh Denard

Contents

H. Denard
Department of Classics – School of Histories and Humanities, Trinity College Dublin, Dublin, Ireland
e-mail: hdenard@gmail.com

The *London Charter* (2009) is today widely recognised as the de facto benchmark to which heritage visualisation processes and outputs should be held accountable. At the time of going to print, the London Charter website includes versions in English, Bosnian, Chinese, Farsi, German, Italian, Japanese, Polish and Spanish, and additional translations, including into French and Russian, are in progress. The Charter has won formal endorsement from national and international bodies including adoption as an official guideline by the Italian Ministry of Culture. It has also been the subject of and widely cited in numerous publications, not least *Paradata and Transparency in Virtual Heritage* (Bentkowska-Kafel et al. 2012), for which it is the fundamental text. The Charter was a key result of the EPOCH European Network of Excellence in Open Cultural Heritage (http://www.epoch-net.org) and currently forms an important part of the activities of V-MUST: Virtual Museum Transnational Network (http://www.v-must.net/). The Charter has also inspired the creation of the *Principles of Seville: International Principles of Virtual Archaeology*, which 'aim to increase the conditions of applicability of the *London Charter* in order to improve its implementation specifically in the field of archaeological heritage' (International Forum of Virtual Archaeology 2010; see Chap. 16 by López-Menchero in this volume).

The Charter arose out of a recognition that the use of three-dimensional computer modelling in the historical, archaeological and broader cultural heritage domain, and particularly the rise of

C. Corsi et al. (eds.), *Good Practice in Archaeological Diagnostics*, Natural Science in Archaeology,
DOI 10.1007/978-3-319-01784-6_15, © Springer International Publishing Switzerland 2013

hyperrealism, risked introducing a dangerous lack of intellectual transparency, a lack which could critically compromise the actual and perceived validity of such research. Computer models of historical sites, monuments and artefacts began to be published but often without sufficient accompanying information to enable viewers to determine their historical validity. Such publications were doubly perilous because of their aesthetic persuasiveness: an image could seem to indicate a degree of certainty about the past which, in fact, the evidence did not support.

The potential for visualisations to mislead their viewers was identified as a burning issue in numerous publications from the mid-1990s on (e.g. Ryan 1996; Roberts and Ryan 1997; Niccolucci 1999; Goodrick and Gillings 2000; Frischer et al. 2002; Jablonka et al. 2003; Roussou and Drettakis 2003; Zuk et al. 2003; Denard 2005; Hermon and Niccolucci 2006; see also Mudge 2012). Many scholars agreed that there was a pressing need to reconcile digital visualisation methods with the professional norms of academic research, which demand, for example, that the relationship between evidence and hypothesis should be made clear. However, there was no clear consensus on how this intellectual transparency should be achieved in the case of digital visualisations or on the criteria by which such work could be evaluated.

In 2006, at a symposium on 'Making 3D Visual Research Outcomes Transparent', I proposed that the assembled experts should establish a common framework of principles and that these should be published in the form of a 'Charter' which we could recommend to our various disciplines and professional bodies as a shared basis for using and evaluating digital visualisations of cultural heritage. Such a Charter, I argued, would provide a significant boost, internationally, to understandings of and compliance with best practice. At an Expert Seminar immediately following the symposium, we created the first draft of the London Charter (Beacham et al. 2006), which has since been revised into the current version (2.1, 2009). Through extensive consultation with expert communities since 2006, the London Charter has succeeded in establishing international consensus on the principles that should inform best practice in heritage visualisation across numerous disciplines. The Charter not only improves methodological rigour but also offers a means of achieving significant efficiencies in teaching and training as well as in the research and communication of cultural heritage.

For the reader's convenience, we reproduce the current main text of the *London Charter* below (omitting only its Glossary), after which I will discuss some aspects of it in greater detail to give fuller explanations of the reasoning behind certain key points and to flesh out some of their implications for those seeking to implement it.

* * * * *

15.1 The London Charter for the Computer-Based Visualisation of Cultural Heritage (Version 2.1, 2009)

15.1.1 Preamble

While computer-based visualisation methods are now employed in a wide range of contexts to assist in the research, communication and preservation of cultural heritage, a set of principles is needed that will ensure that digital heritage visualisation is, and is seen to be, at least as intellectually and technically rigorous as longer established cultural heritage research and communication methods. At the same time, such principles must reflect the distinctive properties of computer-based visualisation technologies and methods.

Numerous articles, documents, including the AHDS Guides to Good Practice for CAD (2002) and Virtual Reality (2002) and initiatives, including the Virtual Archaeology Special Interest Group (VASIG) and the Cultural Virtual Reality Organisation (CVRO) and others have underlined the importance of ensuring both that computer-based visualisation methods are applied with scholarly rigour, and that the outcomes of research that include computer-based visualisation should accurately convey to users the status of the knowledge that they represent, such as distinctions between evidence and hypothesis, and between different levels of probability.

The London Charter seeks to capture, and to build, a consensus on these and related issues in a way that demands wide recognition and an expectation of compliance within relevant subject communities. In doing so, the Charter aims to enhance the rigour with which computer-based visualisation methods and outcomes are used and evaluated in heritage contexts, thereby promoting understanding and recognition of such methods and outcomes.

The Charter defines principles for the use of computer-based visualisation methods in relation to intellectual integrity, reliability, documentation, sustainability and access.

The Charter recognises that the range of available computer-based visualisation methods is constantly increasing, and that these methods can be applied to address an equally expanding range of research aims. The Charter therefore does not seek to prescribe specific aims or methods, but rather establishes those broad principles for the use, in research and communication of cultural heritage, of computer-based visualisation upon which the intellectual integrity of such methods and outcomes depend.

The Charter is concerned with the research and dissemination of cultural heritage across academic, educational, curatorial and commercial domains. It has relevance, therefore, for those aspects of the entertainment industry involving the reconstruction or evocation of cultural heritage, but not for the use of computer-based visualisation in, for example, contemporary art, fashion, or design. As the aims that motivate the use of visualisation methods vary widely from domain to domain, Principle 1: "Implementation", signals the importance of devising detailed guidelines appropriate to each community of practice.

15.1.2 Objectives

The London Charter seeks to establish principles for the use of computer-based visualisation methods and outcomes in the research and communication of cultural heritage in order to:

- Provide a benchmark having widespread recognition among stakeholders.
- Promote intellectual and technical rigour in digital heritage visualisation.
- Ensure that computer-based visualisation processes and outcomes can be properly understood and evaluated by users
- Enable computer-based visualisation authoritatively to contribute to the study, interpretation and management of cultural heritage assets.
- Ensure access and sustainability strategies are determined and applied.
- Offer a robust foundation upon which communities of practice can build detailed London Charter Implementation Guidelines.

15.1.3 Principles

15.1.3.1 Principle 1: Implementation

The principles of the London Charter are valid wherever computer-based visualisation is applied to the research or dissemination of cultural heritage.

1.1 Each community of practice, whether academic, educational, curatorial or commercial, should develop London Charter Implementation Guidelines that cohere with its own aims, objectives and methods.

1.2 Every computer-based visualisation heritage activity should develop, and monitor the application of, a London Charter Implementation Strategy.

1.3 In collaborative activities, all participants whose role involves either directly or indirectly contributing to the visualisation process should be made aware of the principles of the London Charter, together with relevant Charter Implementation Guidelines, and to assess their implications for the planning, documentation and dissemination of the project as a whole.

1.4 The costs of implementing such a strategy should be considered in relation to the added intellectual, explanatory and/or economic value of producing outputs that demonstrate a high level of intellectual integrity.

15.1.3.2 Principle 2: Aims and Methods

A computer-based visualisation method should normally be used only when it is the most appropriate available method for that purpose.

2.1 It should not be assumed that computer-based visualisation is the most appropriate means of addressing all cultural heritage research or communication aims.

2.2 A systematic, documented evaluation of the suitability of each method to each aim should be carried out, in order to ascertain what, if any, type of computer-based visualisation is likely to prove most appropriate.

2.3 While it is recognised that, particularly in innovative or complex activities, it may not always be possible to determine, *a priori,* the most appropriate method, the choice of computer-based visualisation method (e.g. more or less photo-realistic, impressionistic or schematic; representation of hypotheses or of the available evidence; dynamic or static) or the decision to develop a new method, should be based on an evaluation of the likely success of each approach in addressing each aim.

15.1.3.3 Principle 3: Research Sources

In order to ensure the intellectual integrity of computer-based visualisation methods and outcomes, relevant research sources should be identified and evaluated in a structured and documented way.

3.1 In the context of the Charter, research sources are defined as all information, digital and non-digital, considered during, or directly influencing, the creation of computer-based visualisation outcomes.

3.2 Research sources should be selected, analysed and evaluated with reference to current understandings and best practice within communities of practice.

3.3 Particular attention should be given to the way in which visual sources may be affected by ideological, historical, social, religious and aesthetic and other such factors.

15.1.3.4 Principle 4: Documentation

Sufficient information should be documented and disseminated to allow computer-based visualisation methods and outcomes to be understood and evaluated in relation to the contexts and purposes for which they are deployed.

Enhancing Practice

4.1 Documentation strategies should be designed and resourced in such a way that they actively enhance the visualisation activity by encouraging, and helping to structure, thoughtful practice.

4.2 Documentation strategies should be designed to enable rigorous, comparative analysis and evaluation of computer-based visualisations, and to facilitate the recognition and addressing of issues that visualisation activities reveal.

4.3 Documentation strategies may assist in the management of Intellectual Property Rights or privileged information.

Documentation of Knowledge Claims

4.4 It should be made clear to users what a computer-based visualisation seeks to represent, for example the existing state, an evidence-based restoration or an hypothetical reconstruction of a cultural heritage object or site, and the extent and nature of any factual uncertainty.

Documentation of Research Sources

4.5 A complete list of research sources used and their provenance should be disseminated.

Documentation of Process (Paradata)

4.6 Documentation of the evaluative, analytical, deductive, interpretative and creative decisions made in the course of computer-based visualisation should be disseminated in such a way that the relationship between research sources, implicit knowledge, explicit reasoning, and visualisation-based outcomes can be understood.

Documentation of Methods

4.7 The rationale for choosing a computer-based visualisation method, and for rejecting other methods, should be documented and disseminated to allow the activity's methodology to be evaluated and to inform subsequent activities.

4.8 A description of the visualisation methods should be disseminated if these are not likely to be widely understood within relevant communities of practice.

4.9 Where computer-based visualisation methods are used in interdisciplinary contexts that lack a common set of understandings about the nature of research questions, methods and outcomes, project documentation should be undertaken in such a way that it assists in articulating such implicit knowledge and in identifying the different lexica of participating members from diverse subject communities.

Documentation of Dependency Relationships

4.10 Computer-based visualisation outcomes should be disseminated in such a way that the nature and importance of significant, hypothetical dependency relationships between elements can be clearly identified by users and the reasoning underlying such hypotheses understood.

Documentation Formats and Standards

4.11 Documentation should be disseminated using the most effective available media, including graphical, textual, video, audio, numerical or combinations of the above.

4.12 Documentation should be disseminated sustainably with reference to relevant standards and ontologies according to best practice in relevant communities of practice and in such a way that facilitates its inclusion in relevant citation indexes.

15.1.3.5 Principle 5: Sustainability

Strategies should be planned and implemented to ensure the long-term sustainability of cultural heritage-related computer-based visualisation outcomes and documentation, in order to avoid loss of this growing part of human intellectual, social, economic and cultural heritage.

5.1 The most reliable and sustainable available form of archiving computer-based visualisation outcomes, whether analogue or digital, should be identified and implemented.

5.2 Digital preservation strategies should aim to preserve the computer-based visualisation data, rather than the medium on which they were originally stored, and also information sufficient to enable their use in the future, for example through migration to different formats or software emulation.

5.3 Where digital archiving is not the most reliable means of ensuring the long-term survival of a computer-based visualisation outcome, a partial, two-dimensional record of a computer-based visualisation output, evoking as far as possible the scope and properties of the original output, should be preferred to the absence of a record.

5.4 Documentation strategies should be designed to be sustainable in relation to available resources and prevailing working practices.

15.1.3.6 Principle 6: Access

The creation and dissemination of computer-based visualisation should be planned in such a way as to ensure that maximum possible benefits are achieved for the study, understanding, interpretation, preservation and management of cultural heritage.

6.1 The aims, methods and dissemination plans of computer-based visualisation should reflect consideration of how such work can enhance access to cultural heritage that is otherwise inaccessible due to health and safety, disability, economic, political, or environmental reasons, or because the object of the visualisation is lost, endangered, dispersed, or has been destroyed, restored or reconstructed.

6.2 Projects should take cognizance of the types and degrees of access that computer-based visualisation can uniquely provide to cultural heritage stakeholders, including the study of change over time, magnification, modification, manipulation of virtual objects, embedding of datasets, instantaneous global distribution.

* * * * *

15.2 Commentary on the London Charter

The following discussion closely follows 'A New Introduction to the London Charter' (Denard 2012a, 2012b), with edits and updates appropriate to the current volume.

15.2.1 Preamble and Objectives

The Preamble's notion of demonstrable parity with traditional scholarly and professional standards ('a set of principles [...] that will ensure that digital heritage visualization is, and is seen to be, at least as intellectually and technically rigorous as longer established cultural heritage research and communication methods') guides much of the thinking underlying the Charter. Previous approaches to this problem were similar in this respect, for example, likening the process of archaeological visualisation to that of philological textual analysis (e.g. Frischer et al. 2002).

The challenge of the scholarly validation of heritage visualisation can most simply be illustrated by considering how one evaluates scholarly print publications: authors are expected, as a minimum, to situate their questions and arguments in relation to prior scholarship; to present and assess their sources, duly referenced in footnotes and bibliographies; and to remain within a range of currently acceptable logical and stylistic norms. These criteria draw upon continually evolving conventions that reflect the prevailing assumptions, at any given moment, about (inter alia) the nature and purpose of scholarship as defined by the various disciplines. The inherently linear nature of textual exegesis renders the author's very process of interpretation visible to the community, which, in turn, allied with the listing of sources, allows the argument to be evaluated.

Consider, by contrast, a visualisation. Some subjects, and arguments, do not so readily lend themselves to verbal expression, not only because they may rely on intensive, detailed reference to visual or spatial materials that resist textual description but because the subject matter and understandings that the author wishes to communicate are inherently non-linear or synthetic – for example, a building or event. If communicating by means of text, the author must force a synchronic perception into a diachronic narrative, that is, must strip the object down into a sequence of layers, to be presented sequentially rather than simultaneously, thereby negating the very cognitive experience that the author ultimately wishes to evoke. In such cases, a visualisation – whether static image, real-time model or printed object – might well become the expressive medium of choice, conveying, all at once, the complete, synthetic image of the author's idea.

However, although for certain purposes visualisation can exceed text in expressive power, its explanatory value may be less. No matter how thoughtfully a research question is posed in relation to the existing field of knowledge, how painstakingly available sources are researched and interpreted and how discerningly or creatively an argument is elaborated visually to the viewer, a finished image alone does not reveal the process by which it was created. Even a real-time model, while it allows the user to explore a space in linear time, if it lacks an account of the evaluation of sources or of the process of interpretation, does not, in itself, render the research process visible to the visitor and thus fails to allow the viewer to assess it as part of an argument. Such visualisations tend to be viewed as relatively superfluous, if attractive, impressions, rather than as essential means of constructing and contributing to historical debate. The empirical opacity of the synchronous image, then, is the crisis that threatens to isolate visualisations from meaningful disciplinary dialogue. For a heritage visualisation to match the rigour of conventional research, its rigour must be visible. That is why at the heart of the London Charter is the principle that heritage visualisations 'should accurately convey to users the status of the knowledge that they represent, such as distinctions between evidence and hypothesis, and between different levels of probability'.

Visibility is also at the centre of the Charter initiative itself: by calling itself a 'Charter' (rather than, say, a co-authored article), the document seeks to draw a certain kind of attention to itself, so that its encapsulation of expert thought on best practice in heritage visualisation will be recognised as a common reference point spanning nations and academic and professional contexts. Agreement on an internationally agreed benchmark that insisted on intellectual accountability, it has been realised, could be the single most effective means of enabling heritage visualisation methods to become an equal and valued part of communities engaged in the research and communication of human culture. With this in mind, the principles of the Charter are, for the most part, deliberately conservative – no more than a consolidation of existing consensus so that the document might legitimately demand 'wide recognition and an expectation of compliance within relevant subject communities'. Once accepted, its value would be to provide a set of principles that, if adhered to, would virtually guarantee the methodological integrity of a heritage visualisation project and act as a common yardstick for professional evaluation – an essential tool enabling the integration of visualisation efforts into the process of mainstream peer evaluation and, thereby, scholarly debate.

Between March 2006 and February 2009, *The London Charter for the Use of 3D Visualisation in the Research and Communication of Cultural Heritage* (Draft 1) evolved into *The London Charter for the Computer-Based Visualisation of Cultural Heritage* (2.1), reflecting a broadening of scope and ambition. Draft 1's strict focus on '3D visualisation' was expanded to encompass 'computer-based visualisation' – embracing 2D, 3D, 4D and even hard-copy printouts or computer-generated physical objects such as replicas of museum artefacts. In addition, at a meeting of the Advisory Board in Brighton in November 2007, following much vigorous debate, it was agreed that the Charter should aspire to influence the use of visualisations not only in academic and curatorial contexts but also in 'those aspects of the entertainment industry involving the reconstruction or evocation of cultural heritage' although omitting 'the use of computer-based visualisation in [...] contemporary art, fashion, or design' (see Preamble). Computer-generated imagery, it was argued, plays an increasingly influential role in shaping public perceptions of the past even when, as so often, it is factually erroneous. Audiences understand that studios are bound to take creative licence. However, such is the persuasive power of visualisations, especially when photorealism is used, that alongside their entertainment value, they also often lend an unjustifiable impression of historical accuracy. If we agree – as those with a professional interest in cultural heritage are surely bound to do – that it is a matter of no small importance whether or not a generation's impressions of the past conform to the contours of historical understanding, then it follows that we should take an active role in encouraging and urging the commercial sector to make available, through documentaries and other formats, sufficient information to enable their audiences to distinguish between fact and fiction. An awareness-raising effort of this kind, aiming to raise producers' and audiences' expectations of historical accountability, including an emphasis on the additional frisson that integrity brings to our encounters with the past, would require significant resources and further research. But even in the absence (to date) of a concerted attempt to realise this ambition, this broadening of the scope of the Charter's validity raises numerous questions which are likely to become more, rather than less, publicly and energetically explored in time.

This development also highlights the importance of writing the Charter in a style that is accessible to the widest possible audience, spanning not only a variety of professional and disciplinary contexts but also all levels of expertise, from the seasoned expert to the general public. This is appropriate given that the Charter addresses issues that affect such diverse stakeholders, from journalists to researchers and from museum curators to international organisations. This stylistic accessibility is possible because the Charter, rather than making highly specific technical recommendations, addresses methodological issues at quite an abstract level. Having said that, communication across disciplines and languages has presented formidable challenges.

A comparative study of the translation issues that the London Charter has encountered, which I shall not attempt here, could deliver fascinating insights into the diversity of cultural approaches that an international Charter relating to cultural heritage must support. Certain key terms such as 'accessibility' and 'sources', for example, mean different things in different academic contexts, and equally crucial concepts, such as 'cultural heritage', 'intellectual transparency' and 'subject communities', have no obvious linguistic or cultural equivalent in different languages. Several of the terminological changes from Draft 1.1 to Version 2.1 of the Charter are the result of translators and community members drawing our attention to such issues. Indeed, Principle 3 was renamed 'Research Sources' because of the different connotations the term 'sources' had in humanities disciplines from those in computer science and the other sciences. The effects of this learning process are also registered in Principle 4.9, which draws attention to the importance of making explicit, in an interdisciplinary project, the 'implicit knowledge' that each expert community holds, as well as of 'identifying the different lexica of participating members from diverse subject communities'. The visualisation community is deeply indebted to the care and integrity of those who have acted as editors, translators and correspondents in the collaborative search for successful linguistic and cultural analogues for the Charter.

15.2.1.1 Principle 1: Implementation

In order to retain cross-context relevance, the Charter addresses only fundamental methodological principles that will, by their nature, remain valid even as technologies or technical standards evolve. For this reason, the Principle 1.1 of the Charter acknowledges the need for more detailed, discipline- and technology-specific Implementation Guidelines that map out the technical implications of these methodological principles. One consequence of this is that one should be able to envisage, in the not too distant future, a definitive text of the Charter. Implementation Guidelines, however, will continue to be needed and revised on an ongoing basis.

Principle 1.2 ('every computer-based visualisation heritage activity should develop, and monitor the application of, a London Charter Implementation Strategy') reflects extensive expert experience of developing and helping others to develop Charter implementation plans, which has unfailingly demonstrated the Charter's value as a coherent conceptual framework that is quickly and easily understandable by people from diverse professional and disciplinary contexts. When used as a set of prompts in the project planning phase, the Charter saves significant amounts of time in designing a robust, structured visualisation methodology and subsequently, during the life of the project, provides a clear and concise agreed reference point for project participants, guaranteeing that the project both is, and is seen to be, methodologically rigorous.

15.2.1.2 Principle 2: Aims and Methods

The principle that 'a computer-based visualisation method should normally be used only when it is the most appropriate available method for that purpose', while it articulates an enduring and fundamental methodological tenet, can also be seen as the expression of a specific moment of epistemological crisis precipitated by technological change. The London Charter initiative arose as a response to the increasingly widely recognised need to ensure the ever-increasing expressive power of computer graphics becomes accountable to the rigorous standards of historical research. Those archaeologists, classicists and historians of various denominations who had perceived the enormous potential of digital 3D visualisation for fashioning intuitively understandable representations of spatial arguments also, as mentioned above, from the mid-1990s, began to signal, each in their own disciplinary contexts and terms, concern about how this means of representation could so beguilingly elide distinctions between information and speculation.

According to the conventional narrative of the 'evolution' of computer graphics, we were destined to develop increasingly lifelike synthetic simulacra of reality which would, in turn, be experienced in increasingly 'immersive' ways.

Indeed, the capacity of 3D computer visualisations to conjoin geometrical information with 'textures' that suggested, or even photographically reproduced, the surface appearance of actual objects seemed one of the most compelling arguments for the application of computer graphics to the representation of cultural heritage: one day, we dreamed, we might create complete experiences of long-lost places and periods that would be virtually indistinguishable in appearance from actuality. We may at times struggle to distance ourselves from the assumption that methodological advancement is an inevitable sequitur of technological progress, particularly when both our quotidian and our professional lives increasingly rely on and constantly anticipate the emergence of digital technologies. It was a key realisation, therefore, that the headlong career towards hyperrealism, photorealism, virtual reality and other such graphical innovations might actually be taking us further away from, not nearer to, accountable knowledge representation. What was needed was a forceful interruption of the flow of teleological assumptions. This realisation therefore gave rise to Principle 2.3, that 'the choice of computer-based visualisation method (e.g. more or less photo-realistic, impressionistic or schematic; representation of hypotheses or of the available evidence; dynamic or static) or the decision to develop a new method, should be based on an evaluation of the likely success of each approach in addressing each aim'.

A common misconception is that the Charter prescribes absolute precepts governing which particular method or approach should be used in each given circumstance. Nothing could be further from the truth: the Charter consistently, and insistently, throws the ball back into the court of those about to undertake computer-based heritage visualisation, asking them to articulate the particular aims and requirements of each strand of each project and to make decisions appropriate to each of their specific requirements. Thus, for instance, recognising that there is a potentially infinite range of valid aims and methods, Principle 2.2 asks only that 'a systematic, documented evaluation of the suitability of each method to each aim should be carried out, in order to ascertain what, if any, type of computer-based visualisation is likely to prove most appropriate'.

15.2.1.3 Principle 3: Research Sources

In Principle 3, the comparison with conventional textual scholarship is most in evidence in the stipulation that 'in order to ensure the intellectual integrity of computer-based visualisation methods and outcomes, relevant research sources should be identified and evaluated in a structured and documented way'. Visual resources pose particular challenges for precisely the same reason that visualisations themselves do: they are inherently non-linear entities that, although they may, as the saying goes, be worth a thousand words, do not tell their own story (see Pletinckx 2012). Thus, Principle 3.3 requires that 'particular attention should be given to the way in which visual sources may be affected by ideological, historical, social, religious and aesthetic and other such factors'.

15.2.1.4 Principle 4: Documentation

Principle 4 ('sufficient information should be documented and disseminated to allow computer-based visualisation methods and outcomes to be understood and evaluated in relation to the contexts and purposes for which they are deployed') in many ways carries the central conceptual burden of the Charter. The phrase 'in relation to' is crucial: recognising that heritage visualisation efforts vary widely in scale and purpose, this principle refuses to prescribe a fixed list of information types to be documented, much less to dictate one particular level of detail of documentation. Rather, it requires the practitioner to make a contextually appropriate determination of what type and quantity of documentation is required to ensure that their visualisation processes and outcomes will be intellectually transparent.

The Charter goes on to elucidate the various areas in which documentation should operate in order to (Principle 4.2) 'enable rigorous, comparative analysis and evaluation of computer-based visualisations'; (Principle 4.4) ensure that audiences can understand what each visualisation seeks to represent; (Principle 4.5) publish research sources; and (Principle 4.7) explain the methodological rationale of a visualisation. It also makes recommendations on documentation formats and standards (Principle 4.11–4.12).

Principle 4.6, the Documentation of Process ('Paradata'), merits particular attention. Consistent with other principles, its assertion that 'documentation of the evaluative, analytical, deductive, interpretative and creative decisions made in the course of computer-based visualisation should be disseminated in such a way that the relationship between research sources, implicit knowledge, explicit reasoning, and visualisation-based outcomes can be understood' necessitates a flexible and relational, rather than rigid and absolute, approach. Again, we see that the Charter does not provide a checklist of tasks, but rather, in effect, a structured set of prompts that asks us to determine what specific measures are appropriate for each individual project relative to its own particular aims and circumstances. The level of documentation that is appropriate may, for instance, be proportionate to the quality of sources upon which a visualisation is based and the weight of importance that visualisation has within an argument. Thus, a minimal record may suffice for a speculative visualisation based on very limited evidence that aims to do no more than give a sense of the possible approximate size of an artefact, structure or site; whereas a model designed to carry the burden of a detailed reconstruction hypothesis, and which is based on extensive, precise site measurements and a weighty corpus of *comparanda*, will require meticulous documentation of both its sources and interpretative processes if others are to be able to understand and evaluate the quality of the underlying empirical analysis and argument being advanced.

However, documenting 'the evaluative, analytical, deductive, interpretative and creative decisions' is a tall order. As Drew Baker (2012) has shown, no visualisation is completely 'objectively' true to fact; despite the aspiration that a 3D model's geometry should be an accurate record of the cultural artefact, it is even truer of computer-based visualisations than of photographs that digital surrogates, within the constraints of a particular technology, *represent*, rather than accurately *reproduce*, some aspect of reality. Interpretation is ineluctably involved at every stage: 3D scanning and processing convert the infinitely granular surface of artefacts into point clouds or polygon meshes; digital cameras and video recorders 'capture' analogue waves as binary sequences; and display surfaces – monitors, screens and print illustrations – attempt to convey on two-dimensional planes the impression of three-dimensional space.

Visualisation creation involves and visualisation outputs invite multiple perspectives: technological, optical (including the human eye), cultural, aesthetic (including connoisseurship) and epistemological (within disciplines, but increasingly also within 'intra-disciplines' such as digital humanities or archaeological computing), right down to the real-time model user's choice of path, viewing point or interaction. In short, visualisations are technical, personal and cultural memory structures, with all the instability that implies, upon which we stage our narratives of the past. Indeed, the whole London Charter is predicated upon the absence of objectivity: tracking the interpretative trail, consensus around methods and representational conventions, is only necessary and meaningful because of the inescapable elusiveness of pure fact.

We are forcibly confronted with this elusiveness when we consider the challenge of documenting dependency relationships. The Charter's glossary defines a dependency relationship as:

> A dependent relationship between the properties of elements within digital models, such that a change in one property will necessitate change in the dependent properties. (For instance, a change in the height of a door will necessitate a corresponding change in the height of the doorframe.)

A visualisation is essentially a complex set of dependency relationships, and it is this which makes them at once such powerful empirical instruments – a means of exploring what the implications of each piece of knowledge might be for each other piece of knowledge – and so very difficult to render amenable to intellectual accountability. It is for this reason that Principle 4.10 proposes that 'computer-based visualisation outcomes should be disseminated in such a way that the nature and importance of significant, hypothetical dependency relationships between elements can be clearly identified by users and the reasoning underlying such hypotheses understood'. When this kind of approach is adopted, computer-generated graphics, by enabling us systematically, iteratively and precisely to explore and record the reciprocal interpretative implications of pieces of evidence and hypotheses, remain a uniquely enabling means of constructing knowledge. The spatial specificity that digital visualisations demand, when exploited intelligently and rigorously, far from being a fatal, Siren seduction, becomes a fertile ground of enriched understandings.

Principle 4.1's proposal that 'documentation strategies should be designed and resourced in such a way that they actively enhance the visualisation activity by encouraging, and helping to structure, thoughtful practice' reflects an awareness that the requirement to document process has an inevitable impact on working practice. Recording our process in a lab book, for example, may play an essential role in providing the detailed record of the research activity which will ultimately provide the necessary basis of publication, but it can also interrupt the creative flow of visual interpretation.

When absorbed in the visualisation process, we make an infinite number of moment-by-moment decisions, each shaped by a host of factors, not least the deep, acquired subject and technical knowledge that at any given moment may represent itself to us as little more than 'instinct' – a feeling for what is and what isn't right. As visualisers we need to be as aware as possible of the kinds of decisions that we make on 'instinct' so that we can monitor their validity, but we must also be realistic about the level of detail about the interpretative process that it is possible, or even appropriate, to capture. By the same token, as consumers, we must be sophisticated enough to recognise, and accept, the unavoidable role of subjectivity in influencing the style, aesthetics and interpretative choices that we find manifested in heritage visualisations.

A simple reading could view the London Charter as a limiting scope for creativity by seeking to tie everything down to the most minute detail. But that would be to forget that heritage visualisation is, above all, a hypothesis machine. One may know, for example, that a certain object was part of a greater structure but lack sufficient information to determine precisely how it fitted into the whole. In such cases, documented visualisation in fact gives greater liberty to try out possibilities, because when a hypothesis is published along with its rationale and evidence base, it acquires a recognisable standard of *methodological* validity. This remains the case even if the visualisation output takes the form of an interactive tool rather than a static image or fixed model: one that allows others dynamically to test hypotheses by altering variables within a digital environment. The London Charter encourages manifold interpretative interventions, each being just one of multiple possible stories we might tell about the past. Documentation, while it does not legitimise outcomes or conclusions as historical hypotheses, nevertheless allows us to present even highly speculative experiments and entitles us to expect that they be evaluated on their own terms. Taken further, documented visualisation is not only a hypothesis engine, it is an epistemological engine: one that, by licencing ludic intervention as a means of producing knowledge, could ultimately contain the potential to affect our assumptions about the nature, aims and methods of research and communication in the heritage domain.

15.2.1.5 Principle 5: Sustainability

The Charter, being designed to facilitate the integrity of digital visualisation-enabled research and communication of cultural heritage, is at its most fundamental level an ethical document. Its essentially ethical character becomes most explicit in Principles 5 and 6. Principle 5, 'Sustainability', draws attention to the fact that computer-based visualisations constitute, in their own right, part of our common 'human intellectual, social, economic and cultural heritage'. Considerable resources, often drawn – directly or indirectly – from the public purse, flow into the creation of heritage visualisations. It therefore behoves the visualisation community to behave as good stewards of that investment, both through ensuring that this work is preserved and that it reaches those public audiences to whom it may have genuine value. This provides the context for Principle 5 ('strategies should be planned and implemented to ensure the long-term sustainability of cultural heritage-related computer-based visualisation outcomes and documentation').

Digital preservation is a vast and rapidly evolving area spanning policy and resources, intellectual property rights, technical development, ethical and epistemological debates and practical methodological and project workflow considerations. Significant initiatives in this area include the USA's National Digital Information Infrastructure and Preservation Program National Digital Information Infrastructure and Preservation Program (USA). http://www.digitalpreservation.gov/about/index.html. Accessed 2 July 2013 and the UK-based Digital Preservation Alliance Digital Preservation Alliance (UK). http://www.dpconline.org. Accessed 2 July 2013. The UK's Archaeology Data Service has set the London Charter as a benchmark for the deposit of digital visualisation; their Collections Policy (4th edition, section 2.2.8, 'Visualisation') states that:

> 3-D reconstructions, including computer-generated solid models, VRML, and other visualisations will be collected where it is feasible to maintain them and where they are considered to be capable of reuse and restudy or are seen as being of importance for the history of the discipline in accordance with the procedures defined in the AHDS Guide to Good Practice for Virtual Reality. In general the ADS will also preserve the data from which the model is derived, and sufficient metadata in accordance with the principles of the London Charter (2009) Archaeology Data Service Collections Policy. http://archaeologydataservice.ac.uk/advice/collectionsPolicy. Accessed 2 July 2013.

Big issues in digital preservation include the sustainability of complex objects such as games, virtual worlds, interactive and mixed-media and mixed reality, such as Augmented Reality applications or motion sensor-based installations (see Anderson et al. 2012). It's important to note that in many cases, the technical and resource challenges are such that documentation may be the only way in which such complex and ephemeral entities can be preserved. The status of documentation as a historical 'object' in its own right becomes clear, and as such it is also manifest that documentation merits a level of investment in its preservation that is commensurate with its potential cultural and historical significance.

Extensive effort is being made to develop data models that will enable the integration of 3D content into digital repositories. However, digital preservation is still a relatively new field, and, as Principle 5.4 alludes to, the active maintenance of digital assets over time requires costly infrastructure (digital assets will not survive if simply neglected). The long-term sustainability of digital heritage visualisation outputs is therefore far from guaranteed. Consequently, Principle 5.1 recommends that physical as well as digital formats should be considered, depending on which has the best prospects of being successfully sustained.

An idea which the Charter does not (yet) explicitly acknowledge, but which deserves consideration, is that the social and cultural processes that accompany archaeological investigation and digital visualisation may also constitute complex, ephemeral and valuable histories. The myriad factors, from weather and food to human behaviours and relationships, and all the mundane and extraordinary events that surround the research and communication environment, although they may significantly shape our experience or influence our interpretative instincts, are generally written out of our academic narratives. Their often unstructured and anecdotal character, or the tacit or seemingly 'marginal' nature of the kind of knowledge they supply, do not lend themselves well to articulation within the disciplines of academic or professional documentation. Consequently, both we and the readers/users of our research outputs are left with an impoverished record, one which, ironically, eradicates traces of the very strata of human experience – physical, psychological, emotional and even spiritual – that most potently move us to explore and ascribe value to human cultural heritage in the first place. We might, in future, consider whether a project, or our audiences, might benefit from a more holistic approach to documentation that would capture the intangible, experiential, environmental and social dimensions of our processes, as well as their more formally scientific aspects.

15.2.1.6 Principle 6: Access

Principle 6's advice that 'the creation and dissemination of computer-based visualisation should be planned in such a way as to ensure that maximum possible benefits are achieved for the study, understanding, interpretation, preservation and management of cultural heritage' continues the overtly 'ethical' appeal of Principle 5 and, like Principle 5, raises all kinds of challenging issues for the public-funding models that frequently underwrite heritage visualisation initiatives. While national-level public funding bodies tend to concentrate on content creation and transnational bodies such as the EC invest in the development of new technologies, there remains a critical shortage of funding for promoting the visibility of, access to, and deployment of visual, digital heritage assets in diverse contexts. Principle 6.1's exhortation to stakeholders to consider 'how such work can enhance access to cultural heritage that is otherwise inaccessible due to health and safety, disability, economic, political, or environmental reasons, or because the object of the visualisation is lost, endangered, dispersed, or has been destroyed, restored or reconstructed' gives only a hint of the vast unlocked potential that a well-wrought heritage visualisation can have if made available for use in new contexts, including uses beyond those envisaged by its creators.

This emphasis, at the close of the Charter, on the issue of 'access' begins to bring us full circle. The London Charter was, in part, born out of an anxiety that serious heritage visualisation was suffering, in prestige and perceived integrity, by its superficial similarities to computer-generated imagery seen in a historical popular games and films. We now find ourselves, however, gravitating towards a, perhaps wiser, recognition of interconnectedness: an observation that computer-based visualisations are valuable in part precisely because they collapse boundaries between the mysteries of rarefied academic research and popular understanding. A visualisation may at once both embody deep and complex specialist knowledge and at the same time make the contours of that knowledge intuitively accessible to a non-expert audience in a way that a text-based publication never could.

We need collectively to think through the implications of these new parameters, in relation to wider debates about the characteristics and prospects of a 'digital society' and the ways in which the language of 'impact' and 'knowledge transfer' has altered, and continues to transform, our professional environments. Our challenge is to shape these exchanges and transactions, and the language we use, in such a way that they serve actively to enhance the integrity of deep and rigorous scholarly enquiry, including through dialogue with more diverse kinds of stakeholders and audiences than heretofore. Heritage visualisation, as an engine of intensely demanding interdisciplinary research and of lively public engagement, can make a persuasive contribution to the cultivation of popular understanding of the essential role that cultural heritage plays in generating a healthy, changing and self-aware culture. High-integrity, computer-based heritage visualisations can be focal points, equally accessible to all, around which we aggregate debates about what is at stake in the images, experiences and narratives we construct about the past, and present, of human culture.

As the Charter's methodological principles become increasingly commonly understood and adopted, sustainability and access are likely to become increasingly viewed as the central, burning issues in heritage visualisation. Methodological rigour will ensure that heritage visualisations have excellent prospects of being of enduring intrinsic quality; however, it remains for us to take the difficult steps needed to secure their survival and fully to realise their value in a shared future.

References

Anderson D, Delve J, Dobrerva M, Baker D, Billenness C, Konstantelos L (eds) (2012) The preservation of complex objects, vol. 1: visualisations and simulations. University of Portsmouth, Portsmouth

Baker D (2012) Defining paradata in heritage visualization. In: Bentkowska-Kafel A, Baker D, Denard H (eds) Paradata and transparency in virtual heritage. Ashgate, Farnham, pp 163–175

Beacham R, Denard H, Niccolucci F (2006) An introduction to the London Charter. In: Ioannides M, Arnold D, Niccolucci F, Mania K (eds) The e-volution of information communication and technology in cultural heritage: proceedings of VAST 2006. Archaeolingua, Budapest, pp 263–269

Bentkowska-Kafel A, Baker D, Denard H (eds) (2012) Paradata and transparency in virtual heritage. Ashgate, Farnham

Denard H (2005) At the foot of Pompey's statue: reconceiving Rome's *Theatrum Lapideum*. In: Bowman AK, Brady M (eds) Images and artefacts of the ancient world. Oxford University Press, Oxford, pp 69–76

Denard H (2012a) A new introduction to the London Charter. In: Bentkowska-Kafel A, Baker D, Denard H (eds) Paradata and transparency in virtual heritage. Ashgate, Farnham, pp 57–71

Denard H (2012b) Ecologies of research and performance: preservation challenges in the London Charter. In: Anderson D, Delve J, Dobrerva M, Baker D, Billenness C, Konstantelos L (eds) The preservation of complex objects, vol. 1: visualisations and simulations. University of Portsmouth, Portsmouth, pp 87–102

Frischer B, Niccolucci F, Ryan N, Barcelò JA (2002) From CVR to CVRO: the past, present, and future of cultural virtual reality. In: Franco Niccolucci (ed) Virtual archaeology: proceedings of the VAST Euroconference,Arezzo, 24–25 November 2000. Bar International Series 1075. Oxford: Archaeopress, pp 7–18

Goodrick G, Gillings M (2000) Constructs, simulations and hyperreal worlds: the role of virtual reality (VR) in archaeological research. In: Lock G, Brown K (eds) On the theory and practice of archaeological computing. Oxford University School of Archaeology, Oxford, pp 41–58

Hermon S, Niccolucci F (2006) A fuzzy logic approach to reliability in archaeological virtual reconstruction. In: Niccolucci F, Hermon S (eds) Beyond the artefact: digital interpretation of the past. Proceedings of CAA2004. Archaeolingua, Budapest, pp 11–17

International Forum of Virtual Archaeology (2010) Principles of Seville: international principles of virtual archaeology, final draft. http://www.arqueologiavirtual.com/carta/?page_id=437. Accessed 2 July 2013

Jablonka P, Kirchner S, Serangeli J (2003) Troia VR. A virtual reality model of Troy and the Troad. In: Doerr M, Sarris A (eds) The digital heritage of archaeology. Hellenic Ministry of Culture, Athens, pp 13–20

Mudge M (2012) Transparency for empirical data. In: Bentkowska-Kafel A, Baker D, Denard H (eds) Paradata and transparency in virtual heritage. Ashgate, Farnham, pp 203–242

Niccolucci, Franco (1999) Virtual reality in archaeology: a useful tool or a dreadful toy? In: Proceedings of the Medi@terra symposium, Athens, November 1999. 238ff. Athens

Pletinckx D (2012) How to make sustainable visualizations of the past: an EPOCH common infrastructure tool for interpretation management. In: Bentkowska-Kafel A, Baker D, Denard H (eds) Paradata and transparency in virtual heritage. Ashgate, Farnham, pp 203–242

Roberts JC, Nick S. Ryan (1997) Alternative archaeological representations within virtual worlds. In: Bowden R (ed) Proceedings of the 4th UK virtual reality specialist interest group conference, Brunel University, Uxbridge, Middlesex, November 1997, pp 179–188. http://www.cs.kent.ac.uk/people/staff/nsr/arch/vrsig97/vrsig.html. Accessed 2 July 2013

Roussou M, Drettakis G (2003) Photorealism and non-photorealism in virtual heritage representation. In: Arnold D, Chalmers A, Niccolucci F (eds) Proceedings of VAST2003, fourth international symposium on virtual reality, archaeology and intelligent cultural heritage. Eurographics, Aire-La-Ville, pp 51–60

Ryan N (1996) Computer-based visualization of the past: technical "realism" and historical credibility. In Main P, Higgins T, Lang J (eds) Imaging the past: electronic imaging and computer graphics in museums and Archaeology, British Museum occasional papers 114. The British Museum, London, pp 95–108

The London Charter for the Computer-based Visualisation of Cultural Heritage (2009) http://www.londoncharter.org. Accessed 2 July 2013

Zuk T, Carpendale S, Glanzaman WD (2003) Visualizing temporal uncertainty in 3D virtual reconstructions. In: Mudge M, Ryan N, Scopigno R (eds) Proceedings of VAST 2005, sixth international symposium on virtual reality, archaeology and cultural heritage. Eurographics, Aire-La-Ville, pp 99–106

International Guidelines for Virtual Archaeology: The Seville Principles

16

Víctor Manuel López-Menchero Bendicho

Contents

V.M. López-Menchero Bendicho
Spanish Society of Virtual Archaeology (SEAV),
Sevilla, Spain
e-mail: victor.lopezmenchero@uclm.es

16.1 Virtual Archaeology: Definition and Term

Our ancestors, whether individually or as a group, knowingly or unknowingly, left a trail of evidence of their existence, of their time on this planet. This evidence is nothing more than a small set of fragments of an infinitely broader and more complex reality that becomes increasingly more obscure as we go further back in time. So there is far less evidence dating back to over 5,000 years which has survived up to the present day than there is dating back scarcely 200 years. For decades now, the work of archaeologists has been to search for all these traces, regardless of their age, size or geographical location, so that, through study and examination, they can endeavour to rebuild the complex puzzle that is history, to uncover the evolution of past human societies. With this purpose in mind, archaeologists use whatever scientific advances and knowledge are available to them at any given time. So, for example, the latest breakthroughs in chemistry or physics enable us to detect substances in archaeological sites that, until very recently, had gone completely unnoticed. The application of the latest advances in geophysics or remote sensing can reveal the location and shape of underground objects, even before excavation begins. The methodical anthropological studies of human groups in the present make it easier for us to understand human groups in the past, and all this evolves in parallel with a society that is becoming ever more interested in discovering and

C. Corsi et al. (eds.), *Good Practice in Archaeological Diagnostics*, Natural Science in Archaeology,
DOI 10.1007/978-3-319-01784-6_16, © Springer International Publishing Switzerland 2013

understanding the past. It is precisely this new relationship between archaeologists and the society in which they work that has brought about substantial changes in archaeological theory and practices. In this way, the practice of archaeology is no longer confined to the field of research, but is now involved in the conservation and dissemination of its results, by means of its most visible formula: archaeological heritage. The material remains of the past are conserved as irrefutable proof of historic discourse, but they are also held up as teaching tools and for the dissemination of the knowledge generated through archaeologists' work. The need to know, to discover and to better understand what our ancestors were like and how they lived has led archaeologists to refine their methods and techniques. It is no longer enough just to vaguely explain what a Roman city may have looked like or what happened at the Battle of Gaugamela. People want to see what happened with their own eyes while also calling for material remains that have survived the passing of time to be conserved for future generations. Such demands surpass the classical conception of archaeology and open the door to the birth of new scientific disciplines. And so virtual archaeology is born.

It seems correct to say that virtual archaeology is a scientific discipline with its own identity, despite feeding on expertise from different areas of knowledge, as it has its own objectives and methods that are different from other disciplines. This condition does not make it a science or a completely independent field of learning as it is directly dependent on the archaeological discipline, just as archaeology has a direct relationship with historical science and anthropology. To this effect, just as archaeology is either history/anthropology or it is nothing, virtual archaeology is either archaeology or it is nothing.

Precisely for this reason, virtual reconstructions that include a large number of elements that have not been verified either archaeologically or historically cannot be considered as virtual archaeology, but rather as historical narrative, in other words, a genre in which reality and fiction become blurred, in which it is impossible for the viewer or the public to distinguish between the two. The same thing occurs between history books and historical novels or between documentaries and films based on historical facts. Obviously this does not mean that any virtual reconstructions and their applications must have a total degree of certainty, as the study of the past is inherently subjective and partial, but rather that they must be based on empirical hypotheses that are constructed through a detailed study of the past and its remains.

From a purely nominative point of view, the use of the term virtual archaeology is intrinsically wrong if we understand it to be the sum of two already existing words (virtual + archaeology), and it would probably be more correct to say digital archaeology or cyber-archaeology. However, if we choose to approach the term as if it were a single item, its meaning is more or less clear, in so far as it is widely used by the international scientific community. In this case, we understand language to be a social construct that is arranged around words or expressions that are used by a community of speakers, and on the basis of that circumstance, we accept the use of virtual archaeology as a valid term. This does not, however, prevent many experts from preferring to speak of digital archaeology or cyber-archaeology when referring to the same concept.

16.2 Historical Background to the Creation of an International Charter of Virtual Archaeology

All scientific disciplines are evidently characterised by the existence of a community of experts that research, and at the same time disseminate, the results of their studies. In the case of virtual archaeology, this community of experts has been gradually growing since the 1990s, to the point where it is now large enough to have national societies. This is the case of the Spanish Society of Virtual Archaeology (SEAV) which, since it began in the year 2008, has brought together over 23 research groups and 21 private Spanish companies concerned about the future of virtual archaeology. For this reason, one of the first measures that the SEAV has set up is the International Forum of Virtual Archaeology (IFVA), aimed at laying the theoretical foundations for the international future of virtual archaeology. The Forum's primary objective has been to lead the transnational creation of the International Charter of Virtual Archaeology, also known as the Seville Charter, or the Seville Principles. To facilitate this process, the SEAV created the International Meeting of Archaeology and International Meeting on Graphic Archaeology and Informatics, Cultural Heritage and Innovation (ARQUEOLÓGICA 2.0), which was held for the first time in La Rinconada (Seville) in June 2009. At that time, both the meeting and the Forum proposed that it was a primary objective to establish a debate between some of the leading experts worldwide about whether or not it is relevant to create an international charter aimed at adapting the general principles of the London Charter to the field of virtual archaeology, as all scientific knowledge is based on criteria that are accepted by the majority of that scientific community.

With this purpose in mind, a monographic session was held during ARQUEOLÓGICA 2009 entitled "Reflections on the London Charter", in which three of its signatories, Richard Beacham of King's Visualisation Lab., King's College; Sorin Hermon of the Digital Cultural Heritage and Archaeological Sciences of Cyprus; and Juan A. Barceló of the Universidad Autónoma de Barcelona, explicitly showed their support for the SEAV initiative to create a new document that should be capable of achieving broad international agreement. After this monographic session, a plenary session was held, entitled "Bases of Virtual Archaeology", in which 30 prestigious experts and researchers in this field of expertise took part, reaffirming the need to suitably define and give shape to the field of virtual archaeology, taking into account that nothing of this kind had been possible up until that moment, despite the discipline's growing popularity and the frequent use of the term around the world (Fig. 16.1).

The conclusions drawn from the first meeting of the International Forum of Virtual Archaeology left no doubt: there was an urgent need to start work on creating an international charter of

Fig. 16.1 Session entitled "Reflections on the London Charter". First International Meeting on Graphic Archaeology and Informatics, Cultural Heritage and Innovation (ARQUEOLÓGICA 2.0)

virtual archaeology. In the words of Dr. Almagro, "We cannot overlook the need to establish or define certain rules or guidelines – not legal impositions – to attempt to contain the indiscriminate production of 3D models with no basis or criteria whatsoever, which, thanks to the visual attraction and potential of their means of dissemination, can inundate a market demanding this type of product" (Almagro 2008, 43).

Intense efforts then began at the SEAV to produce a first draft of what would go on to be the International Charter of Virtual Archaeology. This work, in which many members of the SEAV collaborated, was presented in June 2010 in La Rinconada (Seville) during the second meeting of the International Forum, within the framework of the 2nd International Meeting on Graphic Archaeology and Informatics, Cultural Heritage and Innovation (ARQUEOLÓGICA 2.0), and it was warmly received by all participants in the Forum.

In parallel to the ARQUEOLÓGICA 2010 meeting and in keeping with one of the main objectives of the International Forum of Virtual Archaeology, the international scientific journal *Virtual Archaeology Review* (VAR) was launched, with the aim of becoming established as a prestigious international journal that would be capable of reaffirming virtual archaeology as an independent and recognised field of research, as all scientific knowledge aims to divulge its research studies via specialist publications (http://varjournal.es). Since 2010, the creation of VAR has been playing an important role in drawing up the Seville Charter. So, for example, issue number four of the journal is exclusively dedicated to dealing with the theoretical aspects of the discipline, something which is essential to be able to set genuine scientific standards.

Furthermore, in order to improve dissemination and the knowledge that the international scientific community has about the process involved in drawing up the Seville Charter, a website has been created: www.arqueologiavirtual.com. This site offers information not only about the International Charter of Virtual Archaeology but also about other activities, such as the International Forum, ARQUEOLÓGICA 2.0 and the Spanish Society of Virtual Archaeology itself.

The third and fourth meetings of the International Forum of Virtual Archaeology, held during ARQUEOLÓGICA 2011 and 2012, respectively, have enabled a basic document to be established, in collaboration with eminent members of the CIPA-ICOMOS such as José Luis Lerma, Ana Almagro and Mario Santana. The experience provided by these researchers and the valuable recommendations offered by eminent members of the scientific community, such as Alonzo C. Addison (UNESCO) or Jean-Louis Luxen (ex-president of ICOMOS), have enabled a first and final international draft to be approved, under the category of Principles, instead of Charter. This document, which has already been put into practice in many countries, will be discussed below.

16.3 Principles of the Charter

The London Charter (http://www.londoncharter.org) is currently the most advanced international document in this direction. Its various updates reveal the overwhelming need to find a document with recommendations that can serve as a basis for designing new projects with greater rigour in the field of cultural heritage but also to propose new recommendations and guidance tailored to the specific needs of each branch of learning and community of experts. For this reason, the objectives set out in the London Charter aim to "offer a robust foundation upon which communities of practice can build detailed London Charter Implementation Guidelines". And we must not forget the immeasurable scope of the concept of cultural heritage, which encompasses such broad areas as monumental, ethnographic, documentary, industrial, artistic, archaeological and oral heritage. The London Charter takes full account of the cultural heritage as a concept and therefore the specific needs of each of its constituent parts. For this reason, the Preamble to the London Charter recognises these needs: "as the aims that motivate the use of visualization methods vary widely from domain to domain, Principle 1: 'Implementation', signals the importance of devising detailed guidelines appropriate to each community of practice". Principle 1.1 recommends, "Each community of practice, whether academic, educational, curatorial or commercial, should develop London Charter Implementation Guidelines that cohere with its own aims, objectives and methods". It therefore seems obvious that given the importance of archaeological heritage as part of cultural heritage, and since many recognise the existence of a community of experts who focus specifically on the concept of virtual archaeology, consideration must be given to the preparation of guidelines, documents and recommendations that, following the general guidelines established by the London Charter, take into account the specific nature of virtual archaeology. The principles discussed below aim to increase the conditions of applicability of the London Charter in order to improve its implementation specifically in the field of archaeological heritage, including industrial archaeological heritage, simplifying and organising its bases sequentially while at the same time offering new recommendations, taking into account the specific nature of archaeological heritage in relation to cultural heritage.

16.3.1 Principle 1: Interdisciplinarity

"Any project involving the use of new technologies, linked to computer-based visualisation in the field of archaeological heritage, whether for research, documentation, conservation or dissemination, must be supported by a team of professionals from different branches of knowledge". "Given the complex nature of computer-based visualisation of archaeological heritage, it can not be addressed only by a single type of expert but needs the cooperation of a large number of specialists (archaeologists, computer scientists, historians, architects, engineers etc.)" (1.1). Under a traditional classification of scientific knowledge, virtual archaeology is a blend of social/human sciences (anthropology, history, didactics, etc.) and natural/exact sciences (geography, biology, chemistry, geology, IT, engineering, etc.). This hybrid nature, a result of the overlap between many different existing sciences, is typical of the astounding growth produced by scientific knowledge throughout the twentieth and early twenty-first centuries, which has led to numerous sciences being born out of the juxtaposition of segments that had already been established by previous sciences. We need look no further than the case of biochemistry, which arose out of the union between chemistry and biology, or electrochemistry, a result of electricity and chemistry coming together. Roughly speaking, we could say that virtual archaeology is a result of the union between archaeology and IT, although it relies on collaboration from many other scientific disciplines.

"A truly interdisciplinary work involves the regular and fluid exchange of ideas and views among specialists from different fields. Work divided into watertight compartments can never be considered interdisciplinary even with the participation of experts from different disciplines" (1.2). The dialogue and ideas that arise in any attempt to share information between professionals from different fields of expertise are always more rewarding and enriching than a mere sum of isolated ideas, as they promote critical thinking and a diversity of perspectives.

"Among the experts who must collaborate in this interdisciplinary model, it is essential to ensure the specific presence of archaeologists and historians, preferably those who are or were responsible for the scientific management of the excavation work or archaeological remains to be reconstructed" (1.3). The reasoning that justifies and recommends the presence of the archaeologists who took part in the excavation process is related to the paradox of archaeological destruction/investigation (Wheeler 1979, p. 9; Carandini 1997, p. 256), according to which any excavation is synonymous with destruction as it is the same as "burning the pages of the only existing copy of a book, immediately after reading it" (Carandini 1997, p. 256). Excavating entails selecting, rejecting, destroying, conserving and establishing hierarchies and priorities of certain details over others. During the excavation process, many details are not recorded in the corresponding reports, photographs or drawings, simply because it is impossible to document everything. Nevertheless, such details always remain in the excavator's mind. This is why there is no more comprehensive report or more detailed documentation about an excavation than that which is captured in the memory of the archaeologist in charge of that excavation. That information can be priceless, for example, when carrying out a virtual reconstruction.

16.3.2 Principle 2: Purpose

"Prior to the development of any computer-based visualisation, the ultimate purpose or goal of our work must always be completely clear. Therefore, different levels of detail, resolutions and accuracies might be required". "Any proposed computer-based visualisation will always aim to improve aspects related to the research, conservation or dissemination of archaeological heritage. The overall aim of the project must be encompassed within one of these categories (research, conservation and/or dissemination). The category concerning dissemination includes both educational projects, whether formal or informal education, and recreational projects (cultural tourism)" (2.1). Establishing in advance what the main objective of our intervention is may represent significant time and money savings; for example, the level of detail required to consider a virtual reconstruction for research purposes will never be the same as that required for a virtual reconstruction for the purposes of recreation. In the first case, referring to research, the most important thing is to generate working hypotheses that are particularly in line with reality, i.e. accurate and precise, without worrying too much about the superficial quality of the generated image, while in turn offering the archaeologist the chance to move freely around the recreated virtual setting in order to verify or reject the interpretative model. In contrast, in the second case referring to dissemination, even respecting the principle of historical rigour, we would need to work much harder on the finishes to achieve the most realistic image possible of the past, generating credible virtual shots that are easily understood by an audience that is not specialised in this subject. In this second case, it is probably not necessary for the viewer to have the chance to move around the virtual space, as the reconstructions and recreations will generally be used in non-interactive media, such as documentaries, fixed panels and leaflets (López-Menchero 2011).

"In addition to clarifying the main purpose of computer-based visualisation, more specific objectives must always be defined in order to obtain more precise knowledge of the problem or problems to be resolved" (2.2). Breaking the principal objective down into smaller, more accessible objectives will enable the work on the project to be arranged in order and hierarchy more efficiently, while at the same time making it easier to carry out a subsequent assessment of the project once it has been completed, as it will be easier to see, in detail, whether or not the proposed objectives have been met.

"Computer-based visualisation must be always at the service of archaeological heritage rather than archaeological heritage being at the service of computer-based visualisation. The main objective of applying new technologies in the comprehensive management of archaeological heritage must be to satisfy the real needs of archaeologists, curators, restorers, museographers, managers and/or other professionals in the field of heritage and not vice versa" (2.3).

"Ultimately, the main purpose of virtual archaeology will always be to serve society as a whole and contribute to increase the human knowledge" (2.4). Virtual archaeology only makes sense if it is developed to improve people's quality of life through culture: firstly, for ethical reasons, but secondly for practical reasons, as most virtual archaeology projects are funded by public financing. If people do not perceive any benefit in their everyday life, they will stop financing this type of project; if, however, they consider the project to be something useful and valuable for societies' progress and development, they will increase the funds available for this purpose.

16.3.3 Principle 3: Complementarity

"The application of computer-based visualisation for the comprehensive management of archaeological heritage must be treated as a complementary and not alternative tool to other more traditional but equally effective management instruments". To this effect, "Computer-based visualisation should not aspire to replace other methods and techniques employed for the comprehensive management of archaeological heritage (e.g., virtual restoration should not aspire to replace real restoration, just as virtual visits should not aspire to replace real visits)" (3.1). A marvellous example of complementarity can be found in the restoration of the famous fountain in the Court of the Lions in the Alhambra (Granada, Spain), where 3D digitalisation was firstly carried out with a laser scanner covering the entire complex and was later used to analyse the most damaged areas and their possible restoration on the computer. The restorers did not start on the real restoration until the virtual restoration of the complex was completed. This method avoided subsequent complications and gave the restoration team great confidence when it came to tackling the real restoration, as they had prior experience of having carried out the same work in a virtual way (Cano et al. 2010).

"Computer-based visualisation should seek forms of collaboration with other methods and techniques of a different nature to help improve current archaeological heritage

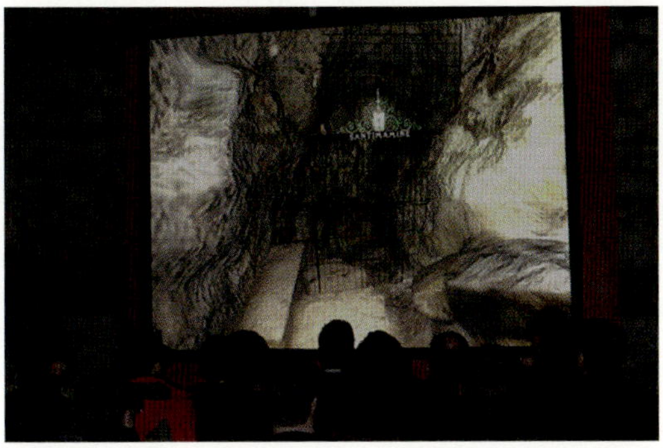

Fig. 16.2 Santimamiñe virtual cave. Developed by the Spanish company Virtualware – www.virtualwaregroup.com

research, conservation and dissemination processes. To do so, compliance with "Principle 1: Interdisciplinarity" will be fundamental" (3.2). "Nevertheless, computer-based visualisations might be an alternative approach when original archaeological remains have been destroyed (e.g., due to the construction of large infrastructures), are placed in areas with difficult accessibility (e.g., without roads) or at risk of deterioration due to the huge influx of tourists (e.g., rock paintings)" (3.3). In these exceptional circumstances, the virtual solution is the only possible answer, whether to conserve the archaeological heritage (3D digitalisation guarantees the heritage is conserved, albeit digitally) or for its dissemination (see the case of virtual caves, such as Santimamiñe: Barrera and Baeza 2010) (Fig. 16.2).

16.3.4 Principle 4: Authenticity

"Computer-based visualisation normally reconstructs or recreates historical buildings, artifacts and environments as we believe they were in the past. For that reason, it should always be possible to distinguish what is real, genuine or authentic from what is not. In this sense, authenticity must be a permanent operational concept in any virtual archaeology project". However, we must always bear in mind that the methods used to increase the 3D models' levels of scientific transparency must differ according to the sector in question. So, for example, the method used for conventional users, i.e. the general public, must be simple, fast and intuitive, as is the case with the system developed by the "Troia VR" project. In contrast, when the end recipients are other researchers, the level of precision must be much higher, providing as much information as possible.

"Since archaeology is complex and not an exact and irrefutable science, it must be openly committed to making alternative virtual interpretations provided they afford the same scientific validity. When that equality does not exist, only the main hypothesis will be endorsed" (4.1). Unfortunately, in many 3D visualisations aimed at the public, a monolithic, almost positivistic idea of archaeological knowledge is conveyed, without leaving any room for alternative interpretations that, in many cases, afford the same scientific validity as the principal hypothesis. This attitude breaks with the principles of authenticity, rigour and transparency that any scientific research study must uphold, as it prevents the visitor from understanding the complexity and scope of archaeological research (San Martín 1994: 15). In the Roman town of Treignes (Viroinval, Belgium), one of the explanatory panels shows two possible reconstructive hypotheses about the appearance that the town's main façade may have had, awarding equal validity to both possibilities. In this way, the visitor can discover the real status of the research, with all its certainties but also with its uncertainties.

"When performing virtual restorations or reconstructions, these must explicitly or through additional interpretations show the different levels of accuracy on which the restoration or reconstruction is based" (4.2). One of the best systems of scientific transparency (levels of veracity) for researchers that has been developed to date is that of the Vendicari Tower (Sicily, Italy). This system was developed by the Italian company NoReal, under the direction of the architect Davide Borra (2009), and considers its primary premise that any virtual 3D reconstruction must be based on a set of historical and archaeological hypotheses (Fig. 16.3). Those hypotheses are gradually configured, architectural element by

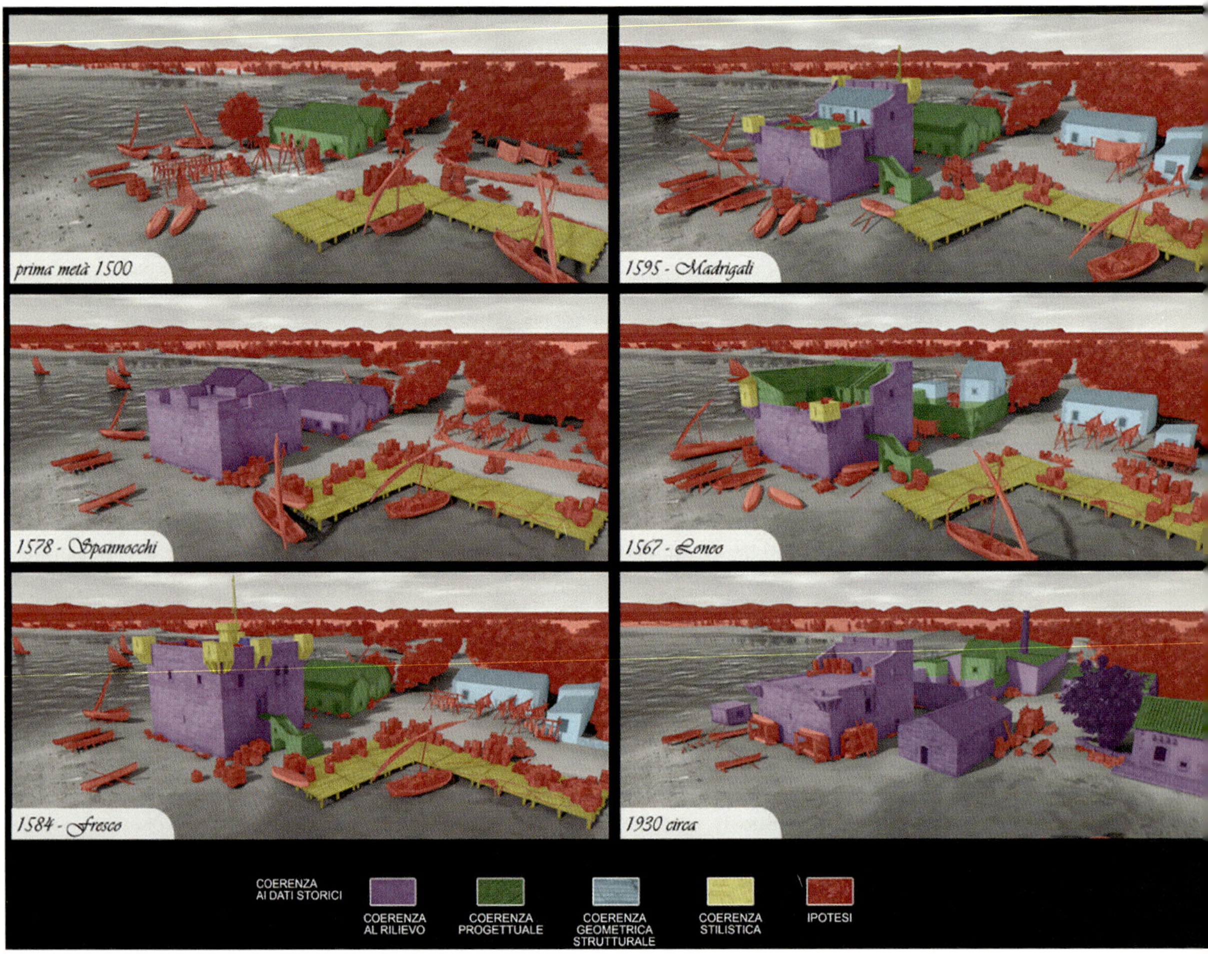

Fig. 16.3 Davide Borra system of scientific transparency (levels of veracity). Vendicari Tower (Sicily, Italy). Developed by the Italian company NoReal

architectural element, object by object, stone by stone, until they finally reach the complete virtual hypothesis, which is habitually known as a virtual reconstruction. In order to discern the 3D model's degree of authenticity, five levels or degrees of certainty are established, each represented by a different colour. For this purpose, a colour scale is defined in which the cold colours represent those elements that have a greater degree of certainty, while the warm colours refer to those elements about which there is less historical or archaeological information available. The colour representation on the 3D model itself enables the degree of certainty to be quickly and intuitively identified, without losing depth and relevance in the information provided. The five levels proposed by Borra are:

Relief (purple). This represents those parts of the 3D model based on archaeological or historical remains that are still visible or reproduced in "objective" iconographic sources.

Project or design coherence (green). This represents those parts of the 3D model that, even if not based on archaeological or historical remains that are still visible or reproduced in "objective" iconographic sources, may be deduced on the basis of those iconographic sources or remains or by studying written sources.

Geometric/structural coherence (blue). This represents those parts of the 3D model whose existence can be felt using the principle of geometric and structural continuity, with regard to the remains that are still visible or reproduced in "objective" iconographic sources.

Stylistic coherence (yellow). This represents those parts of the 3D model whose existence is presupposed out of a similarity with other structures or objects that have been found in similar archaeological or historical contexts.

Hypothesis (red). This represents those parts of the 3D model whose existence in the past is merely conjecture that cannot be defended through the study of written, iconographic or archaeological sources, but which can, however, be reasonably considered to have existed, by sheer common sense.

"In so far as many archaeological remains have been and are being restored or reconstructed, computer-based visualisation should really help both professionals and the public to differentiate clearly among: remains that have been conserved "in situ"; remains that have been returned to their original position (real anastylosis); areas that have been partially or completely rebuilt on the original remains; and finally, areas that have been virtually restored or reconstructed" (4.3). Digital 3D documentation through the use, for example, of a laser scanner enables us to differentiate between what is currently visible and a possible reconstructive hypothesis; however, what we can see today may have undergone major changes as a result of restorers' work, as they often reconstruct or reincorporate lost parts in the archaeological artefacts and structures. In such cases, for the benefit of authenticity, the 3D models need to provide additional information about which zones have been physically reconstructed and which have been conserved just as they were found during the excavation process.

16.3.5 Principle 5: Historical Rigour

"To achieve optimum levels of historical rigour and veracity, any form of computer-based visualisation of the past must be supported by solid research, and historical and archaeological documentation". "The historical rigour of any computer-based visualisation of the past will depend on both the rigour with which prior archaeological research has been performed and the rigour with which that information is used to create the virtual model" (5.1). As Marcelo Martín (2003, p. 21) has pointed out, we must always bear in mind that "conservation alone means a museum without a public, and dissemination on its own means advertising", as it is research that gives purpose and meaning to the field of archaeological heritage in so far as it is responsible for generating the necessary contents to be able to proceed with its restoration and dissemination. In this sense, investment in archaeological research is fundamental in ensuring the rigour and veracity of any virtual archaeology project. Insufficient funding of research moves virtual archaeology away from science and closer to the world of fiction and entertainment.

"All historical phases recorded during archaeological research are extremely valuable. Thus, a rigorous approach would not be one that shows only the time of splendour of reconstructed or recreated archaeological remains but rather one that shows all the phases, including periods of decline. Nor should it display an idyllic image of the past with seemingly newly constructed buildings, people who look like models, etc., but rather a real image, i.e., with buildings in varying states of conservation, people of different sizes and weights, etc." (5.2). Respect for the value of all the phases and additional elements of a monument or other heritage has been defended for decades now by numerous international documents (Venice Charter, art. 11; Burra Charter, art. 15.4; Ename Charter, art. 3.2 & 3.3). Restricting virtual reconstructions to moments of "maximum splendour" detracts from historical reality, offering a still image of the past that does not correspond to the truth; for if anything characterises human societies, it is precisely their capacity for constant transformation (Figs. 16.4).

Fig. 16.4 Virtual reconstruction of the *Carnuntum* landscape in both summer (*top*) and winter (*bottom*). Developed by the Austrian company 7Reasons Medien GmbH

Fig. 16.5 This installation in the Ename Church (Belgium) is a fantastic example of how to develop economically and technologically sustainable systems. Developed by the Belgian company Visual Dimension

"The environment, landscape or context associated with archaeological remains is as important as the ruin itself. Charcoal, paleobotanical, paleozoological and physical paleoanthropological research must serve as a basis for conducting rigorous virtual recreations of landscape and context. They cannot systematically show lifeless cities, lonely buildings or dead landscapes, because this is an historical falsehood" (5.3). On this point, it is worth recalling the words of Sir Mortimer Wheeler (1979 p. 7), the founder of modern archaeology, when he claimed, "the archaeologist is digging up, not things, but people. If the fragments and pieces with which he works are not alive for him, it would be better if he had looked for another occupation […] Dead archaeology is the driest dust that blows".

"Archaeological heritage recording is extremely important not only for archiving, documentation, analyses and dissemination but for management. New techniques such as photogrammetry or laser scanners can be used to increase the quality of the scientific documentation. In this way, the better the metric documentation of archaeological heritage is carried out, the greater will be the chance to monitor and obtain historically and valuable replicas" (5.4).

16.3.6 Principle 6: Efficiency

"The concept of efficiency applied to the field of virtual archaeology depends inexorably on achieving appropriate economic and technological sustainability. Using fewer resources to achieve steadily more and better results is the key to efficiency". If the new technological means used are excessively complicated, heavy or expensive to run, the archaeological heritage managers and the archaeologists themselves will reject them and keep to their traditional methods. This is currently one of the main challenges facing new technologies, including computer-based visualisation, to make its way in the field of archaeological heritage (Fig. 16.5).

"Any project that involves the use of computer-based visualisation in the field of archaeological heritage must pre-screen the economic and technological maintenance needs that will be generated once installed and operative" (6.1). "Priority must be given to systems that may initially require high investments but yield long term profit, with minimum maintenance cost and high veracity, i.e., low-consumption, resistant, easy to repair or modify systems will be preferred" (6.2). If we are talking about research or conservation, the means used must be as inexpensive and uncomplicated as possible, as they will, to a great extent, have to be totally or partially transported to the excavation site. Likewise, the information generated in a given programme or format must be able to be easily extrapolated to another more modern programme to prevent a definitive loss of that information, as is often the case when information gets trapped in obsolete formats (Howell 2007). On this point, it is advisable to follow the guidelines set by the UNESCO Charter on the preservation of digital heritage (2003).

"Whenever possible, draw on the results obtained by previous visualisation projects, avoiding duplicity, i.e., performing the same work twice" (6.3). Constantly wanting to reinvent the wheel is not only absurd but an unnecessary expense. Logically, in order to make the most of the results obtained in previous projects, those projects must meet some minimum requisites of scientific transparency, as specified in Principle 7 of this document.

16.3.7 Principle 7: Scientific Transparency

"All computer-based visualisation must be essentially transparent, i.e., testable by other researchers or professionals, since the validity, and therefore the scope, of the conclusions produced by such visualisation will depend largely on the ability of others to confirm or refute the results obtained".

"It is clear that all computer-based visualisation involves a large amount of scientific research. Consequently, to achieve scientific and academic rigour in virtual archaeology projects it is essential to prepare documentary bases in which to gather and present transparently the entire work process: objectives, methodology, techniques, reasoning, origin and characteristics of the sources of research, results and conclusions" (7.1). The more exhaustive the report, the greater the scientific transparency, which will make the results easier to reuse in the future. For the specific case of virtual reconstructions, the recording method put forward by Daniel Pletinckx (2007) could be very useful.

"Without prejudice to the creation of such databases it is essential to promote the publication of the results of virtual archaeological projects in journals, books, reports and editorial media, both scientific and popular science, for information, review and consultation by the international scientific community and society in general" (7.2). Virtual archaeology will find it hard to reach the status of scientific discipline if it does not pay attention to scientific publications, as, at this moment in time, they enable us to assess the quality and impact of a researcher's work. Furthermore, given that most of the funding for virtual archaeology projects comes from public administrations, the recipients of that funding must take on an ethical obligation to the society that allows them to carry out their work. This obligation must include publishing articles in journals that are accessible to the general public.

"The incorporation of metadata and paradata is crucial to ensure scientific transparency of any virtual archaeology project. Paradata and metadata should be clear, concise and easily available. In addition, it should provide as much information as possible. The scientific community should contribute with international standardization of metadata and paradata" (7.3) (Bentkowska-Kafel et al. 2012). While this standardisation is in process, "in general, the registration and organisation of all documentation relating to virtual archaeological projects will be based on the Principles for the recording of monuments, groups of buildings and sites ratified by the 11th ICOMOS General Assembly in 1996" (7.4).

"In the interests of scientific transparency, it is necessary to create a large globally-accessible database with projects that offer optimum levels of quality (Art 8.4), without undermining the creation of national or regional databases of this type" (7.5). The compilation of good practices can considerably help the discipline's progress, which is often marred by the abundance of ongoing projects and the inability to discern their quality.

16.3.8 Principle 8: Training and Evaluation

"Virtual archaeology is a scientific discipline related to the comprehensive management of archaeological heritage that has its own specific language and techniques. Like any other academic discipline, it requires specific training and evaluation programmes". The training programmes that have been run to date are clearly not enough, both due to the lack of programmes and to the number of training hours offered. Among the most interesting initiatives are the International Summer School course "3D modelling in archaeology and cultural heritage" organised by Dr. Fabio Remondino since 2008 in different locations: Ascona (Switzerland), Trento (Italy), Durham (UK) and Grosseto (Italy); the Italian School of Virtual Archaeology (Scuola Italiana di Archeologia Virtuale) organised by the CNR ITABC since 2009; and the Specialisation

Fig. 16.6 Specialisation Course in Virtual Heritage and Archaeology run by Professor Alfredo Grande since 2009 at the Centre for Virtual Archaeology Research and Development (CIDAV) in La Rinconada (Spain)

Course in Virtual Heritage and Archaeology run by Professor Alfredo Grande since 2009 at the Centre for Virtual Archaeology Research and Development (CIDAV) in La Rinconada (Spain) (Fig. 16.6).

To counteract this deficit, "high-level post-graduate training programmes must be promoted to strengthen training and specialisation of a sufficient number of qualified professionals in this field" (8.1). The launch of a training programme under the European project V-Must (www.v-must.net) and the ambitious programme started up by the Spanish Society of Virtual Archaeology (www.seavtraining.com) are aiming to remedy this situation, although there is clearly still a long way to go.

"When computer-based visualisations are designed as instruments for edutainment and knowledge of the general public, the most appropriate method of evaluation will be visitors' studies" (8.2).

"When computer-based visualisations are intended to serve as an instrument for archaeological research and conservation, the most appropriate archaeological evaluation method will be testing by a representative number of end users, i.e. professionals" (8.3). It makes no sense to develop programmes and products for professional groups, such as archaeologists or restorers, without knowing these groups' real needs. The end user must always be the one who determines the problem to be solved. Besides, it is particularly advisable to involve the end user in the process of creating the solution.

"The final quality of any computer-based visualisation must be evaluated based on the rigour of the measures and not the spectacularity of its results. Compliance with all the principles will determine whether the end result of a computer-based visualisation can be considered "top quality" or not" (8.4).

16.4 Definitions

To conclude this chapter, we have felt it appropriate to pause for a moment to look at the question of terminology, as recent decades have seen the appearance of a specialist jargon in the field of virtual archaeology. This new technical language has grown and evolved alongside the new technological and heritage realities, often without its community of speakers even fully realising the process. The main problem with this reality is the manifold meanings that have arisen around many words (Abejón et al. 2006: 471–472). A better understanding of the language we use in the present is crucial to encouraging the progress of any scientific discipline in the future, as words intrinsically possess a performative value that helps to build realities. Probably one of the first international documents to take on board this statement was the Australia ICOMOS Charter for cultural significance sites (the Burra Charter) that set 17 definitions in its first article. Given that the London Charter already has an extensive glossary of terms, the Seville Principles only highlight those concepts that are specifically associated with virtual archaeology, endeavouring to find a direct correlation between terms and definitions that already exist in other international documents dealing with archaeological or architectural heritage and the terms and definitions inherent in the virtual discipline. So, for example, the definition of virtual anastylosis is closely related to the definition of anastylosis given by the Venice Charter in 1964.

The terms defined by the Seville Principles are as follows:

Virtual archaeology: the scientific discipline that seeks to research and develop ways of using computer-based visualisation for the comprehensive management of archaeological heritage.

Archaeological heritage: the set of tangible assets, both movable and immovable, irrespective of whether they have been extracted or not and whether they are on the surface or underground or on land or in water, which together with their context, which will also be considered a part of archaeological heritage, serve as a historical source of knowledge on the history of humankind. The distinguishing feature of these elements, which were or have been abandoned by the cultures that produced them, is that they may be studied, recovered or located using archaeological methodology as the primary method of research and using mainly excavation and surveying or prospection techniques, without compromising the possibility of using other complementary methods for knowledge.

Comprehensive management: this includes inventories, surveys, excavation work, documentation, research, maintenance, conservation, preservation, restoration, interpretation, presentation, access and public use of the material remains of the past.

Virtual restoration: this involves using a virtual model to reorder available material remains in order to visually recreate something that existed in the past. Thus, virtual restoration includes virtual anastylosis.

Virtual anastylosis: this involves restructuring existing but dismembered parts in a virtual model.

Virtual reconstruction: this involves using a virtual model to visually recover a building or object made by humans at a given moment in the past from available physical evidence of these buildings or objects, scientifically reasonable comparative inferences and in general all studies carried out by archaeologists and other experts in relation to archaeological and historical science.

Virtual recreation: this involves using a virtual model to visually recover an archaeological site at a given moment in the past, including material culture (movable and immovable heritage), environment, landscape, customs and general cultural significance.

Acknowledgements "The research leading to these results is partly funded by the EU Community's FP7 ICT under the V-MusT.net Project (Grant Agreement 270404). The publication reflects only the author's views and the Community is not liable for any use that may be made of the information contained therein. Neither the V-MusT.net consortium as a whole, nor any participant in the V-MusT.net consortium, warrants that the information contained in this document is capable of use, nor that use of the information is free from risk, or accepts any liability for loss or damage suffered by any person using this information".

References

Abejón T, Maldonado Á, Florián L (2006) Contribución al estudio de la terminología de la gestión del patrimonio cultural y del turismo cultural a través de los datos del proyecto ESLEE. In: La terminología en el siglo XXI: contribución a la cultura de la paz, la diversidad y la sostenibilidad. Actas del IX Simposio Iberoamericano de Terminología RITERM2004, AA.VV. Institut Universitari de Lingüística Aplicada. Universitat Pompeu Fabra, Barcelona, pp 471–482

Almagro Vidal A (2008) El concepto de espacio en la arquitectura palatina andalusí. Un análisis perceptivo a través de la infografía. CSIC, Madrid

Barrera S, Baeza U (2010) La Realidad Virtual aplicada a la explotación sostenible del Patrimonio Arqueológico. Un caso éxito:la Cueva de Santimamiñe. Virtual Archaeol Rev 1:69–72

Bentkowska-Kafel A, Denard H, Baker D (eds) (2012) Paradata and transparency in virtual heritage. Ashgate, Farnham

Borra D (2009) L'anastilosi virtuale della torre di vendicari. In: Susan G (ed) La torre di vendicari. Un'architettura che nasce e si sviluppa dal mare. Lombardi, Siracusa, pp 173–202

Cano P, Lamolda F, Torres JC, del Mar Villafranca M (2010) Uso de escáner láser 3D para el registro del estado previo a la intervención de la Fuente de los Leones de La Alhambra. Virtual Archaeol Rev 2: 89–94

Carandini A (1997) Historias en la Tierra. Ed. Crítica, Barcelona

Howell A (2007) Preserving digital information: challenges and solutions. National Library of Australia. http://nla.gov.au/nla.arc-49633. Accessed 10 Mar 2009

López-Menchero VM (2011) Propuesta para profundizar en la Carta de Londres y mejorar su aplicabilidad en el campo del patrimonio arqueológico. Virtual Archaeol Rev 4:65–69

López-Menchero VM, Grande A (2011) Hacia una carta internacional de arqueología virtual. El borrador SEAV. Virtual Archaeol Rev 4:71–75

Martín M (2003) La gestión del patrimonio se sustenta sobre tres grandes tareas:investigar, conservar y difundir. Boletín de Interpretación 8:21–22

Pletinckx D (2007) Interpretation management: how to make sustainable visualisations of the past. EPOCH knowhow book. The Interactive Institute, Kista

San Martín Montilla C (1994) La protección del Patrimonio Arqueológico desde el museo II. Criterios de difusión. Boletín Informativo del Instituto Andaluz del Patrimonio Histórico 8:14–16

Wheeler M (1979) Arqueología de campo, 2nd edn. Fondo de Cultura Económica, México

Reconstructing Past Landscapes for Virtual Museums

17

Ivana Cerato and Sofia Pescarin

Contents

I. Cerato (✉) • S. Pescarin
Virtual Heritage Lab, CNR-ITABC,
Via Salaria, km 29, Monterotondo St., Roma, Italy
e-mail: ivana.cerato@itabc.cnr.it;
sofia.pescarin@itabc.cnr.it

17.1 Introduction

The reconstruction of ancient landscape is a challenging research activity that implies the management of a high level of uncertainty, on one side, and, on the other, requires the cooperation of several different disciplines.

The traditional two-dimensional mapping output from GIS was not enough to represent landscape complexity and dynamics. Static visualisation, moreover, could not fully answer the request of updating interpretations and simulations. For these reasons three-dimensional modelling and interactive applications have been more and more adopted by the scientific community, coupling the growing interest of not expert users for virtual reconstructions and interactive/immersive virtual museums.

No matter how many disciplines are involved and no matter how much information is acquired on the field, in the archives, etc., the whole picture would never be complete and the number of variables being always too high.

Although landscape has significantly changed (or completely disappeared), we could still use and integrate within a GIS environment available information from very different sources, such as archaeology, cartography, historical geography, soil science, geophysical surveying, geomorphology, landscape and historical ecology.

Through landscape analysis, remote sensing and cross-discipline analysis, we are able to highlight a full range of aspects of past landscapes; through interactive or not interactive

C. Corsi et al. (eds.), *Good Practice in Archaeological Diagnostics*, Natural Science in Archaeology,
DOI 10.1007/978-3-319-01784-6_17, © Springer International Publishing Switzerland 2013

Table 17.1 Workflow of the landscape reconstruction

Data collection	Data acquisition	GIS	Data post-processing	Landscape reconstruction
1. Geophysical surveys; 2. Active aerial photography; 3. Geomorphological surveys; 4. Archaeological field survey and excavation 5. Historical map collection	1. Range image modelling; 2. Based image modelling; 3. Photogrammetry 4. UAV	A GIS-based integration of all survey data, maps and re-studied legacy data has procured a formidable database for the computer aided digital 3D mapping and interpretation of these complex ancient sites	Understanding of the ancient settlement dynamic and the interactive human-environment relationship	Utilised and transformed into three-dimensional simulations in an attempt to visualise, research and understand past experiences

visual applications, it becomes easier to communicate them to wider audiences, in most cases far away from the landscape under study.

The archaeological landscape, as we can observe today, is the result of a long process of transformation, and it is very important to store diachronic relations, describing as territory and site changes over time, to propose a reliable interpretation of the ecological context. The digital reconstruction of the archaeological landscape is a very complex process, which includes considering many kinds of data and setting up several activities, in a multidisciplinary approach. Environment and archaeological structures can be reconstructed through different techniques and data sources integrated in a coherent methodology of processing and communication: such as cartography, remote sensing, photo-interpretation, field survey, laser scanner and photogrammetry. Each technique/source is selected according to the type of structure, its characteristic and to the required information (Table 17.1). The management of these data is an essential task for the use, analysis and communication of the information gathered.

Digital and spatial technologies are changing how archaeology and related disciplines approach the past, in relation to the contemporary world. These technologies provide alternative ways to study and understand past and present.

17.2 Uncertainty Towards Transparency

One of the most important research challenge in landscape digital reconstruction is the definition of how digital models may be used to convey uncertainty and different hypotheses of how buildings were constructed or used and which relation they have with the territory, to examine relationships among different phases of the same site and to explore visibility inside and outside structures. On the other hand, one of the critics commonly raised by archaeologists is that 3D models are a "closed box", with no possibility of evaluation and often without a particular aim, the emphasis being on computer graphics and artistic aspects, rather than on the attempt to solve a particular scientific problem. 3D reconstructions (reality based or not reality based: Remondino and El-Hakim 2006) have to declare the methodology and the type of data from which they have been obtained, so to allow discussion and critical awareness. One of the most significant consequences of the introduction of digital 3D modelling in the Cultural Heritage field is the possibility to use 3D models as highly effective and intuitive means of communication as well as interface to share and visualise information collected in databases (Manferdini and Remondino 2010).

For the present work, we have considered two types of users: common users (mainly not expert on the content) and expert users (belonging to the scientific community). We have assumed that the main aim for common users is to understand the past, to live an experience and to build "affinity, empathy" (Zeki 2009) that can help in finding joy and motivation to increase their knowledge (Forte et al. 2006). This process was also described by Thomas Mann: "for a significant intellectual product to make a broad and deep immediate appeal, there must be a hidden affinity, indeed a congruence, between the personal destiny of the author and the wider destiny of his generation" (Mann 1925).

On the other side, expert users have other requirements: they need to read a reconstruction and propose alternative possibilities of sites and landscape reconstruction. This represents a shift from past research with a significant impact on how material culture is documented and understood.

It is therefore essential for the scientific community to avoid "closed boxes" and to approach digital reconstructions in a more "transparent" way (Principles 3 and 4 of London Charter: London Charter 299. See Chap. 15 by Denard in this volume). But this requirement is not just limited to expert users. Common users should also have the possibility, at least on request and in accordance with the type of media, to access further explanatory information on how reconstructions were made and which sources were used. This is particularly true with Virtual Museums (V-MUST DEL 2.1) where the focus is communication that should not just be aimed at demonstrating technical or artistic skills, but rather to attain a degree of "visual fidelity" and accuracy.

Therefore, a "transparent" approach enables a better evaluation of 3D visualisations and a more efficient way to share information within the scientific community, thus avoiding the risk of duplicating works, resources, etc., starting each time from the beginning. It offers potentially a rapid advance of the research. Metadata are now taken as a practical way to follow this approach (Felicetti and Lorenzini 2011). Are metadata useful to make the process transparent also to final users (Gockel et al. 2013)?

Normally data used for the reconstruction remain hidden to the users, leaving them a vague perception of what there is "behind the scene" and which is its reliability. How could expert users handle and improve the transparency, keeping track of reliability and uncertainty in virtual archaeology (VA) (Forte and Williams 2003) project? Should data used in the reconstruction process be transparent to final users? How can they serve to improve users' understanding of the past?

We have tried to understand how these issues are considered important in scientific publications, presented in the last couple of years at conferences, and in virtual archaeology (VA) applications or virtual museums (VM) and how these issues have been solved. We have finally tried to understand the trend in the research domain and the gaps that need further developments. We have therefore carried out two surveys: one (see 17.2.1) related to VM and VA applications, among those presented from 2006 to 2012 at Archeovirtual international exhibition (www.archeovirtual.it) (Pescarin et al. 2012a, b) and those analysed by V-MUST.net (www.v-must.net; see V-MUST DEL 2.3), and a second one (see 17.2.2) related to papers published in the last 2 years (2011–2012).

In the first survey we wanted to obtain information regarding how reconstructions are communicated to a public (Q1) and in the second one how the scientists communicate the results of their surveys and interpretation to the community of researchers (Q2).

We have considered projects focused on archaeological landscape (Forte and Ryan Williams 2003) and ancient potential landscape (Forte and Williams 2003; Pescarin 2007). To widen the research, we have also included beyond landscapes sites (medium/small range) and urban reconstructions (urban landscapes).

We have finally taken into consideration projects related with "reconstructions", whose concept included the entire process, from acquisition and interpretation to reconstruction and visual representation of results (interactive or not interactive).

17.2.1 Virtual Museums and Virtual Archaeology Survey

In the first survey, we collected information on 112 VM and VA applications:

- Fifty-seven applications/demo presented at Archeovirtual
- Fifty-five virtual museums surveyed by V-MUST.net

Among these applications, 42 deal with a wider concept of landscape/site/city reconstruction (both "restitution" and "reconstruction"), and only 34 regard strictly "reconstructions" not necessarily based on spatial GIS-based data. Among these last, 20 applications deal with spatial digital assets. Even if a not web-GIS based is more oriented to narration instead of the Gis based one, what came out is that 50% of GIS-based projects use a narrative approach regarding maetadata and trasparency.

They were taken into account only by the 5 % of most recent projects (2011–2012) (2/42, among those who deal with a wider concept of landscape reconstruction). When we consider all virtual museums surveyed within V-MUST and Archeovirtual, the percentage is higher: 10 % (11/112). Nevertheless in 38 % of cases (16/42), although we cannot talk about metadata, there is an attempt to propose different ways to provide extra information, although often not in structured way. If we consider projects with narrative approach, we see that all 5 % are not interested in narration, while 19 % (8/16) are interested.

It is clear therefore how metadata are in most cases developed for expert users, although there is a good percentage of projects that provide anyhow not-structured extra information (more multimedia), as part of their communication strategy.

Some examples of the projects that provide metadata are *Locus Imaginis* (Fig. 17.1a; *Locus Imaginis*) and Behind Livia's Villa (Fig. 17.1b) presented during Archeovirtual 2012, while examples of projects that present not-structured extra information are Virtual Rome (Fig. 17.2a) and the Virtual Museum of Ancient Via Flaminia (Fig. 17.2b).

Fig. 17.1 (**a**) *Locus Imaginis*: http://www.map.archi.fr/ldl/Locus/Locus_Imaginis/. (**b**) Behind Livia's villa

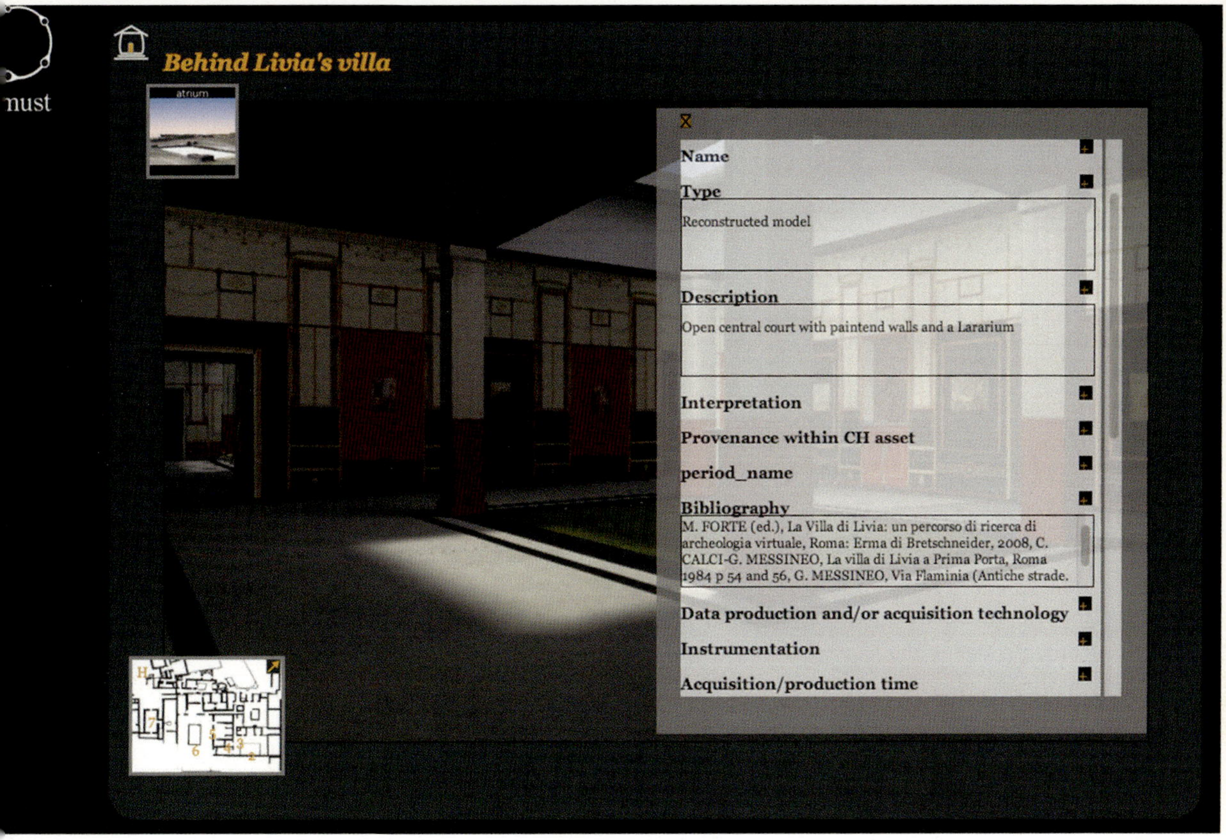

Fig. 17.1 (continued)

In the *Aquae Patavinae* project (Fleury and Medeleine 2012), we tried to overcome the problem of the transparency, offering to the user a virtual reality navigation system on-line and different layers of exploration and tools to get into more depth the archaeological informations (Fig. 17.3).

17.2.2 Survey on Published Works

In the second survey we analysed abstracts or, when already printed, papers of most important conferences with topics regarding 3D reconstruction, multimedia and virtual reality and case studies:

- Fourth ISPRS International Workshop 3D-ARCH 2011: "3D Virtual Reconstruction and Visualization of Complex Architectures" (http://www.3d-arch.org/)
- Virtual Retrospect 2009 (http://archeovision.cnrs.fr/spip.php?article 144)
- VAST2012: The 13th International Symposium on Virtual Reality, Archaeology and Cultural Heritage (http://www.vast2012.org/)
- Eighteenth International Conference on Virtual Systems and Multimedia – VSMM 2012 (http://www.vsmm2012.org/)
- The Computer Applications and Quantitative Methods in Archaeology (CAA) 2012 (http://caaconference.org/)
- International Conference on Cultural Heritage, Oct. 29–Nov. 3, 2012, Lemesos, Cyprus (Euromed) 2012 (http://www.euromed2012.eu/)

At the end of this analysis, we evaluated 686 publications (Fig. 17.4). Among these, 149

(27 %) projects deal generally with archaeological reconstructions, of which only 5 % (29/686) strictly with landscape reconstruction. The majority of these describe the acquisition system, data presentation and visualisation methods. In many cases the focus is on the presentation of new ways to visualise virtual reconstructions and landscape evolution. The goal of "Urban Archaeology" project, for instance (Capone 2011?), is to highlight issues; to propose possible but not unique solutions, stimulating cultural debate; and to make the cultural message understandable to a broad audience. For this reasons they have chosen a case study particularly complex, necessary to test a methodological path appropriate to the content to communicate. Another example is focused on the analysis of the use of remote sensing from space-based and airborne platforms to study ancient land management strategies at the ancient Greek agricultural territory (*chora*) of Metaponto in southern Italy (Trelogan et al. 2012). The paper (Shaw 2012), which focuses on the Stonehenge World Heritage site, details work carried out as part of a masters dissertation which looked at how smartphones and applications can be used to aid the ergonomics of landscape study, analysis and interpretation within archaeology.

Only six projects (1 %), presented in the conferences, describe the use of metadata and suggest possible approaches to integrate 3D modelling into the archaeological research methodology, by describing some validation methods of the models. The ArcSeer project (Lynam 2011), for instance, presents the initial results of the prototype tool QueryArch3D (Fig. 17.5). The goal is to create a Web-based tool that allows interactive visualisation and queries of multiresolution Cultural Heritage 3D models. The visualisation front end allows the user to navigate interactively in a virtual environment, where existing structures can be explored and queried, at different levels of detail. It should be possible to distinguish (e.g. "switch" on and off) real structures from virtually reconstructed ones. Another example (Richards-Rissetto et al. 2012)

Fig. 17.2 (**a**) Virtual Rome: www.virtualrome.it. (**b**) Flaminia: http://www.vhlab.itabc.cnr.it/flaminia/

Fig. 17.2 (continued)

explores the potential of using Microsoft's Kinect to create a low-cost and portable system to virtually navigate, through a prototype 3D GIS, the digitally reconstructed ancient Maya city of Copan in Honduras (Fig. 17.6). As users move their "bodies" through the VR environment, the ability to click on 3D models and acquire archaeological information with a simple hand gesture maintains the continuity of the experience in the VR environment. In the "Plan the Rome" project (Felicetti and Lorenzini 2011), the expected results are a digital 3D model of Rome as it was in the fourth century AD, a digital 3D model of the principal machinery used in the Roman world, and links for each digital model to the body of ancient source material (documentary, archaeological and iconographic).

Fig. 17.3 *Aquae Patavinae*: http://www.aquaepatavinae.lettere.unipd.it/portale/?page_id=2174

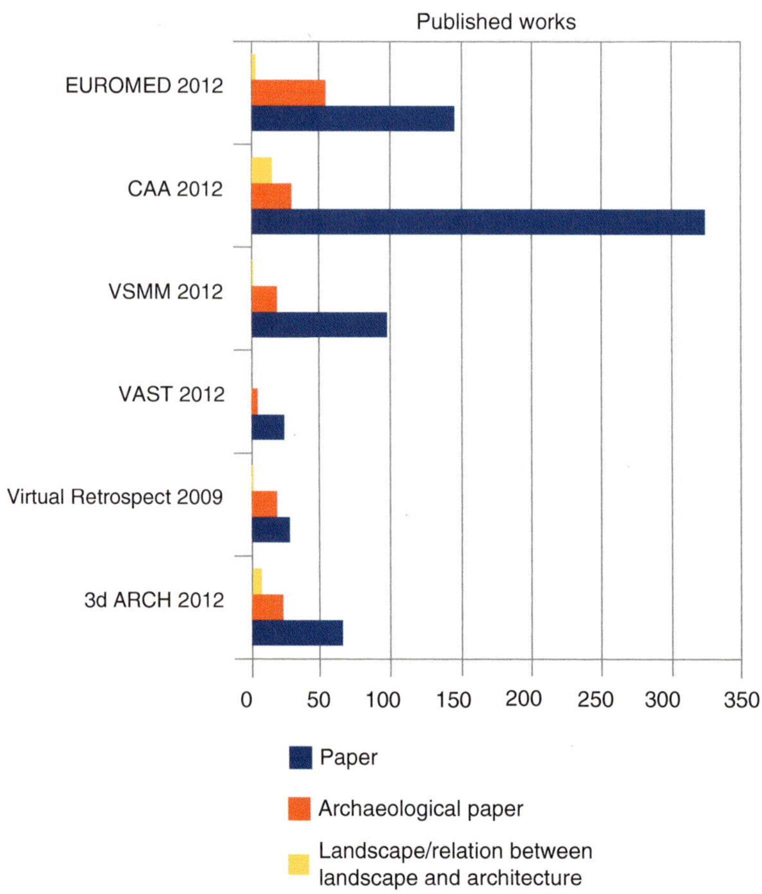

Fig. 17.4 Graph of the published works

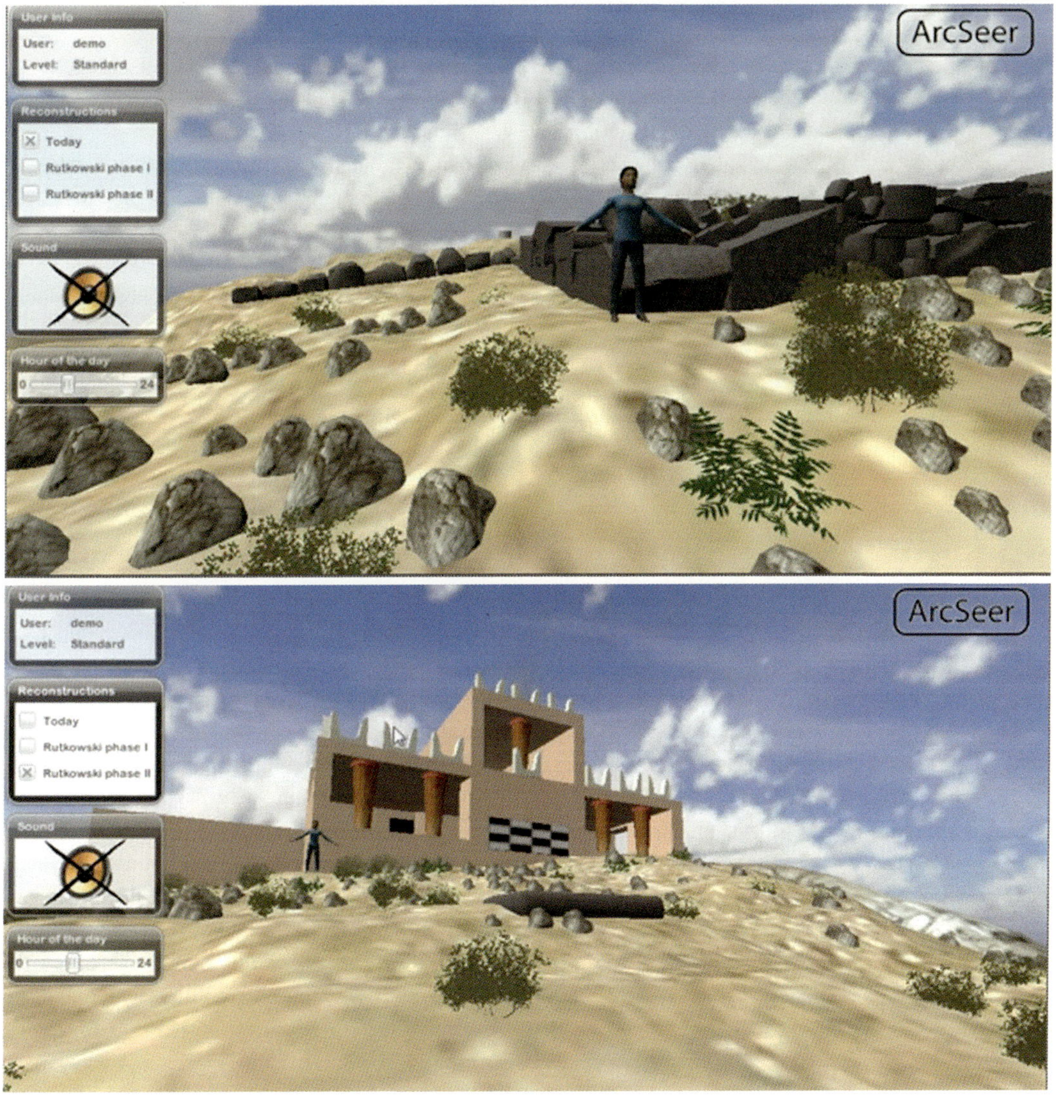

Fig. 17.5 ArcSeer (www.arcseer.com)

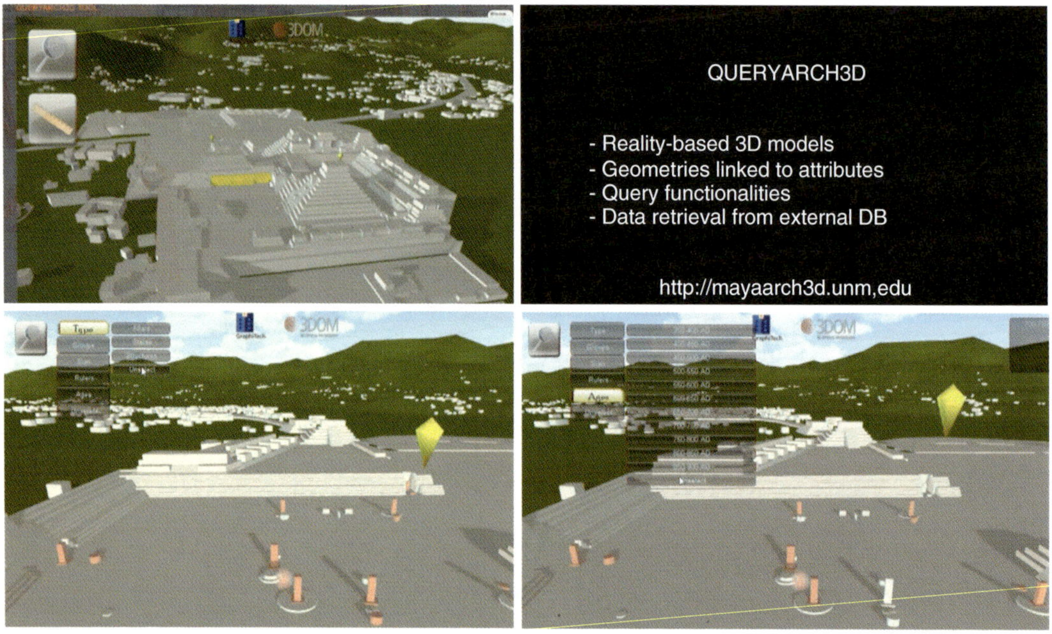

Fig. 17.6 Kinect and 3D GIS in archaeology. A Web-based interactive tool for multi-resolution 3D model access and visualisation (http://mayaarch3d.unm.edu/)

Conclusion

Virtual reconstructions of archaeological sites, artefacts, and architecture play an important role, supporting scientific discussions among experts and bringing the past to broad audiences, through virtual museums.

Regarding the initial questions of the current work – Q1 (How do we communicate reconstructions to non-expert users?) and Q2 (How do we communicate the results of our interpretation to the community of researchers?) – these are some of the conclusions:

In the case of Q1: extra information is provided in a not-structured way (no database/metadata behind). There is an increasing interest in providing information regarding the level of uncertainty and reliability, in some cases connecting models to sources. The continuity of the experience is preferred to discontinuity, preferring a reconstruction of a full realistic ecosystem, obtained from raster representations derived from interpolation (colour maps). The majority of the projects are focused on reality-based reconstruction (or restitution) more than on reconstruction of the past. A narrative approach is in most cases the preferred communication style.

Regarding Q2, maps are often used to communicate results, with a preference of vector representation, stressing discontinuity and therefore just reliable data. Raster continuous visualisations are used especially for spatial analysis. Metadata are used increasingly, although their adoption is still at a very early stage.

Since the level of uncertainty is a challenge, transparency is essential to understand and build research hypotheses and conclusions, particularly in areas where data is questionable, incomplete or conflicting.

Some approaches used to distinguish what is certain from what is not certain (transparency effects, different colours, models without maps or with modern maps) or to access extra information in a not-structured way are unsatisfactory for expert users, although they seem to be a good compromise for common users. In complex multiphase models, some of these approaches can be unclear to users. If metadata could be the answer for researchers, there are still two open issues: how metadata could be easily used and integrated with reconstruction and which semantic model to use? Evaluation results (Lucci Baldassari et al. 2013) highlight Virtual Museum solutions edited for not expert users imply metadata in more communicative and efficient way. How could they make users aware of the process, of used sources and of different possible hypothesis?

One of the possible way to solve this issue is to create interactive 3D environments where users could choose to live a narrative experience or to explore these reconstructed landscapes in expert mode, querying and accessing sources, texts, multimedia material, pictures, bibliography, etc. (Fleury and Medeleine 2012). Metadata access needs to be improved and their use adapted to landscape issues. From the User Interface and User Interaction perspective, a lot still need to be done. A final remark regards the necessity to improve scientific publications on reconstructions, including evidence of results in an explicit theoretical approach. There is a high number of publications with no visible or accessible results connected with sources and methods used. Is this lack connected with not appropriate publication medium or evaluation system?

References

Capone M (2011) Urban archaeology. How to represent uncertainty: the archaeological excavations of Piazza Bovio. DisegnareCon 4(7):68–80. http://disegnarecon.unibo.it/issue/view/256/showToc

Felicetti A, Lorenzini M (2011) Metadata and tools for integration and preservation of cultural heritage 3D information. In: Proceedings of the 23rd international CIPA symposium, Prague, 12–16 September 2011. CIPA

Fleury P, Medeleine S (2012) Reviving ancient Rome: virtual reality at the service of cultural heritage. In: Ioannides M et al (eds) Progress in cultural heritage preservation. Proceedings of the 4th international conference, EuroMed 2012, Limassol, October 29–November 3, 2012. Springer, Berlin/Heidelberg, pp 159–169

Forte M, Williams PR (eds) (2003) The reconstruction of archaeological landscapes through digital technologies. Italy-United states workshop, Boston, November, 1–3, 2001, BAR International series S1151. Archaeopress, Oxford

Forte M, Pescarin S, Pujol Tost L (2006) VR applications, new devices and museums: public's feedback and learning. A preliminary report. In: Proceedings of the 7th international symposium on virtual reality, archaeology and cultural heritage, VAST 2006, Nicosia, 30 October–4 November 2006, Short papers volume. Eurographics, Geneva, pp 64–69

Gockel B, Graf H, Pagano A, Pescarin S, Eriksson J (2013) VMUXE: an approach to user experience evaluation for virtual museums. In: Proceedings of the HCI international 2013, Las Vegas, 21–26 July 2013, LNCS. Springer, Heidelberg

London Charter for the Computer-Based Visualisation of Cultural Heritage (Version 2.1, 2009). http://www.londoncharter.org/downloads.html. Accessed 23 June 2012

Lucci Baldassari G, Demetrescu E, Pescarin S, Eriksson J, Graf H (2013) Behind Livia's Villa. A case study for the devolution of large scale interactive "in-site" to "on-line" application. In: Proceedings of the HCI international 2013, Las Vegas, 21–26 July 2013, LNCS. Springer, Heidelberg

Lynam F (2011) ArcRange and ArcSeer: presenting a new approach to archaeological data management and representation. Paper presented at the 3D-ARCH international workshop, Trento, 2–5 March 2011. Portal of architectural image-based-modeling. http://www.map.archi.fr/aibm/Portal_of_Architectural_Image-Based-Modeling/Article-Lynam.html

Manferdini AM, Remondino F (2010) Reality-based 3D modeling, segmentation and web-based visualization. In: Digital heritage. Proceedings of the third international conference, EuroMed 2010, Lemessos, 8–13 November 2010, LNCS Vol. 6436. Springer, Heidelberg, pp 110–124

Mann T (1925) Death in Venice [and other stories]. Alfred A. Knopf, New York

Pescarin S (2007) Reconstructing archaeological landscape. Interpretation and integration in spatial and real-time open system. In: Keeler D, Salisbury RB (eds) Space – archaeology's final frontier? An intercontinental approach. Cambridge Scholars Publishing, Newcastle, pp 69–93

Pescarin S, Fanini B, Ferdani D, Lucci Baldassari G (2012a) Archeologia virtuale a Montegrotto Terme. In: Bassani M, Bressan M, Ghedini F (eds) Aquae Patavinae. Montegrotto Terme e il Termalismo in Italia. Aggiornamenti e nuove prospettive di valorizzazione, atti del convegno nazionale, Padova 14–15 giugno 2011, Antenor Quaderni 26. Padova University Press, Padova, pp 305–322

Pescarin S, Pagano A, Wallergård M, Hupperetz W, Ray C (2012b) Archeovirtual 2011: an evaluation approach to virtual museums. In: Guidi G, Addison AC (eds) 18th international conference on virtual systems and multimedia. Proceedings of the VSMM 2012, virtual systems in the information society, Milan, 2–5 September 2012. IEEE, Los Alamitos, pp 25–32

Remondino F, El-Hakim S (2006) Image-based 3D modelling: a review. Photogramm Rec 21(115):269–291

Richards-Rissetto H, Robertsson J, Remondino F, Agugiaro G, Girardi G, von Schwerin J (2012) Kinect and 3D GIS in archaeology. In: Guidi G, Addison AC (eds) 18th international conference on virtual systems and multimedia. Proceedings of the VSMM 2012, virtual systems in the information society, Milan, 2–5 September 2012. IEEE, Los Alamitos, pp 331–337

Shaw L (2012) There's an App for that: how can smartphones improve the ergonomics of landscape study, analysis and interpretation? In: The computer applications and quantitative methods in archaeology conference, CAA 2012, Southampton, 26–29 March 2012

Trelogan J, Rizzo A, Moscatelli E (2012) Landscape change at Metaponto: a tale of two DEMs. In: The computer applications and quantitative methods in archaeology conference, CAA 2012, Southampton, 26–29 March 2012

[V-MUST DEL 2.1]. Consortium of the V-MusT.net: terminology, definitions and type for virtual museums. Deliverable report: V-MusT.net – D 2.Version 3 (2013)

[V-Must DEL. 2.3]. Consortium of the V-MusT.net: state of the art on virtual museums in Europe and outside Europe. Deliverable report: V-MusT.net – D 2.Version 1.6 (2013)

Zeki S (2009) Splendors and miseries of the brain: love, creativity, and the quest for human happiness. Wiley-Blackwell, Chichester

18 Computer-Aided 3D Visualisation of Roman Towns

Michael Klein

Contents

M. Klein
7reasons Medien GmbH, Vienna, Austria
e-mail: michael.klein@7reasons.at

18.1 Introduction

The scientific discipline that seeks to research and develop ways of using computer-based visualisation for the comprehensive management of archaeological heritage is called *virtual archaeology*. It includes virtual reconstruction, recreation, anastylosis and restoration of archaeological finds aiming to produce highly detailed images of the past (Principles of Seville 2012, 3; see Chap. 16 by López-Menchero in this volume).

Most recently, a main issue has become the visualisation of the geophysical survey results and their interpretation. The visualisation of sites whose archaeological remains lie mainly still buried can be approached by referencing the existing data with better preserved sites of the region comparing similar structures and dimensions, aiming to preserve architectural local features and details of decoration.

Digital elevation models, geophysical survey results, 3D laser and LiDAR (light detection and ranging) scans are taken into account to build the ancient terrain features, where the architectural 3D reconstruction is located.

The methods involved in such virtual reconstructions are multifarious and have to be adapted to the special characteristics of each site. Procedures which had been developed and tested over the past years by several teams are applied and refined in each project, while other new techniques have to be developed case by case to suit the necessity of the specific project.

C. Corsi et al. (eds.), *Good Practice in Archaeological Diagnostics,* Natural Science in Archaeology, DOI 10.1007/978-3-319-01784-6_18, © Springer International Publishing Switzerland 2013

Furthermore, in order to achieve high-quality reconstructions of a complex site, different approaches need to be developed to aid the process of communication between the researchers working in the framework of the project.

The ideal workflow and the ranges of activities undertaken in these frameworks will be further described in each step.

18.2 Workflow

18.2.1 Data Acquisition

The interpretation of archaeological datasets from complex urban environments and their subsequent transformation into three-dimensional representations of the interpretation usually requires a time-consuming process of re-evaluation of scientific acquisitions and discussion. Data from excavation records often presents only a limited window into an archaeological site, and we will therefore focus this paper on the more extensive datasets derived from geophysical survey and remote sensing.

Typically, the data used during the reconstruction process is provided by the scientific teams involved in the project: documents, maps, graphics and images from archives, excavation reports, aerial photography, LiDAR, GPR (Ground Penetrating Radar), ground magnetic survey data, etc. and their interpretations by geophysicists and archaeologists.

Innovative methods of cost-effective geometrisation SFM (Structure From Motion) can be introduced to digitise standing remains and finds, which are subsequently used to verify measurement carried out in the reconstruction. This process is achieved by matching image pairs and calculating their perspective offset, respectively, to each other in order to estimate their distance in a 3D coordinate system and resulting in a 3D geometry.

These 3D models can be used to obtain surface information like 3D point clouds, to reconstruct surfaces and their textures. Often, this procedure results in a large dataset which can only be implemented in the animation process if these models are simplified to a great extent. The task is being achieved by reproducing the dense surface by a coarse geometry which resides on the exact position of its reference geometry.

The best available technology to output a highly realistic terrain model is a fractal approach. The detail of the geometry is adapted depending on the distance to the virtual camera. With this method, replicas of millions of items (boulders, trees, houses, etc.) can be spread over the landscape.

Certain terrain features like streets and roads, urban boundaries, agricultural land-use features or waterways can be defined by simple maps possessing RGB and greyscale values, which are called splat maps. The easiness of using these splat maps is one of their advantages allowing the dissemination of the information as a simple image which can be edited by the noncomputer graphic specialist members of the project team.

When attempting the reconstruction of ancient landscapes, LiDAR data can be used to filter and discard various modern features and elements like vegetation growth and urban construction.

Apart from archaeological interpretation, aerial images taken from airplanes or drones can be used to build geometry without the cost-intensive and laborious procedure of a LiDAR scan with the use of photogrammetry.

Fluid simulation can be applied to visualise realistic water but also to simulate river flows and evolution of coastlines.

18.2.2 GIS Analysis and Processing

To combine data used in large-scale reconstruction like the remodelling of urban settlements, GIS systems are used. Various georeferenced cartographic and topographic information are collected and assembled to construct a database which is used in the particular task of reconstruction.

Topographic data like DEM (Digital Elevation Model) and DTM (Digital Terrain Model) are needed for generating the virtual environment, which is further modified to achieve an approximation according to the situation of the desired period. To accomplish these tasks, specialised terrain simulators and editors can be used to calculate the geomorphological processes such as erosion and landfills. The results are displayed in fractal geometry, producing a highly realistic and natural appearance. A further task is the insertion of data from historical and archaeological maps which can be used to approximate various historical landscape features like river courses, roads and settlements (Fig. 18.1).

Mainly, when referring to objects, such as furniture, pottery or tools, other approaches of digitalisation include photogrammetric geometrisation. With the aid of current software and a series of images taken from several sides of an object, the creation of accurate geometry is possible without applying markers or circumstantial measurements. This procedure can be utilised in case of artefacts of any size but has been very recently also applied to aerial photos (Fig. 18.2).

18.2.3 Archiving and Data Preprocessing

Raw data from photogrammetric surveys (images) and scans (3D point clouds) are processed with different software: the extraction of textured models with different geometric resolution is usually one of the main issues, depending on the environment, where the model is used or placed. One special procedure was established to suit the need of a large online database, where high-resolution scanned objects had to be reduced to 1 % of their original density, while still preserving all of their visual qualities.

18.2.4 Interpretation and Reconstruction

The interpretation of the raw data serves as a basis for an urban infrastructure model. Therefore, a street grid needs to be determined on the basis of what is known of the town. The main house boundaries are defined by locating obvious open areas such as courtyards, atriums and public squares. With scientific guidance and through comparative studies, building typologies are developed and then integrated into the existing terrain.

The first approach to an architectural layout is made with a set of modularised house parts which are assembled in an electronic "library", within a

Fig. 18.1 Terrain editor with various stages of the terrain construction process

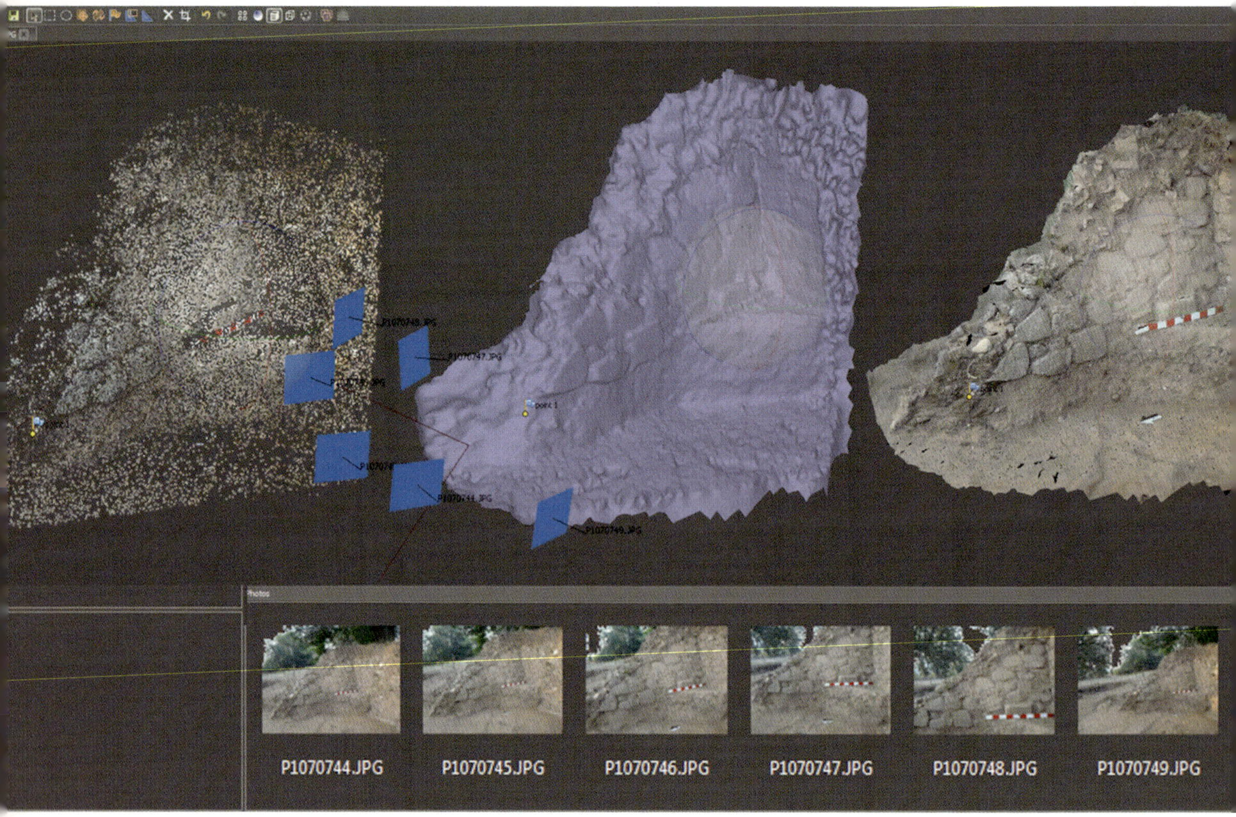

Fig. 18.2 Production of a 3D reconstruction with Agisoft

real-time editor from where the operator can choose the desired items. The latter are placed on the digital terrain model which can be plotted on the basis of what was mapped with geophysical results and interpretations (Fig. 18.3). This procedure, however, may encounter difficulties due to the large amount of single modules which have to be set in order to construct the individual houses. One option is therefore to produce larger units of these modules to speed up the process. This modular approach turns out to be very effective even for operators who are not trained in computer-aided 3D drawing.

At this point, already a crude 3D model of the settlement exists, which is inserted into a real-time engine, to allow modification and interactive display during further sessions of discussion with the scientists. This point is crucial in every project, as several questions are characteristically unsolved yet: the model and the incorporated data like maps, orography and excavation plans can meet the needs for further interaction between specialists of different disciplines. When satisfactory answers have been found, and the layout thus modified, the houses in the scene are textured, to give them a realistic appearance, and also the landscape, roads and other features are modelled with more detail.

Various methods of production are implemented depending on the use for the final format. Accuracy and authenticity is crucial to sustain a high-quality product. It is also essential to attach the documentation of the resulting objects. When the desired level of detail is achieved, the process of texturing (mapping images onto the surfaces) is applied.

A wide range of software is used for the construction of different items and this operation demands highly skilled personal to ensure a high-quality output.

Sometimes, the decision is made to show a close-up of certain objects, like a single house or

Fig. 18.3 Real-time construction editor with the library of modularised building blocks

compound. In this case, a model with higher degree of geometric resolution has to be created, showing also the interior organisation, like rooms, stairs and corridors. To complete the scene, correct building materials, interior decoration and furniture have to be shown. This is a task which is of similar importance as the quality of the general layout, if the overall result should be scientifically correct.

Special attention in the reconstruction work has obviously to be given to the prominent public buildings, as they are normally the best known from archaeological research and are more easily framed into comparative research.

As an example, we will present here the case of the forum of *Ammaia*, the Roman town in central *Lusitania* which has been one of the targets of the research carried out in the framework of the EU-funded project Radio-Past (one of the so-called open labs; Corsi and Vermeulen 2012, 3–5; Van Roode et al. 2012). This complex can be classified as a typical "tripartite" forum, meaning that the central square is bordered by a portico, and the "sacred" area with the temple occupies one of the sides, while opposite is the basilica (Vermeulen et al. 2012). Detailed reconstruction of the structure of the forum temple is based on excavations, geophysical research and comparative research. Among similar monuments, those of the nearby towns of *Conimbriga* (Frova 1990; Correia 2009) and Evora (Hauschield 1992) have been considered as the closest examples (Fig. 18.4).

Excavations and geophysical research provided the basic outline of the shape and dimensions of the temple: the Flavian forum temple of *Conimbriga*, with almost identical dimensions in the plan, reaches a total height of approximately 18 m (60 Roman feet, from base to rooftop), whereas the columns used in the *Ammaia* reconstruction were set to an idealistic 30 Roman feet (9 m) height, reaching the perfect proportion for

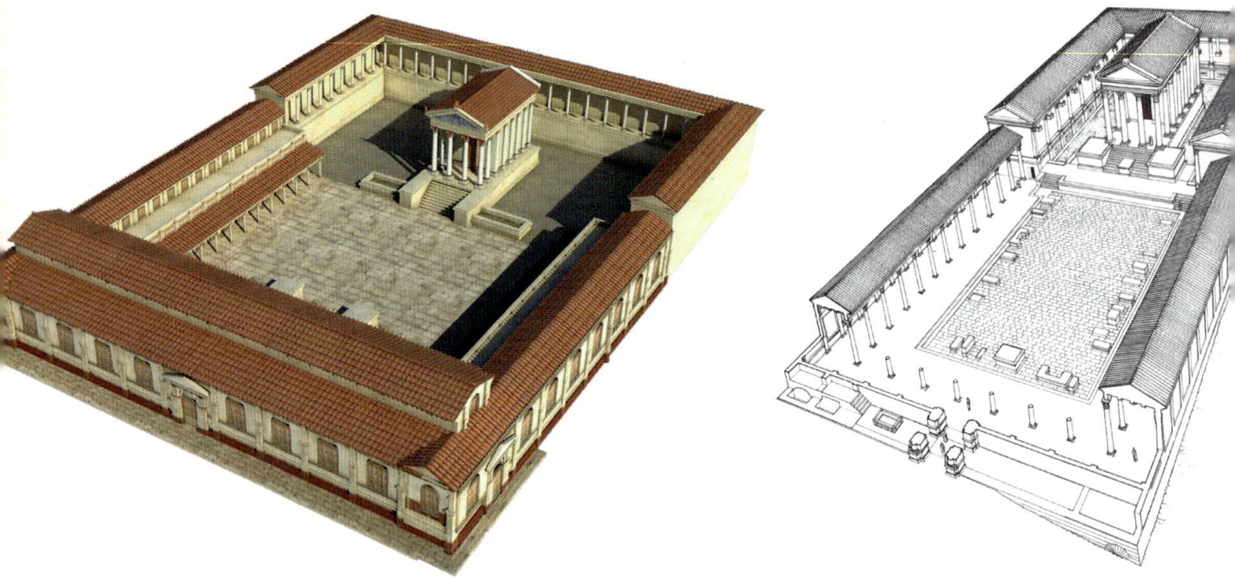

Fig. 18.4 *Left*: Reconstruction of the forum in *Ammaia. Right*: The Flavian forum of *Conimbriga* (Correia 2009)

a "tetrastylos" temple, resulting in a 1:2 ratio for its ground plan/outline. The colouring of the structure is based on the fact that the Roman citizens had a more or less uniform culture during the Early Empire, and according to that, frescoes from *Pompeii* and its surroundings were used as a starting point. An interesting example for comparison was also the reconstruction of the colouring of the *Ara Pacis* in Rome (Foresta 2011).

It is not known which type of columns was used for the forum temple; they could be either Ionic or Corinthian. Therefore, two versions of the temple reconstruction were made, showing the mentioned orders (Fig. 18.5).

The Corinthian decorations were first made in high-resolution mesh – high poly in 3ds Max and imported into a specialised 3D program. There they were retopologised and appropriately painted. The surface information of the detailed model was projected onto the reduced one, providing a high visual quality with low computational cost. With some appropriate minor alternations, these columns were also remade into temple pilasters.

Simple Ionic capitals are a typical find on the site of *Ammaia*, but it is still possible that the forum temple was decorated with more elaborated architectural elements. A more idealistic version of this capital was modelled using the same procedure as described above. Corner capitals were also constructed.

Another example of the procedures adopted for the reconstruction of typical Roman urban monuments is the southern gate of *Ammaia*. This complex is partially preserved above ground and excavations focused on the area for several campaigns (Corsi and Vermeulen 2012: 8–9). The gate, in Portuguese defined as *Porta Sul*, consists of round twin towers, the gate itself and a square occupying the inner side. There was an arch which was removed in the eighteenth century to be integrated in the city walls of a nearby town of Castelo de Vide (where it was known as *Arco de Aramenha*) and only its base remains in situ (Fig. 18.6, left). While the base gives exact measurements of its width, taking into account a rather decorative purpose of the *Porta* and the city walls (rather than a fortification purpose), a certain height was added to support architraves and decorative elements above and around the gate, whereas the dimension of the surrounding building could be defined by the visible remains.

These measures were used to design the other elements like the city wall and the round twin towers which were built in Flavian times. The diameter of each tower is 6.30 m and their most probable height deducted from comparison should range around 6–7 m, little higher than the city wall.

Fig. 18.5 Two versions of the forum temple with Corinthian/Ionic order

Fig. 18.6 *Left*: The *Arco de Aramenha* (the original monumental southern gate to the town of *Ammaia*) inserted in the walls of the nearby town of Castelo de Vide. *Right*: The *square* at the entrance of the southern gate of *Ammaia* paved with granite blocks, where *round marks* are still visible

Behind the gate is an open space, crossed in the middle by one of the main urban axes (the so-called *cardo maximus*), *c*. 4 m wide. The square is paved with massive (approximately 1×1m) squared granite slabs forming a rectangular area of *c*. 23×12 m on each side of the *cardo*, thus reaching a total of 50×23 m.

It is not clear which was the use of this open space, but taking into account the presence on the eastern side of the square of a structure that could be interpreted as a *macellum*, it is possible that this square functioned as market. In this sense, the clearly visible round marks (holes of approximately 7 cm diameter, displayed in regular intervals 3 m from the surrounding walls; Fig. 18.6, right) could witness the presence of provisional covers, like tends or wooden roofs, used as shelters together with a possible portico which would have surrounded the square.

Fig. 18.7 *Ammaia*. Reconstructed *Porta Sul*

The reconstruction is based in full on the visual remains, documentation, excavations, 3D scan and comparative research (Porta Miñá, Lugo, Spain; *Portus Adurni*, Portchester, Great Britain; South Gate, Tiberias, Israel; *Porta di Venere*, Spello, Italy; *Porta Montanara*, Rimini, Italy) and modelled according to the already described procedure (Fig. 18.7).

All of the produced content was prepared for the rendering of the different scenes integrating them into a 3D environment. At this stage in order to achieve realism, lightning and thousands of botanical features as well as a realistic terrain surface have to be set and optimised to gain considerable render quality in balance with the computing time needed.

When attempting "technical reconstructions", for instance, if reconstructing operating cycles, effectivity and ergology of ancient tools, machinery and their use, functional presentations are devised and simulations are carried out (Fig. 18.8).

From ergological studies concerning Iron Age salt-mining procedures, transportation (wagons and chariots), over Roman water supply systems, pumps, lifting devices and building machinery and different issues are examined. These functional studies are indispensable for correct interpretation and validation of certain archaeological documentation for visualisation.

18.3 3D Animations and Movies

Since, special programs are used to achieve realistic results and breathe life into the scenes. Sophisticated hard- and software like an in-house motion-capture system developed at 7Reasons (Humer et al. 2010, 2011) are used to drive the animation of computer-generated people. This in-house motion-capture system consists of a magnetic suit, which allows the actor to perform freely in every environment. This can be used to drive a virtual character constraining him/her to the movement of a specialist or an actor performing in this suit. Optical systems are used to reproduce facial and complete body movements as well as gesture recognition.

Data cleaning and preparation must be done in conjunction to the recording to assure smooth transitions in movement and behaviour. Complete capture sessions can be applied but also smaller parts of these sessions can be used to create clips which then are utilised in turn to create a storyline similar to common film editing. These systems are applied to characters within 3D sceneries for movie production and real-time applications.

Fig. 18.8 Screenshots of the interactive real-time application

This system ensures correct movements while keeping the production costs feasible. Specialised render algorithms will enable the creation of terrain features, like plants, stones and boulders, as well as populating the scenes with animated characters.

Humanoid and non-humanoid 3D models are created with a toolset specialised for organic construction. Depending on the needs, this process can start up from a skeletal phase building up volume by applying muscular systems and ending at the texturing process.

When the desired shape of a model is reached, a kinematic bone structure is inserted with a mesh-binding which allows animation of various parts to proceed. The results can be used for highly detailed visuals as well as crowd animations to enliven reconstructed sceneries.

In the case of a short film format, a camera path is created to define the sceneries. According to the camera's clipping, scenes can be further refined and a storyline can be established (Fig. 18.9). To give a reference to human proportions and to provide a more interesting backdrop, animated human figures and artefacts characteristic for the given time are inserted.

On the other hand, the material can be used to produce interactive 3D real-time applications, allowing the user to freely move around the scenes and access information on demand.

In a standard production, the movie starts with the localisation of the site from maps, satellite images and orthographic photos leading the spectator down to the area of the reconstruction.

Then, the journey through Roman times begins with a progressive approach to the settlement, leading the viewer to the city walls and inviting him to stroll through the streets and eminent places of the town. At certain locations, real footage of the standing remains is presented to

Fig. 18.9 Screenshot of the resulting video

Fig. 18.10 Screenshots of the interactive real-time application

show what has been added through the reconstruction process.

The classical short film production will always be a major part of the media market due to its advantage of linear storytelling.

18.4 Real-Time Applications

Additionally, a real-time interactive application can be attached to these media; it allows the user to fly and walk through various scenes and inspect different objects of interest in the Roman town as well as in a virtual museum, which usually reproduces what is displayed in the existing on-site museums (Fig. 18.10).

Real-time applications are made for almost all platforms and hardware devices, and as we saw, they recreate interactive scenery where the spectator is invited to discover reconstructed sites and acquire information.

The long experience gained in these fields in the course of the last decennium enables us to apply innovative solutions like archaeological sandbox systems functional to aid communication between scientists and artists. This system can also be used to display ideas of a project team to a broader audience.

The increasing quality of the current game engines will allow us to produce in near future not only real-time but also film footage without the need of rendering.[1]

18.5 Documentation

The decisions taken during the reconstruction process are documented and commented in online blogs, citing used material, comparative sources, etc. This ensures not only a good communication within the project team but also transparency, retraceability and a scholarly approach to the topics in question. With this proper base, the evolution of reconstruction concepts can be published later on in scientific papers. Without these measures, most of the work conducted would be rendered meaningless, as the main goal cannot only be to produce attractive images of historical objects and scenery but to gain new insights and knowledge.

[1] Various productions have been made handling large datasets like in the project of the Germano-Raetian limes, where over 170 km of scanned terrain data was inserted. Other successful productions like *Carnuntum*, Marvão, Caerleon and the Austrian limes show that the new trend of real-time media is very well received by the market. See Humer et al. 2011.

Conclusion

The real-time construction system has demonstrated to be a valuable tool for the rapid construction of the archaeological sites since it enables non-trained users to sketch out their reconstruction ideas in a very short time. Therefore, the development of this tool will be carried on to refine various features and will be released when it reaches a mature state.

In most products, the overall quality of the production can be evaluated of very high level, although there is some uncertainty in various aspects of reconstructions (mainly about single monuments, and especially private houses, and decoration styles), leaving room for further discussions. For these reasons, in order to include other possible interpretations, the raw data should be made publicly available.

References

Correia VH (2009) O forum de Conimbriga e a evolução do centro urbano. In: Bassarate TN (ed) Cidade e Foro na Lusitânia Romana, Studia Lusitana 4. Museo Nacional de Arte Romano, Merida, pp 89–105

Corsi C, Vermeulen F (eds) (2012) Ammaia I: the survey. A Romano-Lusitanian townscape revealed (ARGU 8). Academia Press, Ghent

Foresta S (2011) La policromia dell'Ara Pacis e i colori del Campo Marzio settentrionale. In: Rossi M (ed) Colore e Colorimetria. Contributi Multidisciplinari, vol VII A. Maggioli Editore, Milano, pp 333–340

Frova A (ed) (1990) Il Capitolium di Brescia. La Ciità nell'Italia settentrionale in étà romana. Morfologia, strutture e funzionamento dei centri urbani delle Regiones X e XI. Atti del convegno di Trieste (13–15 marzo 1987). École Française de Rome, Rome, pp 341–363

Hauschield T (1992) El templo romano de Évora. Cuadernos de arquitectura romana 1:107–117

Humer F, Pregesbauer M, Vermeulen F, Corsi C, Klein M (2010) 3D culture database Carnuntum. In: Ioannides M, Fellner D, Georgopoulos A, Hadjimitsis DG (eds) Digital heritage: proceedings of the 3th international conference dedicated to digital heritage, Limassol, 8th – 13th November 2010, Archaeolingua, Budapest, pp 91–93

Humer F, Gugl C, Pregesbauer M, Vermeulen F, Corsi C, Klein M (2011) Current productions: Carnuntum, German limes and radiopast. Virtual Archaeol Rev 2(4):131–137. http://varjournal.es/doc/varj02_004_16.pdf

Principles of Seville (2012) SEAV. http://www.arqueologiavirtual.com/carta/wp-content/uploads/2012/03/BORRADOR-FINAL-FINAL-DRAFT.pdf. Accessed 1 Jan 2013 van Roode S, Corsi C, Klein M, Vermeulen F, Weinlinger G (2012) Radiography of a townscape. Understanding, visualising and managing a Roman townsite. In: Kluiving SJ, Guttmann-Bond E (eds) Landscape archaeology between art and science: from a multi- to an interdisciplinary approach. Proceedings of the 1st landscape archaeology conference (LAC2010, Amsterdam 26th – 28th January 2010). Amsterdam University Press, Amsterdam, pp 429–441

Vermeulen F, Corsi C, De Dapper M (2012) Surveying the townscape of Roman Ammaia in Portugal: an integrated geoarchaeological investigation of the forum area. Geoarchaeology 27(2):123–139

Preservation of Virtual Reconstructions

19

Daniel Pletinckx

Contents

D. Pletinckx
Visual Dimension bvba, Ename, Belgium
e-mail: daniel.pletinckx@visualdimension.be

19.1 Six Issues to Deal with

The London Charter provides a very complete and well-structured framework to carry out documented 3D visualisation of complex cultural heritage (CH) objects, such as objects of art, man-made structures and historical landscapes. When focusing however on the practical implementation of the preservation of such digital visualisations, we need to deal with six major issues, which are:

- Lack of methodology to document and exchange 3D CH objects
- Lack of communication methodology
- Lack of stimuli to document and preserve
- Lack of long-term storage and digital preservation strategies
- Lack of business models for reuse and exchange
- Lack of updating methodology

The *first issue* is still very basic: *we still do not have any tradition or adopted methodology or standard for the way we document the creation of a 3D visualisation.* Although the London Charter (http://www.londoncharter.org/, see Chap. 15 by Denard in this volume) outlines very well the principles, we need a more practical methodology that can be adopted by the majority of people involved in 3D visualisation. There are already some initial guidelines, for example, on implementing heritage visualisation in Second Life (The London Charter in Second Life http://iu.di.unipi.it/sl/london/) or on general documentation of interpretation processes (paradata) in

C. Corsi et al. (eds.), *Good Practice in Archaeological Diagnostics*, Natural Science in Archaeology, DOI 10.1007/978-3-319-01784-6_19, © Springer International Publishing Switzerland 2013

Fig. 19.1 The archaeological remains of Monte Michele tomb 5

3D visualisation (Interpretation Management 2010) developed within the EPOCH Network of Excellence. But we need more good examples and best practices on how to take on such documentation activity, on what tools to use and on the workflow to follow. We need major involvement by the community to reach consensus that 3D visualisation and its documentation are a normal part of cultural heritage practice.

But these guidelines need to deal not only with the lonely researcher that creates such 3D visualisations of complex cultural heritage but also with geographically distributed multidisciplinary teams. In other words, *exchange* of such documentation and methodology to *collaborate* are essential elements in the practical implementation of a documentation and preservation strategy. This is clearly linked to the capabilities of the tools used. The InMan methodology (Interpretation Management 2010) as developed within EPOCH used a wiki as medium for the documentation process because of its discussion and versioning capabilities. Recent experiments for the 3D visualisation of the Abbey Theatre in Dublin (Abbey Theatre Blog: http://blog.oldabbeytheatre.net/) and the Etruscan Regolini-Galassi tomb used a blog to record the interpretation process and related visualisation issues and to stimulate discussion and consensus creation amongst the group of 3D experts and a wide variety of cultural heritage experts.[1] Although the idea of a peer review process of 3D models was coined already several years ago by Bernard Frischer in the SAVE concept (Serving and Archiving Virtual Environments: http://vwhl.clas.virginia.edu/save.html), very little experience is already available on how exactly to implement such a review process, including source assessment, evaluation of 3D models, discussion amongst peers and improvement of the 3D model (Journal for Digital Application in Archaeology and Cultural Heritage: http://www.journals.elsevier.com/digital-applications-in-archaeology-and-cultural-heritage/). Moreover, a peer review process should be based upon the London Charter and use metrics that reflect the London Charter principles. This still needs to be defined and a sufficient degree of consensus needs to be built on the methodology and implementation.

For example, the reconstruction of an Etruscan funeral chariot from tomb 5 of the Monte Michele necropolis (Fig. 19.1) in Veio (Italy) that we are currently doing needs a wide range of experts. The reconstruction work is performed by a restoration expert,[2] in close cooperation with the

[1] Blog on the 3D visualisation of the Etruscan Regolini-Galassi tomb: http://regolinigalassi.wordpress.com/

[2] Blog on digital restoration: http://worldwidemuseum.wordpress.com/

Fig. 19.2 The virtual reconstruction of the funeral chariot – conceptual model

archaeologist that excavated the site and an expert on Etruscan funeral rituals, who studied in detail funeral chariots (the so-called Tomba del principe sabino: http://www.principisabini.it/). Together with the virtual reconstruction of the chariot (Fig. 19.2), digital restoration of the bronze coverings of the chariot is made. The result of the virtual reconstruction and the digital restoration is passed on to an expert of serious games who turns them into an interactive museum setup with real-time 3D visualisation, see blog on the virtual reconstruction of the Etruscan Monte Michele tomb 5 (http://montemichele.wordpress.com/).

Practice shows that several details, which are crucial for virtual reconstruction and 3D visualisation, are not well recorded by archaeologists. Documenting the interpretation and reconstruction process in detail can show archaeologists which archaeological information is required, so that archaeologists can improve and optimise their way of recording.

This brings us to the *second issue*, which is the *communication of 3D models of cultural heritage objects for collaboration, scientific publication and public use*. We need to make a clear distinction between these three goals, as they have different dynamics and requirements.

Within *collaborative research*, 3D models and their linked paradata need to be passed on from one expert to the other, for study, review and adaption. In practice, this turns out to be quite difficult not only for technical reasons (ownership and knowledge of tools and 3D software) but also for organisational and psychological reasons. Our experience is that it works much better if one central person or team deals with the 3D models and paradata, while experts are consulted to contribute in their domain of expertise. This is also how the multidisciplinary team of Robert Vergnieux at Archéotransfert (http://archeotransfert.cnrs.fr/) deals with 3D visualisation projects with great success. Using a blog is a useful instrument in this process, as mentioned above, but our observations are that the blog needs to be private (i.e. limited to the research team), as experts are reluctant to contribute on a public blog as they see this as a kind of publication with final conclusions, while the contributions are ongoing research, of a volatile and progressive nature.

Scientific publication of 3D visualisation projects is quite common, but very few of these publications allow one to inspect the 3D results interactively in 3D. As most publications result in a PDF file, we can use the 3D capabilities of PDF, which have matured significantly over the last 7 years. PDF is an open format, standardised as ISO32000-1 (an update of this ISO standard, i.e. ISO32000-2 aka PDF2.0 is in preparation). When authoring PDF documents, one can easily add 3D models into a publication from a wide range of 3D formats.[3] Other technologies, such as HTML5, are widely available and start to be standardised to publish 3D online. The uptake of this simple approach however is hampered on one hand by the lack of 3D models in archaeological and historical research and by the lack of education in the cultural heritage domain on how to use 3D in research and documentation. Once there is a sufficient amount of 3D models created and used within the cultural heritage domain, there will be much more pressure to deal with proper 3D publication and digital preservation processes.

[3] Annual Report 2010 of the Netherlands Institute for the Near East (http://www.nino-leiden.nl/doc/AnnualReport NINO-NIT 2010_3D.pdf)

The most common purpose however for 3D cultural heritage data is *public use*. This means that a certain body of scientific results is used to show cultural heritage objects to the public in exhibitions, online or on TV. In this *public use* phase, the focus needs to be on the translation of those scientific results into a 3D visualisation that uses a certain medium (website, serious game, video, TV programme, etc.). Each medium has a specific language, and the creation of results for public use from 3D cultural heritage objects needs specialists who master that language. Most cultural heritage experts do not take this fact into account and have 3D visualisations produced that are perfect for scientific communication but not for public presentation in a certain medium (e.g. as an interactive application with storytelling – Etruscanning3D interactive application with natural interaction interface: http://www.youtube.com/watch?v=iiW4dbfo5yU). Visualisation in 3D, based upon scientific research, would gain a lot more credit and support if it would produce appropriate forms of visualisation.

The CARARE project (http://www.carare.eu/) is delivering cultural heritage objects (archaeology and monuments) to Europeana (Europeana digital library on culture: http://europeana.eu/), partially in 3D. All of this 3D content does exist already but needs to go through a publishing cycle in which PDF is used for most content. Although part of the content will use 3D PDF simply as a file format that can be displayed on every computer and operation system, another part will need curated objects into which 3D is integrated in a document, with links from the text or the photographs to the 3D (3D PDF – CARARE: http://carare.eu/eng/Resources/3D-Virtual-Reality).

Many cultural heritage objects need to be reduced in resolution or complexity to be viewable online. Practice shows that too little effort is done to preserve the original high-resolution data, while the low-resolution public version is preserved, probably because it has much more visibility.

Efforts for establishing an *implementation framework* for 3D cultural heritage objects such as the Seville Principles (http://www.arqueologiavirtual.com/carta/, see Chap. 16 by Lopez-Menchero in this volume) need to make a much clearer distinction between these three uses and identify the different processes that are involved. In the research phase, the focus needs to be on collaboration tools to support annotation, discussion and consensus building. In the publication phase, the focus needs to be on optimal communication, linking the argumentation to the 3D models and passing on the 3D models for further research within the cultural heritage community. In the public use phase, the focus needs to be on transferring the 3D models and their relevant paradata to communication specialists and using the right visual language of a certain medium to convey the story that is told by these cultural heritage objects.

The *third issue* is how to *ensure that documentation and preservation of complex cultural heritage are made*. Practice shows that in most projects, over 90 % of the work goes into the analysis and interpretation of the data, while less than 10 % goes into 3D modelling and texturing. Hence, failing to document and preserve the visualisation process results in the loss of at least 90 % of the invested money. For research projects that receive funding, documenting the visualisation process should be compulsory. Other projects within a more commercial context (e.g. commissioned by a museum to a company) can do the same as most or all of the budget for 3D visualisations is public money. In other words, we need to focus today on creating regulations that make documentation and preservation of digitally born cultural heritage objects a condition for funding or commissioning.

The *fourth issue* is how to *ensure that all these 3D visualisations, 3D models and their related paradata are stored for long-term use*. This issue of course deals directly with several technical preservation issues, such as the file format of the 3D models and the documentation of all related files (textures, bump maps, etc.), for which strategies are at hand. But technical preservation issues are only a part of the problem. Although universities and companies can exist for a long

Fig. 19.3 The abbey and village of Ename in 1065 AD

time, research teams and company teams are very often quite transient. This means in practice that universities or companies are not the right place to store complex cultural heritage objects. In our opinion, storage in a repository at the national level should be compulsory (see, e.g. the ArcheoGrid repository (http://archeogrid.in2p3.fr/) at the national level in France). Storage at such a repository should be subject to a selection and prioritisation procedure on what to preserve and what to let go, to limit the cost of registration and storage. Ownership and IPR should be clear at least at the moment of storage.

The *fifth issue* is the *creation of a model for possible reuse*. It is conceivable that 3D cultural heritage objects should be available for free in low resolution to the public (e.g. in Europeana: http://europeana.eu/) but can have a paid use for high-resolution versions. Museums can use and exhibit digital museum objects from other museums, digital publications can incorporate high-resolution 3D digital objects (for which royalties will be paid, just like for professional photographs today), and even film companies and game developers could pay significant fees to use scientifically correct 3D models of historical buildings and objects. The V-MusT.net project (V-MusT.net Network of Excellence: http://v-must.net/) is developing a business model and practical implementation for exchange and reuse of digital museum objects and virtual environments. In most cases, reuse will invoke changes to the 3D asset, as textures could need improvement (for use in film) or geometric complexity could need to be reduced (for use in games) or extra metadata will become available through use in temporary exhibitions.

An additional aspect is that efficient reuse only can happen if the creation and structure of the 3D asset are well documented. For example, the historical landscape reconstruction of Ename (Belgium) in 1065 AD (Fig. 19.3) can only be reused as a real-time interactive system if the simulation of the landscape (through vegetation maps and ecosystems in the software Vue) can be regenerated in a highly optimised way on a serious game platform (such as Unity3D) thanks to the extensive documentation.

Finally, the *sixth issue* is *how to keep 3D visualisations of cultural heritage objects and their related paradata up to date*. Nearly all digitised cultural heritage objects are static as they represent the physical object as truthfully as possible. Practice however shows that 3D visualisations are not static at all, they are based upon sparse data, and hence they change because of the availability of new research, better excavations or new insights in the use and meaning of objects. The ideal situation would be that any 3D visualisation research by a certain team could be taken up by any other team that continues to improve and complement the results of the first team. If that ideal situation were present, updating 3D visualisations would be quite a natural thing to do. However, this ideal situation is still a distant dream. We can bring that ideal situation a bit closer by dealing with all the issues described above. But still, there will be an important issue of cost, as we need to balance the available resources. What is the use of documenting the finest detail of a 3D visualisation project if that documentation is never used again? In other words, we still need to find out what the optimal amount of documentation and preservation should be, so that long-term overall costs are minimised. This is still uncharted territory and needs more research and practice.

Conclusions

The conclusions of this short chapter are quite simple: the London Charter provides an excellent framework for the digital preservation of complex cultural heritage objects, but we need to focus now on putting the principles of the London Charter into practice. This means that we need to collect the *best practices* by analysing existing projects to find out which approaches do work and why.

This also means we need – based upon the conclusions of these best practices – to define *optimal workflows and specific requirements*, so that documentation and preservation of complex cultural heritage objects become an integrated part of cultural heritage practice. Crucial in this process is the definition of *quality* for the documentation, communication and preservation steps, as described above, and taking care of an appropriate balance between resources, results and the impact of those results.

Finally, we need to realise the *uptake* of such workflows and requirements, first of all by integrating them as soon as possible into the *curriculum* of students in the cultural heritage domain such as archaeology, history, anthropology, monument care and museology. Another major step into the uptake of such documentation, communication and preservation strategies can be the *competence centres* that involved European projects such as V-MusT.net (http://v-must.net/) and 3D-COFORM are setting up to support cultural heritage institutions and their partners.

And why not look into *expanding the London Charter* with clear guidance on these processes, based upon a wide consensus in the cultural heritage domain, so that it can acquire the status of a real Charter that governs documentation, communication and preservation of digital cultural heritage objects?

References

Interpretation Management (2010): EPOCH Knowhow book on Interpretation Management. (http://media.digitalheritage.se/2010/07/Interpretation_Managment_TII.pdf). Accessed 23 Sept 2013

The London Charter. http://www.londoncharter.org/. Accessed 23 Sept 2013

The London Charter in Second Life. http://iu.di.unipi.it/sl/london/. Accessed 23 Sept 2013

Best Practises for a Sustainable Management Plan: The Case Study *Ammaia* in the European Context

20

Sigrid M. van Roode

Contents

S.M. van Roode
Past2Present, Woerden, The Netherlands
e-mail: sigridvanroode@yahoo.com

20.1 Introduction

One of the goals of the Radio-Past project has been to develop a site management plan that would ensure the scientific management and preservation of the site, as well as the creation of a solid community basis for the site management plan (van Roode et al. 2012). In this chapter we will, with *Ammaia* as the case study, identify the main aspects that need to be taken into consideration when combining scientific and cultural potential and produce a few practical suggestions when planning for a sustainable site management plan. In these considerations, the focus will be on the aspects that will most likely prove to be most vital for the conservation of the site: the position a heritage site has in the local community, taking the current economical state of affairs in the region into account, and its value as heritage commodity.

20.2 Value-Based Management

Traditionally, archaeological research has been carried out with as the main focus enhancing our knowledge of the past. Sites were regarded and subsequently treated as 'knowledge quarries'; once the excavations were over, the site itself was left as it was. The context of a site was studied in a context of the past, not of the present, and in some regions the local inhabitants were seen as intruders in a scientific expedition rather than as owners of the area. In the last decennia, a gradual change has been set in motion.

C. Corsi et al. (eds.), *Good Practice in Archaeological Diagnostics*, Natural Science in Archaeology,
DOI 10.1007/978-3-319-01784-6_20, © Springer International Publishing Switzerland 2013

As a result of increasing knowledge and understanding of other cultures, archaeological sites are being seen in their contemporary cultural context as well. They not only fulfil a function in archaeological science but have also economical, touristic and traditional values in a community. The question has arisen: 'Who owns the past?' The scientific community, who can interpret the remains of the past in a broader context? Or the local community, who has lived with the remains for sometimes many centuries, spun its own oral traditions around it that are in turn to be seen as intangible heritage?

A heritage policy, even if this is related to one site in particular only, not only should address the management of the physical condition of the site to ensure its scientific potential for the future but also should incorporate the place the site has occupied in the social landscape, its possibilities for educational and touristic purposes and its relation to other important aspects of the region it is located in Bos (2006). Ideally, a site management plan should search, find and solidify the link between the past and the present (Torre 2002).

A research project, executed by the Getty Conservation Institute from 1995 onwards, has clearly shown the need for value-based management. This new approach to heritage management has been accepted worldwide and has been in development ever since, gradually becoming the standard in heritage management. In Dutch archaeology a comprehensive evaluation system is used as part of the archaeological process and integrated in the Quality Norm Dutch Archaeology (Kwaliteitsnorm Nederlandse Archeologie, usually referred to as the KNA). Heritage management of Dutch archaeological sites handled over the past 10 years is mostly based on enumerating and interpreting these listed values (Bosman and van Roode 2007). Portugal also uses an evaluation system that is less refined but nonetheless clarifies the criteria that Portugal recognises as important for the determination of the value of its heritage (Zihão 2004). The importance attached to value-based management by the scientific community is also reflected in the criteria for international heritage as formulated by UNESCO. In order to be included on the World Heritage List, sites must be of *outstanding universal value* and meet at least one out of ten selection criteria. These criteria are explained in the 'Operational Guidelines for the Implementation of the World Heritage Convention' which, besides the text of the Convention, is the main working tool on World Heritage. At the Getty workshop 'Management Planning for Archaeological Sites' held in Corinth, Greece, in 2000, the importance of the identifications of these values was once more stressed (Teutonico and Palumbo 2002). Heritage policies, be it regarding archaeology or broader cultural heritage, are nowadays all based on a valorisation of several pre-established criteria. Many evaluation systems of archaeological heritage have been developed worldwide, and they are used as the basis of heritage policies on different national and international scales. Of course, the perspectives of these systems vary considerably, and there is no unbiased classification available by which heritage may be evaluated. The classifications are usually determined by the management objective of the organisation or field of expertise designing it.

When studying the different evaluation systems used in cultural heritage management, it becomes clear, however, that the perspectives from which these systems are set up vary considerably and differ between countries or even cultural heritage fields. In order to draw a sustainable heritage policy, the different values of a site need to be identified. These can largely be divided into two main groups: scientific values and socio-economic values. Identifying these values however is, though time consuming, not the main challenge of a site management plan: combining the two should be the core of any effort towards a sustainable plan (Avrami et al. 2000).

20.3 Identifying Scientific Values: The Example of *Ammaia* (Alto Alentejo, Portugal)

The archaeological potential of the archaeological site of *Ammaia* is regionally significant but modest in a broader European context. *Ammaia* is not one of the larger or famous cities of the Roman Empire and certainly not the only one in its kind. As it is mainly a green site, however, its research potential for new non-destructive survey techniques is very high (Corsi and Vermeulen 2012). This agrees fully with the Treaty of Valletta, in which is conveyed that preservation in situ is preferred over excavating and thus destroying a site (van Roode 2008). By developing new techniques and refining the existing methods, *Ammaia* has the potential to become the case study for research that is compliant with the aims of the Treaty of Valletta: leaving heritage in situ where possible. The research potential of *Ammaia* has been described in detail in the 2008 Master Plan (Vermeulen and Corsi 2008).

One of the significant aspects of *Ammaia* is its location on a point of intersection of two different transportation routes: roads systems and water systems. This fact places the study of *Ammaia* in the context of the landscape as perceived and used by the Roman community (Vermeulen et al. 2005). The main road connects *Ammaia* with *Augusta Emerita* (Mérida) in Spain and *Olisipo* (Lisbon) in Portugal. As such, *Ammaia* formed an important station on one of the main roads covering the Iberian Peninsula. Many secondary roads have also been discovered, indicating the industrious use of the landscape in antiquity through the use of its natural resources: both stone quarries and gold and silver mines have been discovered.

The water systems in this urban site are of equal importance. Two main rivers, the Sever and the *Tejo* (Tagus), run through the area and form important waterways for transportation and trade. The natural springs, occurring in abundance in the area, made for an ideal settlement location where fresh water was largely available. Both the road systems and the water systems are important scientific values (Oliveira and Balesteros 1989; Roxo and Afonso 2001). They offer the possibility to study an urban site in its socio-economic and natural context, identify the relation between the two and thus provide us with detailed insight on how a Roman town used the landscape and its natural resources both internally and externally and how the town itself was related to other Roman settlement sites in terms of contact, cultural exchange and trade.

20.4 Identifying Socio-Economic Values

An archaeological site can be perceived on two levels of existence: first, there is the physical site, physically located in a certain geographical position. Second, there is the site as community space, connecting archaeologists, inhabitants, visitors and possibly even interested public from around the world that never physically visit the site itself. The community space revolves around the concept of the site as it is shared by both professionals and public alike. In order to create a sustainable management plan, it is of importance to assess the socio-economic factors surrounding the physical site as well as the aspects of the site as community space. First, we will assess the socio-economic factors of the physical site, after which we will explore possible aspects of the site as community space. Following this train of thought, we will also ascertain their possible active function in creating such a community space.

In order to form a clear view of the social context in which to study an archaeological site, it is important not only to address its immediate surroundings, such as the neighbouring settlement of S. Salvador de Aramenha or in other cases inhabitants of the site itself, but also to study the larger economical and political climate in which the site is to be managed. After all, a site is an unalienable element of contemporary society, even when in neglected condition: investigating the larger perspective will offer insight into the factors that determine either success or failure of site management. In order to award an ancient site a distinct place and possibilities, we have to investigate both the legislative context of the site and the region in which it is situated.

The site of *Ammaia* lay within the borders of the municipality of Marvão, in the district of Portalegre, the region of Alto Alentejo.

20.4.1 Legislative and Organisational Context

In order to protect any heritage site, its legislative context is of as much importance as its scientific value. It is pivotal to understand the local and national laws and regulations in order to ensure accurate protection. In addition, the way the organisation surrounding the maintenance of the site is structured is also of importance to be clearly understood before attempting to implement any management plan at all. In the case of *Ammaia*, Portugal has national legislation on archaeological heritage. The country has signed the Treaty of Valletta in 1992, but this was implemented into legislation only after the near destruction of rock art engravings as a result of the planned Coa Dam in 1995. The water reservoir after the construction of the dam would destroy many engravings, and the project was, after years of heated debate over the value and protection of these remains, finally called off in 1997. In the wake of this controversy, the IPA was created: the Portuguese Institute of Archaeology. This has in 2002 been fused into the IPPAR, and finally in 2007 the now existing institute, the IGESPAR, has come into existence.

The site of *Ammaia* has been appointed a national monument in 1949. Due to its location in a natural park, its development will also have to take into account applicable legislation concerning natural and ecological resources. The owner of the land is a foundation, the *Fundação Cidade de Ammaia*, the organisation created to manage the site. The main partners in the Foundation are the University of Évora, the municipality of Marvão and the Natural Park of Serra de São Mamede. The main objectives of the Foundation are the valorisation of the Roman site and the development of cultural tourism. However, with regard to ownership, the owner of the land only owns the topsoil; any archaeological remains automatically fall to the State. Any excavations foreseen on the site will have to be approved by IGESPAR. Although the site is a national monument, this does not imply a restraint on research: as long as the research plans are approved by IGESPAR, excavating is allowed. After the excavations, IGESPAR has to approve the reports of the results.

20.4.2 Regional Attractions

The Alto Alentejo region in itself is visited by specific visitors. The vineyards, medieval towns and splendid nature are the most cited reasons for a visit to this region. Demographically, Alto Alentejo is the least populated region of Portugal and struggles with the current economical crisis even harder.

The landscape of Alentejo is varied and hosts several natural parks, in which outdoor activities such as hiking take place. The hills of Alentejo are home to the bed of the Tejo river and a variety of flora and fauna, transiting graciously in the adjacent deserted plains of Spain. Several lakes and dams provide waterfront flora and fauna and offer recreational possibilities. Interspersed throughout the landscape, Alentejo has a great number of small heritage sites to offer with remains of different historical epochs. Medieval hillfort sites and Stone Age sites are predominant, although also a number of Roman sites are present. The small, secluded villages in the region have largely maintained their historic plan and buildings and show village life in a more traditional way than, for example, at the major tourist destinations such as the Algarve coast. In addition to natural resources, the region has developed a flourishing wine tradition over the last decade. Alentejo wines have conquered a growing segment in Portuguese viticulture.

Visitors to the Alentejo region indicate a set of factors that contributed to their positive experience. Among these are the renowned hospitality and mainly the possibility to rest and recover from work- and environment-induced stress by enjoying the peace and tranquillity. In addition, the slow pace of living and the still very traditional life in Alentejo are also mentioned as a possibility to overcome alienation and loss of 'belonging' caused by modern society and an attempt to reconnect to a 'locus' with its own clear identity. Due to the vastness of the region and the low inhabitancy rate, it is easily negotiable by car, and parking spots are found in abundance. The Alentejo region also has a few drawbacks that need to be considered. Public transport is not optimal, and accessibility of the heritage sites is therefore limited. Heritage sites are often insufficiently prepared as a tourist attraction, lacking complementary services, multilingual information and local and regional road signage for site promotion. In order to visit a more remote site such as *Ammaia*, the visitor needs to know in advance that such a site exists. The cooperation between heritage sites might also be improved: for example, *Ammaia* as a Roman town has no information on the next Roman site, Alter do Chão, which is only 50 min away by car. Information on the next large Roman site in Mérida (Spain) is also absent. Due to the demographic factors in Alentejo, the population is aging and a lack of qualified human resources is a result. Generally for this region there are very few people who speak another language than Portuguese. Even in the tourist industry, it is sometimes difficult to find people who speak English, French or even Spanish. This also reflects in the availability of books and brochures about the history of the region: almost all of these are in Portuguese. Bookstores in museums, larger cities or even the Lisbon international airport provides little to no items in other languages. Obviously, since the majority of visitors are Portuguese, there might not seem to be a need for information in other languages. Offering this extra service however might attract more visitors and help stimulate the economy in this region.

20.4.3 *Ammaia* as Community Space

Currently, the function of *Ammaia* as community space is marginal. The site itself is visited mainly by tourists and school children; local inhabitants of the village rarely visit either site or museum. During the course of the project, several interviews were conducted with local farmers and villagers. These interviews were conducted on the site's premises, either in the museum or on site. From their accounts, it became clear that they had a very different view of the site. They perceived it not so much as part of their community of even village but as separate entity that was more related to scholarly and scientific factors than to their daily lives. For several of the persons interviewed, the occasion of the interview was the first time they had ever set foot in the museum: they had always interpreted the museum as a place for 'learned' people but were pleasantly surprised to find artefacts on display that had an immediate connection to their daily life, such as the Roman olive press or agricultural tools (van Roode 2009).

From a larger perspective, *Ammaia* is directly connected to the hilltop village and castle of Marvão. The medieval city has been built with spolia from the Roman city, and many Roman blocks can be seen in the houses and infrastructure of Marvão. The community living in Marvão is with ca. 150 inhabitants very small and does not experience *Ammaia* as part of their social surroundings. The fortress of the city is used as their community space: festivals and gastronomical and musical events all take place within the fortress. *Ammaia* has incidentally been the backdrop for classical concerts, but the site is not much used to create and bind a community. The infrastructure of the site does allow for this use: the paved square in front of the magnificent southern gate ('Porta Sul') forms almost a natural and intimate podium for community activities such as concerts but also other artistic performances such as recitals or small-scale, local fairs. In the past, this location has been used for small classical concerts, a concept that could be elaborated upon further in the future.

In the non-physical realm, there are several initiatives to create a community around *Ammaia*. On Facebook, the Fundacão has a page, but there is almost no activity on that website. Facebook also has a community page dedicated to *Ammaia*, which is also not maintained. In the context of the Radio-Past project communication programme, a virtual community has been created on Facebook around the scientific aspects of the project, i.e. non-destructive prospection techniques. Through regular updates, this page attracts more traffic but also targets a very specific community. Ideally, social media would present a platform where visitors of *Ammaia* can share experiences, stories, photographs and information for those interested in the site but are unable to visit.

20.5 Creating Sustainable Heritage Management

- A heritage policy serves the following objectives:
- Outline the roles and responsibilities related to the management of the site.
- Create a clear set of guidelines by which to manage the site.
- Ensure the physical protection of this heritage site.
- Create a sustainable environment to vitalise responsible tourism to *Ammaia*.

It is important to realise that a heritage policy by itself cannot set the proper management of the site in motion. It is a management programme, outlining the objectives for the future and providing instruments to realise these. The success of the site management of *Ammaia* will mainly depend on the way the heritage policy is implemented, evaluated and amended when necessary. Therefore, much attention has been paid to the stakeholders of the site and the values they attach to *Ammaia*. For whom are we preserving and managing this site? Many of the management choices will derive directly from the values that the various stakeholders attach to the site. Only if the stakeholders recognise their own values in the management plan, they will be inclined to implement and use it and in doing so transforming it from a paper plan to an active and appreciated working method that serves as an example for comparable sites. As *Ammaia* is largely invisible from the surface, these values will be not only purely archaeological or scientific but also touristic or economical in nature.

In addition to a thorough stakeholder analysis, perhaps of even greater importance is to identify and acknowledge the socio-economic factors within which the heritage policy is to be developed. As has been illustrated earlier in this chapter, the Alto Alentejo region in which *Ammaia* is located has suffered tremendously in the preceding decennia due to a steady decline in jobs and, subsequently, population. This will inevitably have consequences for the scale of perceived actions to be taken for the *Ammaia* site management plan. In the following paragraphs, we will investigate both the scientific and touristic potential of the site, compare those with the current state of affairs and the tangible possibilities for the near future and determine the best course of action to be taken.

20.5.1 Scientific Potential

The research potential of the site, when defined from an archaeological, scientific point of view, revolves around three main points of interest: the road system, the water system and the use of natural resources in the immediate area.

When seen from a technical scientific point of view, the research potential of the site is even larger. Not only has it been proven possible to elaborate and refine methods of non-destructive prospection methods, but it also forms an excellent case study to monitor deterioration of archaeological remains of a complex nature in situ. The subterranean parts of the site are well preserved under a layer of colluvium. The main building materials used are locally quarried schist and granite. Due to the nature of the soil, the expected remains are of hard and solid material such as pottery, glass and metal. Organic materials, such as bone material, wood and textiles, are unlikely to have been preserved. The Roman ruins have been dismantled to be used for the city of Marvão, so the plans revealed by remote sensing are generally showing the ground plans of buildings and streets in the city on foundation level. The climate is extremely dry, especially in summer. In winter some heavy precipitation occurs. It is unclear how much damage has been done by illegal digging in the past centuries (Vermeulen and Themudo Barata 2010). Local stories mention the presence of hidden treasure and some even relate the excavation of gold objects. The museum also owns a set of objects of excellent quality that are said to come from *Ammaia* but were given to the museum as legate by an elderly woman who had possessed the objects that were acquired by her father. Where they come from on the site is unknown. The museum also received a donation of circa 20 Roman coins, which were collected by an old man who as a youth had learned from his father to find coins near a natural spring, where they were retrieved 'much like the gold washing in the

United States'. In addition to the study of the buried remains, a study has been conducted into the preservation of the visible remains. Notably the Porta Sul and the podium on the forum have been studied as well as the many blocks of stone gathered around the museum and the remains visible in the trenches of earlier excavations that have not been backfilled after the research was completed. The study shows that the visible remains suffer from natural erosion and need to be stabilised.

The research potential in *Ammaia* is not limited to the site itself, but extends to the laboratory already existing on site. The laboratory has been installed with the help of the University of Évora and is fully equipped with research tools. Microscopes, hoods and professional sandblaster sets are all present and now finally used, as resources have been raised to hire experienced fellows. If funding will be granted for the future, the laboratory could advance into a regional research centre and even generate a separate stream of revenues.

20.5.2 Current State of Site and Museum

To establish feasible possibilities, it is important to first gain insight into the current state of affairs at both site and museum. The site consists of mainly agricultural land, strewn with trees, intersected by the road to Marvão. Although the site is large, only a few archaeological remains are visible to the visitor (Oliveira et al. 2007; Stilow et al. 2009); the rest is dominated by the beautiful landscape and shady trees. These remains are located some distance apart, so that the visitor has to walk from one view to another. In between the points of interest, the visitor is on his/her own. When on site, several archaeological remains or points of interest can be seen. The most spectacular of these is the *Porta Sul*, where part of the gate towers can be seen along with the paved square with a few columns. In addition some remains of the *thermae* complex can be seen in earlier excavation trenches, and, across the road, the temple podium is visible above ground. None of the elements currently carry explanatory texts, and the interdependence of the archaeological remains and their position in the context of the city is not made clear. The site is therefore interesting but difficult to experience: the visitor has no clear idea of distances, locations and interrelation when he starts out from the museum. Due to the trees and undergrowth, there is no clear view over the site. The walk itself over the site can be somewhat troublesome: there are no clear paths, and the walk itself takes the visitor over trodden grass interspersed with stones and pottery sherds. The crossing of the road to visit the other part of the site where the temple podium can be seen is dangerous and not a clearly indicated route.

The physically challenged will have difficulties experiencing the site; the pathways are not suitable for wheelchairs, and the crossing to the other side of the road is impossible to negotiate for people who are not capable of walking. Since this is the only access point to the other side of the site, disabled visitors can effectively experience only half of the site. The museum is located in a beautiful, large, restored farmhouse. It has a small parking space in front of the building. Although the museum is signposted clearly, its exterior is relatively uninviting. From a distance, one might even state the museum looks uninhabited. The main entrance is in a small courtyard, closed off by an iron gate. The gates are locked during closing hours: the visitor however remains unaware of the opening hours of the museum, since these are only put up at the main door of the museum which is unreachable when the gate is closed. In front of the museum, the remains of a possible funerary monument have been excavated, now visible to the public, but no text is present to 'subtitle' these remains. The courtyard is traditionally undecorated and tidy. The museum itself is a pleasant surprise. It offers a selection in materials in well-lit cases. The selection of glassware found on the site has been transported to the National Museum in Lisbon; of these, only photographs on the wall remain. The museum focuses on finds only: little reconstructions or contextual information is offered. The relation between the museum and the site remains unclear. The explanations are available in Portuguese only, although the staff speaks English as well and is always eager to provide a guided tour over the site and through the museum. The museum has a small gift shop and coffee corner. In the gift shop, small items like replica oil lamps, buttons and T-shirts are sold as well as a limited number of books, all of which are in Portuguese. The Foundation has submitted a request to receive the status of public institution. This will ensure certain benefits in terms of taxes and will make it easier to find new revenue streams.

20.5.3 Established Tourist Destinations in the Vicinity

Ammaia is part of a wider touristic landscape of Alto Alentejo. The region as a whole is not as intensively visited by tourists as the Algarve and the larger cities like Lisbon and Porto. To establish the value of *Ammaia* and its unique selling points, the touristic attractions in the near vicinity have been inventoried in detail. In addition, the presence of comparable Roman sites will be inventoried. After that, an inventory has been made of the touristic amenities in the Alto Alentejo region as a whole. These will be discussed and compared below in order to identify the position of *Ammaia* in this range. How does *Ammaia* connect to these destinations? And how is it different?

In the immediate vicinity of the site are three major tourist attractions of the region. These are the fortresses of Marvão and Castelo de Vide, along with the larger city of Portalegre at a further distance. Marvão is a hilltop fortress that sits on the highest peak of the Serra de S. Mamede. This fortress used to be an important border garrison town. The village itself is small and tucked away in the outer enclosure of the fortress. The population density is low; only 150 residents still live in the village. Tourism is the main source of income. The village offers a museum in a former church and the fortress itself as the main attractions. The museum is a haphazard collection of objects, paintings and photographs, without explanations, chronological order or systematic display. The museum features the reference poster for *Ammaia* on the counter. The fortress itself is a spectacularly restored building with well-kept gardens. In the last few years, the municipality of Marvão has spent great efforts to restore the fortress and to use this location for events and festivals, such as the yearly culinary and artisanal festival which is held in the fortress. The museum in *Ammaia* is significantly better organised and displayed. The historical relationship between the Roman city of *Ammaia* and the medieval city of Marvão could be used, strengthened and made more visible. The cities of Marvão and *Ammaia* share a common past, visible in the spolia from the Roman site that have been used in some of the buildings in Marvão.

Castelo de Vide is a slightly larger town, centred around a fourteenth-century fortress. The town is well signposted and offers several tourist attractions. Within the fortress walls a small village is located that is being restored to create inspiring workshop space for young artists, local handicraft shops and other small businesses. The Jewish quarter just below the fortress is still in the original state and houses one of the oldest synagogues of Portugal. A sixteenth-century marble fountain is also present in this quarter. On the lower parts of the slope, the newer part of town is located, which has been built in the seventeenth and eighteenth centuries. Some Roman remains have been found during excavations in the city, but these are kept in the archaeological storage unit and are not on display.

Portalegre is the largest town in the vicinity. It is centred around a large fortress and has expanded into a medium-sized town during the sixteenth and seventeenth centuries, when the textile industry flourished. Also here, most of the visitors are Portuguese. As for tourist amenities, Portalegre offers a variety of shops, cafes, restaurants, etc. One or two bookstores are available; however, none of them stock books in English. The 'Oficina de Turismo', the Tourist Office, has the reference poster for *Ammaia* on display and upon request a photocopied brochure in black and white. Portalegre can be reached by public transport.

20.5.4 Comparable Roman Sites

Although the main established tourist destinations in the area offer mainly medieval sights, *Ammaia* is not the only Roman heritage site in Alentejo.

The ruins of Alter do Chão are only partly excavated and are part of the former Roman settlement Abelterium. Among them is a large mosaic with a depiction of a scene in the penultimate chapter of Virgil's *Aeneid*. This mosaic is one of 11 with a similar theme known in the Mediterranean. The site is located on the road from *Olisipo* (Lisbon) to *Augusta Emerita* (Mérida) in Spain, and so far part of the *thermae* complex, a Roman villa and a Late Roman necropolis have been excavated. The date of its foundation and its status in Roman times are still unknown. The site is located in the middle of town, which makes further excavations difficult. The site of Alter do Chão is better suited for visitors than *Ammaia*, even though the latter site is significantly larger. The explanations in both Portuguese and English are clear and use a broader perspective: not only this particular site but the situation on the Iberian Peninsula and even the Mediterranean are presented. On the map of other cities in the region during Roman times, the city of *Ammaia* is absent.

Torre de Palma is the site of one of the largest Roman villas in the Iberian Peninsula. This site is famous for its large mosaic of the 9 Muses, now in the National Museum in Lisbon. The site is also located on the road from *Olisipo* (Lisbon) to *Augusta Emerita* (Mérida) in Spain.

The next large Roman site, in both antiquity and nowadays, is Mérida in Spain. *Ammaia* is located only a few kilometres from the Spanish border; Mérida is a 2-h drive from *Ammaia*, roughly equivalent to the distance to Lisbon. Mérida contains splendid Roman ruins among which a combination of an amphitheatre and theatre, several villas, parts of Roman roads, aqueducts and city walls. In addition, the museum in Mérida offers excellent displays.

20.6 Positioning *Ammaia*: Possibilities

Following from the investigation into established tourist destinations and heritage sites in the region, several possible opportunities to position *Ammaia* as a heritage commodity come to light. Here, we will present these possibilities. Heritage presentation and promotion in the region of Alto Alentejo are mainly focused on Megalithic and medieval heritage. *Ammaia* is the perfect addition to this heritage presentation: the site and museum fulfil a natural position as information centre on Roman Lusitania.

In order to position itself as regional information centre, the museum should develop itself not only as museum but also as knowledge institution and networking hub. This will require a strong notion of cooperation, sharing knowledge and building infrastructure. This commitment is vital for a positive development of *Ammaia* as a regional treasure. In the past *Ammaia* was located on a crossroads of transport routes, both on land and by water. This connection not only is of importance from a scientific point of view but also operates as the main framework to establish new connections in the present.

The visitor characteristic of the region is also pivotal for the success of the positioning of the site. Creating a visitor experience that is not suited for or appreciated by the type of visitor in the region will undermine the basis for a site visit instead of strengthening it. Visitors to Alto Alentejo generally have a predilection for heritage and nature and display activities related to those subjects such as walking, biking, enjoying ecotourism and visiting heritage sites.

Following the characteristics of the region and its visitors, focus could be placed on the following themes that integrate both archaeological and regional elements.

20.6.1 Roman Roads: Connecting Roman Sites

In order to establish a more coherent view of the Roman past in Alto Alentejo, *Ammaia* and the other Roman sites should engage in a more active connection with each other. This can be achieved on a principal level by referring to the other sites in the various information centres.

In the museum and information centre of *Ammaia*, the presentation should encompass more than just the history of this site but offer a broader view on the Roman past as well. The visitor to *Ammaia* should be able to gather the main points of the Roman history in Lusitania from the information offered in the museum. This includes a map of Roman Lusitania, preferably in context to the rest of the Roman Empire, with its main cities and infrastructural works indicated. In addition, a concise overview of the history of the region is essential for understanding *Ammaia* and the other Roman sites in their context. In this way, a visit to *Ammaia* not only will be interesting because of the site itself but will also entice the visitors with stories of the Roman past of Lusitania: information that on this moment cannot be found elsewhere.

20.6.2 Borderland: Connecting Countries

The next large settlement in Roman times following the road from *Ammaia* is the capital of Lusitania, *Augusta Emerita*. The splendid ruins of this city can be admired in Mérida, Spain, a 2-h drive from *Ammaia*. Due to the proximity of the Spanish border and the connection with Mérida, this international connection can be elaborated upon. A shared Roman past clarifies similarities and differences in the border regions of two nations and strengthens the ties between the two. Visitors to *Ammaia* can be made aware of the possibilities to visit Mérida and vice versa.

20.6.3 A Tale of Two Cities: Connecting Marvão and *Ammaia*

Not only is *Ammaia* situated within the municipality, but the medieval city of Marvão is the historical pendant of the Roman city of *Ammaia*. The connection between the two should be strengthened and made more visible by including and promoting *Ammaia* more actively in the tourist information centre in Marvão and vice versa. Marvão is spectacularly visible from *Ammaia*, sitting high above the city on a rocky hilltop. This visual connection is a perfect basis to enhance and elaborate upon in two directions. In *Ammaia*, a sign explaining the decline of *Ammaia* and the rise of Marvão could generate more understanding of the interdependent relation between the two cities. When visiting Marvão, the visitor will have learned to look for Roman building material used in constructing houses in Marvão and have a deeper understanding of this local habitation shift.

From the ramparts of Marvão, the location of *Ammaia* in its natural surroundings is extremely well visible. In the information provided in the museum of *Ammaia* regarding the natural surroundings of *Ammaia*, this benefit of a visit to Marvão could be pointed out explicitly. In return, on the ramparts of Marvão, a small telescope can provide a clear view of the site of *Ammaia* so that the site can be enjoyed from above.

20.6.4 En Route: Connecting Visitors

Several cultural routes in Alentejo have been developed, including sometimes parts of Spain as well. By incorporating *Ammaia* actively in these routes, visitors with an interest in the cultural and natural resources of the region will be given a chance to experience the Roman past on this charming site. The history of *Ammaia* as building resource for the medieval hillfort of Marvão fits as background information in cultural routes dedicated to the Middle Ages. Information on the way the citizens of *Ammaia* used the natural resources available can be unlocked in the many natural routes leading through the region.

Pivotal for the success of positioning *Ammaia* as information centre of the Roman history in Alto Alentejo is cooperation with other heritage sites and/or institutions. The connection is essential: trying to create a position without the support and possibilities of cooperation will prove to be a fruitless endeavour. This cooperation should be organised on several levels:

- Cooperation with the municipality of Marvão is essential on an organisational and practical level.
- Cooperation with other Roman sites in the vicinity is vital to strengthen the Roman identity.
- Cooperation with the Parque Natural is essential on an organisational and practical level.

In order to implement this cooperation, a discussion group should be installed that will meet on regular occasions to discuss the progress, difficulties and solutions together. In this discussion platform the main stakeholders of the site should partake: the Fundaçao, the Parque Natural and the municipality of Marvão. In addition, depending on the subject to discuss, other partners should be invited such as the other Roman sites, the tourism agencies of other towns and cities and the developers of touristic routes. The international connection should be investigated by discussing mutual benefits with the museum in Mérida, Spain.

20.7 Implementing the Site Management Strategy

Now that the possibilities for the positioning of the site have been established, it is pivotal to start out in small, realistically feasible steps, taking the current state of affairs into account. Two of the major defining factors are both the financial and carrying capacity of the site. Due to budget constraints, large-scale planning cannot yet be carried through. The carrying capacity of the site is also a matter of further study: when no measures are taken to consolidate the visible remains and to safeguard the subterranean remains, it will be less desirable to increase the amount of visitors and traffic to the site, which in turn will have a negative financial effect.

The first steps should therefore be to create small yet important adjustments on site level, generating an increase in the stream of revenues. Among these first steps could be:

- Creating information leaflets, also in other languages than Portuguese. By offering more and comprehensible information, visitors will be more likely to share their enthusiasm and advise others to visit the site. These leaflets can be easily produced within a low budget.
- Enhancing the visitor experience by small, mainly cosmetic adjustments. Examples are posting the opening hours of the museum on a clearly visible, reachable place and adding information about the museum on the outside of the building. A start has been made with these minor adjustments in 2010, when the signposts on the site have all been repainted.
- Updating and maintaining the *Ammaia* Facebook page. By implementing a social media strategy, it will become possible to generate attention for both site and museum at no extra cost at all. Using the connections of the digital '*Ammaia* community' will greatly enhance the visibility of the site.

Foremost, however, on site level much attention should be paid to the remains themselves. The in situ remains should be safeguarded by clear indications which parts of the site are accessible to visitors and which parts are not. Excavated trenches need to be backfilled in order to protect the remains; visible structures need to be stabilised. For this, new revenue streams must be explored, such as government grants or donations.

On a local level, the search for common goals to create synergy could possibly result in mutually beneficial achievements. The research values into the use of natural resources by the Roman community of *Ammaia* could be beneficial to the goals of the Parque Natural. Visitors to the Parque could also enjoy the site as part of their walk and find common information on land use and natural resources in the Roman times and the present. The municipality of Marvão could benefit not just from one heritage highlight, the fortress, but from the visitors to *Ammaia* as well. Local shop owners in the village of Marvão could offer their traditional, regional wares in the museum shop as well, the gastronomical fair could be expanded to include *Ammaia*, and the museum could function as starting point for a walking tour, renting out equipment such as maps and binoculars and selling sustenance for a long walk. These achievements can however only be realised when the basis for cooperation is solid and the carrying capacity of the site is regarded as of primary importance.

On a regional level, *Ammaia* could be included more expressively in the network of established tourist destinations. The existing cultural routes in the region, such as the Ruta dos Vinhos or the Ruta dos Sabores, could incorporate a visit to the site as well, based on the shared Roman past of viticulture and gastronomy. The cooperation with the other main Roman sites could also lead to a shared success, for example, by creating a new cultural route focused on Roman Lusitania. The research laboratory on site in *Ammaia* can be used as a shared research facility, providing equipment that other sites do not have to obtain or hire themselves. The income generated by this facility could be used to enhance the museum and site further.

Starting out with a simple leaflet can expand into booklets and books. Beginning with a small printed map of the site and its elements can expand into larger adaptations in the museum display and the dissemination of information. Updating and maintaining one Facebook site can generate enough traffic to warrant a larger information portal in collaboration with the other Roman sites. The Radio-Past project has provided several of these building blocks: as a result of 4 years of combined study, there are now several elements at hand that can be used to benefit the site and visitor experience. A plan of the city, insight into the deterioration of the remains and a beautiful reconstruction of Roman *Ammaia* as it once dominated this location in the Severn Valley all provide information on this story in the history of Roman Lusitania. The archaeological scientific values have been compared and combined with the socio-economic values of the region and form a framework in which to operate.

20.8 Best Practices: A Conclusion

With the Roman city of *Ammaia* as case study, we have attempted to outline an approach towards a sustainable heritage management plan.

- In order for a management plan to succeed, care should be taken to assert what legal framework exists and how the protection of the heritage site can be realised within such framework. Into consideration should be taken not only legislation concerning heritage but also laws and regulations that define ownership, legal possibilities of land use and physical planning.
- The nature of the site itself is also of great importance, both intrinsically and in relation to applicable legislation, scientific values and socio-economic values. In the case of sites with visible structures, such as Hadrian's Wall or Mérida, this will be easier to assess than for sites consisting largely of subterranean remains. In this regard, care should also be taken to assess the nature of the remains themselves: subterranean remains such as walls and foundations demand a different approach than remains such as postholes and hearths. The existing legislation may permit forms of land use that are damaging to certain types of subterranean remains, while those particular forms of land use are of greater economical importance than the archaeological site is considered to be. These aspects should be investigated in the early stages of management planning for a heritage site.
- Assessing the scientific value of the heritage site. Research of the site could shed light on a certain aspect of settlement archaeology, provide valuable insight into burial customs or enhance our knowledge of the cultivation of certain plants in a given time. 'Importance' is always relative to a certain point of view. What may be important for one ethnic group or even generation may be looked upon with disdain by another. What may be important on a national level does not necessarily appeal to local inhabitants. Thus, the amount of importance attached to any subject is extremely subjective. In order to construct a reasonably reliable frame to objectify importance, scientific values should be assessed on different levels of knowledge and linked to local carriers of significance. A continuous discussion with local representatives and experts on the local history is pivotal to the success of a balanced research agenda that fills lacunae in knowledge and strengthens the local identity.
- In addition to a thorough stakeholder analysis, perhaps of even greater importance is to identify and acknowledge the socio-economic factors within which the heritage policy is to be developed.

- Last, but not least, the importance of tailoring the efforts of a site management plan to the site-related factors outlined above is often underestimated. Starting out in small, feasible steps, executed by and for local professionals in the first place, creates a growth model in which every step can be achieved, instead of presenting a grandiose but unrealistic end goal. Working together on what may appear as minor adjustments lay the foundation for elaborating upon previous successes and ensures continuing commitment.

References

Avrami E, Mason R, Torre M (2000) Values and heritage conservation. Research report. The Getty Conservation Institute, Los Angeles

Bos JEMF (2006) Position paper Karanis, Egypt. Introductory paper to the site management plan of the Graeco-Roman town of Karanis. ArcheoLogic report 411. Past2Present, Woerden

Bosman AVAJ, van Roode SM (2007) Waarderingvoor het verleden. Selectiebeleid in de praktijk. Archeobrief 3:14–17

Corsi C, Vermeulen F (eds) (2012) Ammaia I: the survey. A Romano-Lusitanian townscape revealed (ARGU 8). Academia Press, Ghent

Oliveira de J, Balesteros C (1989) Levantamento Arqueológico da Barragem da Apertadura, Aramenha-Marvão. Cámara Municipal de Marvão, Marvão

Oliveira de J, Pereira S, Parreira J (2007) Nova carta arqueológica do concelho de Marvão. Ibn Maruán14 (número especial)

Roxo MJ, Afonso do Ó (2001) Driving forces of land use changes in Alentejo and its impact on soil and water. In: RICAMARE workshop on land use changes & cover and water resources in the Mediterranean Region, Médenine, Tunísia, April 18th and 19th, 2001. RICAMARE

Stylow AU, Abascal JM, Cebrián R (2009) Marvão e *Ammaia* ao tempo das guerras peninsulares, Ibn Maruán 16 (número especial)

Teutonico JM, Palumbo G (2002) Management planning for archaeological sites. The Getty Conservation Institute, Los Angeles

Torre dela M (2002) Assessing the values of cultural heritage. Research report. The Getty Conservation Institute, Los Angeles

van Roode SM (2008) Malta in Nederland. Archeologie in de Ruimtelijke Ordening. SDU Uitgevers, Den Haag

Van Roode SM (2009) User Requirement Document RADIO-PAST website. Woerden

van Roode S, Corsi C, Klein M, Vermeulen F, Weinlinger G (2012) Radiography of a townscape. Understanding, visualising and managing a Roman townsite. In: Kluiving SJ, Guttmann-Bond E (eds) Landscape archaeology between art and science: from a multi- to an interdisciplinary approach. Proceedings of the 1st landscape archaeology conference, LAC2010, Amsterdam, 26th–28th January 2010. Amsterdam University Press, Amsterdam, pp 429–441

Vermeulen F, Themudo Barata F (2010) Hidden landscapes and ancient landmarks: some theoretical and methodological issues of studying early historical landscapes in SW-Iberia. In: Orejas A, Mattingly D, Clavel-Leveque M (eds) From present to past through landscape. CSIC, Madrid, pp 69–84

Vermeulen F, Corsi C (2008) Masterplan for scientific research and development of the Roman site of *Ammaia*. CIDEHUS, Universidade de Évora, Évora

Vermeulen F, De Dapper M, Corsi C, Deprez S (2005) Geoarchaeological observations on the Roman town of *Ammaia*. Internet Archaeol 19. http://intarch.ac.uk/journal/issue19/corsi_index.html

Zihão J (2004) Public archaeology and political dynamics in Portugal. Public Archaeol 3(3): 167–183

Part V
Conclusions

Conclusions

21

Simon Keay

Contents

The last 25 years have seen major advances in the application of non-destructive techniques to the study of archaeological sites. They have the advantage of making it possible to more rapidly assess the character and extent of larger sites and landscapes without recourse to large-scale excavation, which is costly and destructive. They have thus made it much easier to develop more incisive strategies for research, management and protection of a range of sites. This is particularly true of the larger settlements and towns of the later Protohistoric and Classical periods in southern Europe, which have traditionally been investigated in piecemeal fashion by small-scale excavations.

The Boeotia survey during the 1980s (Bintliff and Snodgrass 1988) first revealed the potential of surface collection for our understanding of the extent and chronological development of the larger Classical sites and was rapidly followed by a number of other surveys that adapted and further developed non-destructive methodological approaches. The advent of large-scale geophysical survey from the 1990s onwards was a major step here. The combination of surface collection with magnetometry and resistivity made it possible not only to identify buried streets and buildings but also to gain an idea of their chronology. The results of surveys at *Italica* (Rodríguez Hidalgo and Keay 1995), Penaflor (Keay et al. 2000a, b), *Falerii Novi* (Keay et al. 2001), *Portus* (Keay et al. 2005) and *Ephesus* (Groh 2006), amongst other sites, are good examples of this. Furthermore, the effectiveness of non-destructive surveys of

S. Keay
Department of Archaeology,
University of Southampton, Southampton, UK

British School at Rome, Rome, Italy
e-mail: s.j.keay@soton.ac.uk

C. Corsi et al. (eds.), *Good Practice in Archaeological Diagnostics,* Natural Science in Archaeology,
DOI 10.1007/978-3-319-01784-6_21, © Springer International Publishing Switzerland 2013

large Classical sites in Italy such as at *Potentia* (Vermeulen et al. 2006) has been augmented by the addition of aerial photography to the suite of available techniques, successfully building upon earlier work that demonstrated the potential of the technique in a southern Mediterranean context (Guaitoli 2003; Ceraudo 2013). Consequently non-destructive survey is thus now beginning to make a major contribution to our understanding of urbanism in the Classical world, both in the Mediterranean world and beyond (Johnson and Millett 2012; Vermeulen et al. 2012), and there is very much a sense amongst some scholars that the clarity of the results, and the speed and low cost with which they can be achieved, raises fundamental questions over the need to undertake excavations at all.

This is the context within which to understand the Radio-Past project, which was launched in 2009 and concluded in 2013. This EU-funded Marie Curie project has sought to push the methodological agenda one step further by developing "open laboratories for research and experimentation" in non-destructive approaches to the study of buried archaeological sites that involved a range of leading academic and industrial partners involved in this kind of research from across Europe. The principal "open laboratory" was established at the Lusitanian town of *Ammaia* in southern Portugal, and it acted as a test bed for the application and analysis of a full range of non-destructive techniques, both within the site and in the surrounding countryside beyond. The project results have been spectacular (Corsi et al. 2012; Corsi and Vermeulen 2012), as have been those deriving from contingent issues aired at a range of international conferences.

It is against this background that this book, *Good Practice in Archaeological Diagnostics*, needs to be judged. Rather than a unified manual as such, it is a collection of essays by European specialists that address issues germane to the full range of non-destructive techniques used in the Radio-Past project. Thus, while quite a few contributions touch upon the results of the Radio-Past itself, most notably the work at *Ammaia*, others draw upon their own or other published projects. The book is subdivided into five parts, each of which deals with a different aspect of non-destructive approaches to archaeological sites.

The first part of the book concerns *remote sensing*, with Ceraudo discussing the history of vertical and oblique aerial photography followed by a description of the techniques of flying and taking images and interpretation and classification of the different features that give rise to features visible on aerial photographs. Verhoeven et al. discuss new techniques for orthorectification of archaeological aerial photographs with case studies at the Roman towns of *Trea* (Italy), Pitaranha (Portugal) and the Kreutel region in Austria, while Vermeulen addresses related issues; airborne multispectral techniques are described by Cavalli, who advocates an integrated approach to multiplatform and multisensor and frequency data approaches through case studies from Sicily and Italy. The second section focuses upon *geophysical techniques*, with Bevan and Smekalova dealing with magnetic survey, by discussing the choice of instrument, field procedures, data processing and display and interpretation and the report with reference to a range of case studies from Denmark, the Crimea and Egypt. Carreras uses primarily Spanish examples to explain the principles of resistivity survey, while Novo discusses the history of the approach, principles and systems of ground-penetrating radar (GPR), before moving on to discuss basic 3D imaging and the issue of multichannel arrays in surveys. The integration of geophysical survey and a systematic ground truthing of anomalies is touched upon by Meyer, who looks at pitfalls in the interpretation of different kinds of geophysical data and makes suggestions as to what survey reports should contain. The third section covers *topographic and geoarchaeological surveys*, with papers on intra-site survey by Bintliff, who focuses upon surface collection strategies and interpretative frameworks used in the Boeotia survey and reviews those on other large Classical Mediterranean sites. Verhagen assesses the value of coring and test pits in gauging the underlying stratigraphic profile of sites at which surveying takes place. The section includes a paper on the techniques used in the production of digital elevation models (DEMs) and their value by Martínez and Mayoral and a discussion of geomorphological analyses by Fouache. Fourth, and lastly, the volume deals with *interpretation, visualization and site management*. The management of research and heritage is dealt with by Van Roode, while Cerato and Pescarin discuss the issues to be considered by different "users" when attempting to reconstruct past landscapes for virtual museums, and Klein discusses the advantages of real-time approaches to producing high-quality reconstructions of the *Ammaia* project. The London and Seville Charters for ensuring scientifically sound virtual reconstructions are the subject of papers by López Mechero, Pletincx and Denard.

The broad range of techniques covered by these papers provides a very useful summary of the kinds of technique that can be used in non-destructive surveys of larger ancient sites, as well as contingent issues that arise in their integration. While this should greatly help introduce newer practitioners to the potential of integrated non-destructive surveys, it also raises a number of significant issues that should give us pause for reflection. What follows is a personal reflection upon some of these.

21.1 Choice of Techniques

The range of techniques discussed in this book, coupled with inherent differences in the nature of the archaeological record from one site to another, can make it difficult for an archaeologist to decide which technique should be used for the survey of a specific site. Choice should surely be dictated primarily by the nature of the research questions that are posed to a specific site. If the aim is to understand the layout of a town rather than a small group of buildings or a rural site, for example, then it makes most sense to choose the method that will let you cover ground quickly, preferably as cheaply as possible. While this might seem to point to the use of a magnetometer cart, this would make little sense if the ground was uneven or if there were awkward corners to survey and upstanding structures. The choice should also take into account the nature of the building materials at the site (if known), the soils and underlying geology (see, e.g. Slapszak 2013). The best way to ensure that the choice will yield the best result is through close collaboration between archaeologists and geophysicists from inception of the project, beginning with a discussion of the project objectives, to designing the survey strategy, data collection, analysis and, ultimately, publication. Another issue concerns the trend in recent years towards using ever more sensitive and sophisticated survey instruments, suggesting to some that older instruments are outmoded and their results of little value. While these newer instruments do produce spectacular results, they are beyond the financial means of many researchers and institutions. Nevertheless, the choice of the kind of instrument and software should be determined by the resolution of the results needed to answer the research questions posed, and for most situations cheaper and less sophisticated instruments are entirely appropriate.

21.2 Challenges in Comparing Sites and Surveys

While there is a natural tendency to concentrate non-destructive surveys within the built-up areas of towns (walled or otherwise) and rural settlements, there has been a tendency to ignore the "edges" of settlements and therefore to miss the very significant contribution that non-destructive techniques can make to our understanding of urban peripheries or off-site areas. Recent work in the hinterland of *Falerii Novi*, for example, illustrates how much this broader approach can change our perception of the nature of the town and its relationship to its broader hinterland (Hay et al. 2010). Alternatively, extensive surveys of land between sites in Tuscany (Campana 2013; Meyer 2013) have begun to reveal smaller, and hitherto unknown, categories of site that have the potential of transforming our understanding of the ancient countryside. Beyond this, however, there is the even more intractable problem about how to draw comparisons between the results of one survey and another – an issue that is as true of integrated geophysics, surface and aerial surveys as it was of traditional field surveys (Alcock and Cherry 2004). The stumbling block is that more often than not, one survey will have used one technique, or a combination of techniques, while another will have used different ones, and since the choice will have been determined by a consideration of the soil conditions and ancient techniques of building, they may not necessarily be comparable. Problems of comparison also arise when the results of one survey, whether using one or several techniques, are compared with earlier published surveys, even when the same techniques have been used. The problem here lies with the sensitivity of the instruments, the quality of the software used or differences in the spacing of traverses or other aspects of survey strategy – where these are known. Similar kinds of challenge have arisen in the more traditional field surveys (Terrenato 2004), and here the best solution seems to lie in drawing broader fuzzy comparisons between the results of different surveys rather than attempting fine-grained statistically based comparisons. As ever, the survey strategy selected should depend upon the research questions that one asks of the site, and these should always be those that are most appropriate to the resolution, quality and nature of the data sets being analysed.

21.3 The Challenges of Data Integration

One of the major merits of the Radio-Past project has been to subject the surface and hinterland of *Ammaia* to just about every possible kind of non-destructive technique possible. This has provided us with extremely valuable data about the strengths and weaknesses of different kinds of technique, even though these should be tempered by taking into account the particular archaeological and geological reality of the ancient town. While the data derived from each of the techniques in a study of this kind tell us very different things about the soils, buried structures and artefacts in the ground, they are not readily comparable. Moreover, as the instruments and software used on surveys such as this become ever more sophisticated, the challenges of integrating the data from different kinds of technique will become ever greater. There are two potential solutions to this problem. The easier of the two is simply to undertake visual comparisons of the final interpreted results of each technique, draw conclusions from each of them and then consider all of these together. This produces very valuable information, such as in the case of ongoing surveys at *Carnuntum*, *Aquileia* and *Ephesus* (Groh 2013). The tougher challenge, however, lies in finding ways of integrating the digital data drawn from different techniques, with a view to finding the best way of visualizing the end results. The experience of the *Portus Project* suggests that the best way to achieve this is to have a broad strategy of digital integration from the start of a project, with the form that it takes being predicated by the research questions being asked of the site.

21.4 The Limits of Non-destructive Techniques

While the benefits of a manual of good practice such as this are obvious, it needs to be remembered that not all sites are suited to geophysics, aerial photography or other kinds of non-destructive technique. Furthermore, while one particular technique, or suite of techniques, might work well for one site, it does not necessarily follow that it will work for other neighbouring sites. Thus, while the magnetometry results from *Falerii Novi* were extremely clear (Keay et al. 2000a, b), revealing details of public buildings and houses down to the resolution of individual rooms, those from *Ocriculum* were much less clear (Hay et al. 2013), even though the site was close by and shared the same volcanic soils. Alternatively, for reasons that are not clear, some techniques will work, while others will not. At *Portus*, for example, while magnetometry identified the aqueduct, road and canal between the hexagonal basin and the Tiber, it did not reveal any anomalies belonging to the river port and adjacent villa; aerial photography, by contrast, revealed important details of the river port, villa road, aqueduct and road, but nothing of the canal (Keay et al. 2005, pp. 134–156). Another issue, which is frequently forgotten by archaeologists seeking to interpret published geophysical and aerial photography surveys of sites, is that these are not simply buried structures. Instead, they are anomalies created from differential responses of buried construction materials, fills and other features to sensitive instruments, whose alignment or shape suggests to the practitioner that they are walls, floors or ditches, etc. While it is possible to hazard a guess about the nature of the construction material from the strength or character of a particular anomaly, particularly burnt material, iron or lead, it is still an inexact science. More research into how we can better "read" the signatures of anomalies from geophysical surveys is needed in particular. The best way forward here is to follow up surveys with targeted excavations, so that we can try to characterize certain kinds of anomaly with different kinds of excavated material, such as marble, stone, concrete and mud brick. Ongoing work by the *Portus* Project (E. Richley), for example, is analysing the relationship between GPR anomalies and excavated features. Clearly, however, care needs to be exercised in assuming that the characterization of anomalies at one site will necessarily work for another, even if it is an area with similar geology and at a site using similar kinds of building material. A related issue is that, construction materials aside, our interpretation of the shapes that we identify in geophysical surveys and aerial photographs is very much influenced by the plans of buildings with which we are already familiar, whether from published evidence of excavated buildings or other surveys. While this is to some degree inevitable, it makes it harder for us to identify unusual buildings and can tempt us into conflating different kinds of feature because it seems to resemble something with which we are familiar.

21.5 Integration and Visualization of Results

At one level, unidimensional interpreted plans, in the case of aerial photography, and greyscale or coloured plots, in the case of geophysics, have served us well in presenting survey results. However, as three-dimensional and virtual reality models of these interpretations become more common, we need to find ways of three-dimensionally modelling survey results. This is, of course, particularly important for geophysics but is challenging since each technique responds to different aspects of buried aspects in distinct ways. It has been the subject of several important studies in recent years based on 2D data (notably Kvamme 2006) that have advocated approaches that range from data overlays to digital integration. Another way forward, which is being investigated by members of the *Portus Project* and other practitioners, has been to use the 3D data sets of GPR and ERT combined with excavation data. In the case of the former, for example, experimentation with amplitude slice mapping (time slices) has made it possible to examine reflections at different depths and orientations, three-dimensional iso-surface modelling and the creation of voxel data sets and vector representations of interpretations of volumetric data. In the case of ERT, we have experimented with importing processed data into 3D geophysical processing software (GPR Slice) and viewing data points directly in ArcGIS, generating raster interpolations for horizontal profiles and viewing all the information together (Keay et al. 2013).

A second issue concerns the virtual reality reconstructions of sites whose empirical base is largely comprised of geophysical anomalies. The London Charter and Seville Principles have been very important in helping to establish ground rules for archaeological reconstructions in general, and some of the most recent work has started to attain extraordinarily vivid representations of Classical sites. Indeed, these can now seem so lifelike that they appear to be completely unproblematic reconstructions of what the buildings looked like in antiquity, even though this masks many unknowns, variables and assumptions. There is little doubt that these reconstructions have considerable value in communicating to the general public about the character of a site that would otherwise be very difficult to understand: greyscale images have little ready educational value for the general public. This raises the question as to how "real" these reconstructions should really be, an issue related to the broader question of the purpose that they are meant to serve. Given the ephemeral nature of geophysical data – which means that they need to be supplemented with subjectively selected parallels from better known/excavated sites (see above) – there perhaps needs to be more consideration about the kinds of monument or urban stereotypes that may be being inadvertently propagated. The *Portus* Project approach has been to emphasize the limitations of our evidence, stress the importance of proposing alternative reconstructions and build the archaeological limitation of any reconstruction into the image in a way that it can be readily recreated once new information is available.

21.6 The Continued Role of Excavation

In view of the great advances in non-destructive survey, particularly as regards the quality and the resolution of the results and the speed with which major settlements can be surveyed, it is tempting to think that there is no longer any need for excavation. Indeed, in the case of very shallow and relatively simple sites, there is perhaps a case to be made here. Excavation, after all, is very expensive and time consuming, drawbacks that often lead to excavations never being published, and raises ethical questions about the need to destroy a site in order to understand it. Then there is the challenge of what to do with the site when excavated: restore, backfill or present to the public and how? While all of these are real problems, they can and need to be confronted because excavation still has a key role to play in understanding archaeological sites. Aside from financial issues, the choice of whether to undertake survey or excavation depends very much upon the kinds of questions that are being posed about the site in question. Furthermore, as argued above, the interpretation of geophysical anomalies is not a straightforward process. In the case of magnetometry at complex sites, for example, the anomalies represent a compressed palimpsest of features that are not easy to separate and can be fully understood once they have been excavated. Alternatively, in deeply stratified sites, excavation is still the only way of properly understanding the occupational sequence, since some geophysical techniques, such as magnetometry, do not penetrate deeply enough, while others, such as GPR in which depth can be calculated by considering the speed of return of the electromagnetic pulse, are still an inexact science. Our recently published preliminary interpretation of a large building at *Portus* is an example of some of the difficulties in balancing up these different kinds of approach (Keay et al. 2012a, 2012b). The best approach is surely one that involves an integrated and reflexive strategy of non-destructive techniques and targeted question and answer excavation from the outset.

21.7 Dealing with Data

Surveys of the kind discussed in this book are becoming increasingly common. This, together with ever more sophisticated instruments, is generating increasingly large quantities of digital data and thus raising important issues about the formats in which they should be stored, where they should be deposited and how they can be accessed. There is thus a desperate need for explicit, regional, national and EU-based strategies to deal with all of these, or there is a very real risk that much of this data could be lost, unusable or become inaccessible in the not too distant future. Some UK-based archaeologists running publicly funded survey projects in the UK and abroad are fortunate in that they are obliged to archive their data at the Archaeological Data Service (ADS) (http://archaeologydataservice.ac.uk). This body is funded by the Arts and Humanities Research Council and is served by its Guides to Good Practice and related publications on data management, which offer guidance and exemplars informed by a broad range of archaeological and archival expertise. It also offers specific guidance for different forms of archaeological data, methods of research and recording and best practice in dissemination. Its guidance concerns all stages in the data documentation and curation process. While it does not by any means host all the data from UK survey projects, it offers a way forward in ensuring the sustainability of this kind of material. Many other parts of Europe, however, do not have a resource of this kind, and data is held in universities, research institutions or regional heritage agencies, and finding ways of agreeing regionally driven pan-European standards for data storage and access must surely be a priority for the future.

References

Alcock S, Cherry J (eds) (2004) Side by side survey. Comparative regional studies in the Mediterranean world. Oxbow, Oxford

Bintliff J, Snodgrass A (1988) Mediterranean survey and the city. Antiquity 62:57–71

Bourgeois J (2013) Aerial photography in Flanders: form research to cultural heritage management. In: Vermeulen F, Corsi C (eds) Non-destructive approaches to complex archaeological sites in Europe: a round up, 7. Ghent University, Ghent. E-publication. Downloadable at: http://www.7host.at/radiopast/RP_GC_2013.pdf

Campana S (2013) Exploring the archaeological continuum. New perspectives and actual limitations of very large scale geophysical prospection in Mediterranean landscape archaeology. In: Vermeulen F, Corsi C (eds) Non-destructive approaches to complex archaeological sites in Europe: a round up, 10. Ghent University, Ghent. E-publication: 10. Downloadable at: http://www.7host.at/radiopast/RP_GC_2013.pdf

Ceraudo G (2013) Ancient archaeology and ancient landscapes. Combining traditional methods and advanced technologies. In: Vermeulen F, Corsi C (eds) Non-destructive approaches to complex archaeological sites in Europe: a round up, 11. Ghent University, Ghent. E-publication: 11. . Downloadable at: http://www.7host.at/radiopast/RP_GC_2013.pdf

Corsi C, Vermeulen F (2010) Changing landscapes. the impact of Roman towns in the western Mediterranean. Proceedings of the international colloquium Castelo de Vide – Marvao 15th–17th May 2008. AnteQuem, Bologna

Corsi C, Vermeulen F (eds) (2012) Ammaia I: the survey. A Romano-Lusitanian townscape revealed (ARGU 8). Academia Press, Ghent

Corsi C, Johnson PS, Vermeulem F (2012) A geomagnetic survey of Ammaia: a contribution to understanding Roman urbanism in Lusitania. J Rom Archaeol 25:121–145

Gaffney V, Gaffney C (2000) Non-invasive investigations at Wroxeter at the end of the twentieth century. Archaeol Prospect 7(2):65–68, Special issue

Groh S (2006) Neue Forschungen zur Stadtplanung in Ephesos. Jahresh Österr Archäol Inst 75:47–116

Groh S (2013) Methodological proposals for integrated surveys on site and landscape level. In: Vermeulen F, Corsi C (eds) Non-destructive approaches to complex archaeological sites in Europe: a round up, 16. Ghent University, Ghent. E-publication: 16. Downloadable at: http://www.7host.at/radiopast/RP_GC_2013.pdf

Guaitoli M (2003) Lo sguardo di Icaro. Le collezioni dell'Aerofototeca Nazionale per la conoscenza del territorio. Campisano Editore, Rome

Hay S, Johnson PS, Keay S, Millett M (2010) Falerii Novi: further survey of the northern extramural area. PBSR 78:1–38

Hay S, Keay S, Millett M (2013) Ocriculum (Otricoli, Umbria). An archaeological survey of the Roman Town, vol 22, Monographs of the British School at Rome. BSR, London

Hidalgo R, Manuel J, Keay S, Jordan D et al (1999) La Itálica de Adriano. Resultados de las prospecciones arqueológicas de 1991 y 1992. Arch Esp Arqueol 72:73–97

Johnson PS, Martin M (eds) (2012) Archaeological survey and the city, vol 2, University of Cambridge Museum of Classical Archaeology monograph. Oxbow Books, Oxford

Keay S, Creighton J et al (eds) (2000a) Celti-Penaflor. The archaeology of a Hispano-Roman Town in Baetica. Oxbow Books, Oxford

Keay S, Millett M, Poppy S, Robinson J, Taylor J, Terrenato N (2000b) Falerii Novi: a new survey of the walled area. PBSR 68:1–93

Keay S, Millett M, Paroli L, Strutt K (2005) Portus. An archaeological survey of the port of Imperial Rome, vol 15, Archaeological monographs of the British School at Rome. British School at Rome, London

Keay S, Earl G, Felici F, Copeland P, Cascino R, Kay S, Triantafillou C (2012a) An enigmatic new Trajanic building at Portus. J Rom Archaeol 25:486–512

Keay S, Graeme E, Gareth B, Nathan D, Jessica O, Kris S (2012b) Challenges of port landscapes. Integrating geophysics, open area excavation and computer graphic visualization at Portus and the Isola Sacra. In: Johnson PS, Martin M (eds) Archaeological survey and the city, vol 2, University of Cambridge Museum of Classical Archaeology monograph. Oxbow, Oxford, pp 303–357

Klein M (2013) Current productions in virtual archaeology: Carnuntum, German Limes and Radio Past. In: Vermeulen F, Corsi C (eds) Non-destructive approaches to complex archaeological sites in Europe: a round up, 18. Ghent University, Ghent. E-publication: 18. Downloadable at: http://www.7host.at/radiopast/RP_GC_2013.pdf

Meyer C (2013) Fast, precise and flexible: trends in large scale geophysical prospection for geoarchaeology. In: Vermeulen F, Corsi C (eds) Non-destructive approaches to complex archaeological sites in Europe: a round up, 19. Ghent University, Ghent. E-publication: 19. Downloadable at: http://www.7host.at/radiopast/RP_GC_2013.pdf

Rodríguez Hidalgo JM, Keay S (1995) Recent work at Italica. In: Cunliffe B, Keay S (eds) Social complexity and the development of towns in Iberia. From the copper age to the second century A.D. Symposium, London, 24–25 February 1994, Proceedings of the British academy 86. Oxford University Press, Oxford, pp 395–420

Slapszak B (2013) Structuring non-invasive survey on complex (urban) sites. In: Vermeulen F, Corsi C (eds) Non-destructive approaches to complex archaeological sites in Europe: a round up, 25. Ghent University, Ghent. E-publication: 25. Downloadable at: http://www.7host.at/radiopast/RP_GC_2013.pdf

Terrenato N (2004) Sample size matters! The paradox of global trends and local surveys. In: Alcock S, Cherry J (eds) Side by side survey. Comparative regional studies in the Mediterranean world. Oxbow, Oxford, pp 36–48

Vermeulen F, Hay S, Verhoeven G (2006) Potentia: an integrated survey of a Roman colony on the Adriatic Coast. PBSR 74:203–236

Vermeulen F, Burgers G-J, Keay S, Corsi C (2012) Urban landscape survey in Italy and the Mediterranean. Oxbow, Oxford

Index

C. Corsi et al. (eds.), *Good Practice in Archaeological Diagnostics*, Natural Science in Archaeology,
DOI 10.1007/978-3-319-01784-6, © Springer International Publishing Switzerland 2013

Printed by Printforce, the Netherlands